ENNIS AND NANCY HAM LIBRARY
ROCHESTER COLLEGE
800 WEST AVON ROAD
ROCHESTER HILLS, MI 48307

Nuclear Energy
A Sensible Alternative

Nuclear Energy

A Sensible Alternative

Edited by
Karl O. Ott
Purdue University
West Lafayette, Indiana

and
Bernard I. Spinrad
Iowa State University
Ames, Iowa

Plenum Press • New York and London

Library of Congress Cataloging in Publication Data

Main entry under title:

Nuclear energy.

1. Nuclear industry—United States—Addresses, essays, lectures. I. Ott, Karl O. (Karl Otto), 1925-   . II. Spinrad, Bernard I.
HD9698.U52N815   1985   333.79′24′0973   84-24834
ISBN 0-306-41441-4

©1985 Plenum Press, New York
A Division of Plenum Publishing Corporation
233 Spring Street, New York, N.Y. 10013

All rights reserved

No part of this book may be reproduced, stored in a retrieval system, or transmitted in any form or by any means, electronic, mechanical, photocopying, microfilming, recording, or otherwise, without written permission from the Publisher

Printed in the United States of America

The counters stepped up; the pen started its upward rise. It showed no tendency to level off. A chain reaction was taking place in the pile.

Leona Woods walked up to Fermi and in a voice in which there was no fear she whispered: "when do we become scared?"

Under the ceiling of the balloon, the suicide squad was alert, ready with their liquid cadmium: this was the moment. But nothing much happened. The group watched the recording instruments for 28 minutes. The pile behaved as it should, as they all had hoped it would, as they had feared it would not. . . .

. . . Arthur Compton placed a long distance call to Mr. Conant of the Office of Scientific Research and Development at Harvard.

"The Italian Navigator has reached the New World," said Compton as soon as he got Conant on the line.

"And how did he find the natives?"

"Very friendly."

An account of the first man-made self-sustained chain reaction in the Chicago Pile (CP-1), December 2, 1942, Stagg Field, The University of Chicago.

From *Atoms in the Family* by Laura Fermi
The University of Chicago Press, 1954

Contributors

Bernard L. Cohen • Department of Physics, University of Pittsburgh, Pittsburgh, Pennsylvania 15260

T. J. Connolly • Department of Mechanical Engineering, Stanford University, Stanford, California 94305

Ian A. Forbes • Energy Research Group, Inc., Waltham, Massachusetts 02154

Bertrand Goldschmidt • French Atomic Energy Commission, Paris, France.

V. P. Kenney • Department of Physics, University of Notre Dame, Notre Dame, Indiana 46556

Gerald S. Lellouche • Nuclear Power Division, Electric Power Research Institute, Palo Alto, California 94303

J. W. Lucey • Department of Aerospace and Mechanical Engineering, University of Notre Dame, Notre Dame, Indiana 46556

Karl O. Ott • School of Nuclear Engineering, Purdue University, West Lafayette, Indiana 47907

A. David Rossin • Electric Power Research Institute, Palo Alto, California 94303

Bernard I. Spinrad • Department of Nuclear Engineering, Iowa State University, Ames, Iowa 50011

T. G. Theofanous • School of Nuclear Engineering, Purdue University, West Lafayette, Indiana 47907

Joe C. Turnage • Delian Corporation, Del Mar, California 92014

Robert E. Uhrig • Florida Power and Light Company, Juno Beach, Florida 33408

Chris Whipple • Electric Power Research Institute, Palo Alto, California 94303

Richard Wilson • Department of Physics, Harvard University, Cambridge, Massachusetts 02138

C. Pierre Zaleski • Center for Geopolitics of Energy and Raw Materials, University of Paris–Dauphine, 75775 Paris Cedex 16, France, and University of California, Los Angeles, California 90024

E. L. Zebroski • Electric Power Research Institute, Palo Alto, California 94303

Foreword

E. L. Zebroski

During the 1970s, there was rapid growth of a philosophy that assumes that deindustrialization will result in an Elysian postindustrial society. This view is generally antitechnology; commonly in opposition to large-scale energy sources; and often supportive of high-cost, speculative, or at most, small-scale energy sources. The social and economic costs of policies which would lead to deindustrialization are ignored or considered to be irrelevant.

The development of civilian nuclear energy as a by-product of wartime developments also brings with it an association with the fear of nuclear weapons and with the repugnance for war in general. Many of these views and associations mingle to provide significant political constituencies. These have had considerable impact on party platforms and elections. Also, another important aspect is the conservation viewpoint. This view—correctly—concerns the fact that indefinite increase in per capita energy consumption, coupled with increasing U.S. and world populations, must at some point be restrained by limits on resources as well as by limits arising from environmental effects. All of these concerns have been subject to voluminous analysis, publications, and public discussion. They underlie one of the dominant social movements of the 1970s and 1980s. Indefinite exponential growth of energy production is neither possible nor desirable.

On the other hand, many thoughtful people are also concerned over the rapidly growing risks of both domestic and international tensions and the extent to which limited access to world energy supplies, by countries with inadequate domestic-based energy supplies, contributes to these risks. During the 1970s and 1980s, the United States has been drifting into uncharted territory in respect to the risks of major increases in human misery, social stresses, and possibly wars.

E. L. ZEBROSKI • Electric Power Research Institute, Palo Alto, California 94303.

Energy problems contribute to the potential for domestic social stresses and to international tensions that can lead to wars in the following ways:

1. There is a continued major dependence on imported oil for nearly one-third of all transportation energy, a large part of domestic comfort-heating energy, and a significant percentage of electrical energy. (A large part of the electrical capacity margin available and needed is oil- or gas-fired peaking capacity.)
2. There is an unprecedented and continuing escalation in the cost of all forms of energy, dominated by a more than tenfold increase in the basic price of oil and gas in less than a decade, and prospects for continuing increases in the future. The temporary glut in oil and gas supply, related to the economic recession worldwide in the early 1980s, has restrained the rate of increase in oil prices for a while. This does not solve the long-term problem of overdependence on imported oil in the U.S.—and especially in the developing countries.
3. There is a considerable risk of future major short-term and long-term interruptions in large parts of energy supplies for transportaion, agriculture, home heating, and in the supply of many domestic goods to the extent to which they are dependent on imported oil. In the near-term the risk is mainly from regional conflicts or outright wars. In the longer term the risk is from declining production capacity as fields are depleted, in 10 to 20 years.
4. Decline is certain in the per capita capacity of domestic base-load energy sources (coal and nuclear) by the late 1980s or early 1990s, due to a near-stoppage of new commitments since 1976. There is a continuing poor prospect for new commitments due to a combination of unpredictable regulatory delays which increase costs, and the effect of the limited availability and high cost of capital needed for investment in long-term projects. The high cost of capital forces a short-sighted planning horizon for nearly all industry. (With 20% money, the planning time-horizon shrinks to 2 or 3 years. A project which takes more than 5 years to develop to a productive status, almost without regard to its eventual value, has a negative present worth.) The impact of these circumstances has been a rapid increase in the cost of domestically produced energy, not only from oil and gas sources, but also—to a roughly comparable extent—from coal, nuclear, and other alternate energy sources as well.

   Further dramatic increases in the cost of domestic energy are in prospect for the middle and late 1980s. Some increases have been temporarily cushioned by artificial, but only temporarily effective, constraints on oil and gas prices. (There are continuing constraints on the

Foreword

price of electricity, the allowed prices of which have not fully covered the increasing costs of fuel, labor, and capital.) These circumstances reinforce the overdependence on oil and gas and further limit the ability of the U.S. to increase the capacity of domestic-based energy.

5. The proportion of U.S. oil imports from the OPEC nations has declined (to about 36% in 1983). This seems to provide a smaller sensitivity of U.S. prices and supply to possible disruptions in oil imports—for example, cutoffs arising from the chronic shooting conflicts in the Middle East. The sense of improved security of price and supply is probably false, however, as total imports are still equal to about one-half of domestic production and have started to increase again (1984). Several analysts have noted that in the event of a disruption—for example, closure of the Straits of Hormuz, many of the ships scheduled to move oil to the U.S. would change direction toward those countries which are much more heavily dependent on Middle East oil, and prices would zoom accordingly. The National Petroleum Reserve could be used to cushion the impact on oil supply short-term, but the long-term sensitivity to repeated future "oil-shocks" remains.

6. There is a large potential for social turbulence as the further rapid increases in the cost of energy (and the probable irregularities in the supply of energy) have effects on the fundamentals of society in ways that are unprecedented in American history. Some of these continuing effects include contribution of energy costs to the escalation in the costs of essential amenities—most obviously housing—that have been a traditional part of the American Dream. Most of the meaningful amenities have long been the subject of rising expectations. These rising expectations have been largely fulfilled for most Americans for nearly two centuries, with only occasional depressions or short periods of decline or deprivation. The change from a society of rising expectations to a future with decreasing expectations for large sectors of the population—and a noticeably deteriorating reality for many—is dangerous. In some regions of the world it produces intense social turbulence, increased crime, terrorism, and political instability.

   Other factors besides energy contribute to the lowered prospects for housing. The same combination of factors also contributes to lower prospects of many other amenities and freedoms that have been taken for granted. For example, the freedom of choice of locale that comes from mobility is related to the cost of energy. The constraints of travel mobility—together with constraints in the availability and quality of housing—also trend in the direction of reduced social mobility, and reduced individual opportunities, and ultimately in increased social tensions.

The religious fervor and intransigence of the low-energy ideologues is another danger. Some openly threaten social chaos and terrorism (politely called civil disobedience) if their minority cannot have its way. This fervor also drives the use of litigation for the sake of delay, which balloons the costs of construction, and the great economic waste when nearly completed plants are forced to cancellation.

There are some amenities that can be regarded as frivolous in the broad view of history, and their loss is not of real concern. Others have fundamental human values—for example, improvements in health care, in the quality and availability of foodstuffs, and in the level of public education with respect to food and health management. Increased material and energy resources have contributed to the rapid growth in the effectiveness of health care in the past. However, these now contribute to the high costs of health care, the availability of which is beginning to decline on a per capita basis.

7. The allies of the U.S. in Europe and Asia have even larger dependence on imported oil than does the U.S. The rapid escalation in the cost of oil has been somewhat tempered for Europe by the discovery and production of North Sea oil. Also, European economies are not as transportation-intensive as that of the U.S. High taxes on oil and gasoline have already served to limit demand in Europe for many years, and average distances for transport are relatively short. However, the slowdown in growth of domestic energy capacity and in nuclear energy growth has also limited this means of reducing the overdependence on oil imports. (Major exceptions are found in France, which has proceeded with a large and effective nuclear program, as well as Belgium, Finland, Sweden, Switzerland, Britain, Germany, and Canada—all of which have larger percentages of nuclear capacity in operation than the U.S.)

8. The likelihood of conflict over oil was underscored by the policy statements from both the Carter and the Reagan administrations indicating that the Persian Gulf was to be regarded as a virtual protectorate of the U.S. This has been accompanied by the stationing of naval forces and aircraft in the vicinity. *It is unlikely that the combined environmental impacts of all of the energy and industrial activities of the world pose as large a threat of human misery and death as the continuing wars involving oil supplies.*

9. The effect of oil dependence is more pervasive and fundamental than just higher costs—in time it can escalate to have the effect of losing a war almost without knowing it was happening. This can go through three stages:

- Loss of domestic control of increasing parts of the U.S. economy.*
- Loss of degrees of political control—domestically through loss of economic control, internationally through the more sensitive dependence of our allies on OPEC and the Soviet Union.
- Potential loss of military control—through limitations of domestic resources, a complex of energy-related political factors, and loss of assurance of allies.

The export of equity—$90 billion for oil imports in 1980 alone—has changed the ownership and to some degree affects the managerial direction of large segments of American industry, agriculture, and even housing.

The political sensitivity of threats to oil supply has inhibited domestic policy in both overt and subtle ways. For example, for several years the rates of addition to the national petroleum reserve were below the rates authorized by Congress—in part because of pressures from OPEC.

The potential decline of (conventional) military control is more basic than the hostage situation in Iran in 1980 and the Beirut bombings and withdrawal of 1983–84. In the event of a seriously deteriorating situation in the Middle East, such as a major oil cutoff, there is little the U.S. could—or should—do militarily. Our allies know this and see our military buildup in areas near the Middle East as mostly posturing.

In any event, a "protective conflict" is unmanageable as it would likely destroy the very production capacity and political stability we would like to preserve.

For the foregoing reasons, many thoughtful people feel that the present U.S. inhibitions against the use of any major energy technology—and particularly nuclear energy—that can be based entirely on domestic resources, are dangerous: economically, politically, and militarily.

The esthetic and philosophical preferences against large-scale technologies have a widespread emotional appeal. But the danger signals to society and to the world that can result from following such emotional preferences suggest caution. The relative hazards of a society with inadequate security of energy supplies and prices now must be openly and realistically balanced against the known, estimated, or only speculative hazards of energy production. Because

* The permanent export of equity in the last 10 years, due to the effect of oil imports on the increased negative balance of payments, is approximately equivalent to the total worth of the State of California, or of a half-dozen smaller states.

of the emotional and political appeal of the antitechnology philosophies, a large body of mythology has grown on the presumed social or environmental problems of nuclear energy. There has been much less, and obviously less dramatic, dissemination of valid information. This trend reached its low point in the closing years of the 1970s when the Department of Energy was instructed by the Administration to destroy all printed material which had been developed to provide public information on nuclear energy.

The writing of this book was undertaken by a small group of people—mostly in universities and research institutes—who are concerned that the mythology on nuclear subjects has outrun reason and reality to a dangerous extent. Information on nuclear matters in the media suffers from the scarcity of scientifically trained people among journalists and the relatively high percentage (over 90% in the electronic media) of people who—by their own admission—are captive to the antitechnology ideology, as well as having little or no education in the sciences. The flow of public information (fortunately, with some notable exceptions) has contained more misinformation than fact. There are frequent extreme examples of the personal but unstated ideological biases.

The writing of this book is an attempt to provide in-depth articles on the main issues affecting the use and usefulness of nuclear energy for peaceful domestic purposes. Most of the material should be readily intelligible with little more than a high school level of basic education in science or survey courses in college. The writers have striven for the ideal of scientific objectivity, rather than advocacy. ("Advocacy" implies the selection only of evidence that supports the bias of the advocate; objectivity requires looking at evidence on both or several sides of each topic and evaluating them as evenhandedly as possible.) In some cases, this requires "leaning over backwards" to give the benefit of a doubt even to ideas for which there is already much contrary evidence. For this reason, scientists rarely make good advocates because so many subjects must be taken on the basis of "on the one hand, this is true, and on the other hand, that also might be true." The writers have had the benefit of many years of study and publication on the subjects they treat. They have had the experience of extensive exposure and discussion of the material presented, both in scientific journals and forums and in public forums. Many are in the front ranks of people who have researched, published, and taught university-level courses in the basic science, engineering, or economics of the subjects they cover.

Some of the topics covered are described at times as "moral issues" with the implication that it is unnecessary to understand the factual substance of an issue in order to have a valid and reasoned and morally valid position. However well-meaning may be an adherent to such a view, it also is the very fabric of demagoguery. An alternative view—well articulated in the community of bioethic philosophers (and also by people who specialize in risk analysis)—is that no issue should be treated in isolation. The avoidance of one perceived risk can

Foreword

often bring the unknowing acceptance of a much larger danger (and sometimes also, a shift of locale of some of the larger risk elsewhere in society or in the world). For this reason, the writers believe that there is a moral as well as educational or scientific obligation to understand the basic facts of the technology, its uses and usefulness—and the alternatives. The acceptance and promotion of ideology or conclusions or policies that are based on misunderstanding or mythology—the acceptance of which can result in increased human misery—can hardly be regarded as moral.

The association of energy, in particular its shifting from animal to nonanimal sources, with the way in which people live, is a very close one. No single feature of the industrial revolution of the mid-18th to mid-19th century, for example, is more basic than the development and availability of coal as an abundant, reliable, and cheap energy resource. The steam engine was the technology that converted coal energy to mechanical forms that supplemented or replaced and multiplied the work capability of humans and animals. While acknowledging that man does not live by energy alone, it is clear that an adequate and reliable supply of domestic energy is of vital importance—even to the survival level—of the United States and the civilized world. In the past, some major changes in our energy supply systems have taken place with little public attention. These changes include replacement of coal by oil and natural gas for rail transportation and for residential heating, for the electrification of rural America, and for the labor-saving appliances of the American household, factory, and office.

To be sure, these changes occurred only as a result of many initiatives and decisions on the part of consumers, suppliers, and the effects of the Rural Electrification Act. These decisions were mostly local and decentralized and not made as a part of any overall national energy policy. Today, the changes that are taking place require more public attention. If this attention is to result in better energy policy and better future energy systems, it must be based on some awareness of the world's energy systems and the realistic and timely options. It is against this background that the authors present their views on a number of energy topics.

A special acknowledgement is needed for the contribution of Dr. Karl O. Ott. In addition to his work as coeditor, he initiated the concept of this book as a contribution to the broader understanding of the issue. He brought the various authors together and persevered to see the book to completion.

Contents

Introduction and Overview 1
   T. J. Connolly

I. Energy and Society

1. Energy Futures: A World Study 17
   Bernard I. Spinrad

2. Energy Demand–Energy Supplies 49
   V. P. Kenney and J. W. Lucey

3. Exclusive Paths and Difficult Choices: An Analysis of Hard, Soft, and Moderate Energy Paths 73
   Ian A. Forbes and Joe C. Turnage

4. An Energy-Deficient Society 111
   Robert E. Uhrig

5. Energy Shortages: The Downside Risks 119
   A. David Rossin

II. Economics of Nuclear Power

6. Economics of Light-Water Reactors 125
   A. David Rossin

7. Fast Breeder Reactor Economics 153
   C. Pierre Zaleski

III. Recycling and Proliferation

8. International Cooperation in the Nuclear Field: Past, Present, and Prospects 181
   Bertrand Goldschmidt

9. Nuclear Recycling: Costs, Savings, and Safeguards 195
   Bernard I. Spinrad

10. Alternative Fuels, Fuel Cycles, and Reactors: Are They Useful? Are They Necessary? 207
    Bernard I. Spinrad

11. The Nuclear Weapons Proliferation Issue 223
    E. L. Zebroski and Bernard I. Spinrad

12. Paths to a World with More Reliable Nuclear Safeguards 249
    E. L. Zebroski

13. The Homemade Bomb Issue 265
    Bernard I. Spinrad and E. L. Zebroski

IV. Risk Assessment

14. LWR Risk Assessment 273
    Gerald S. Lellouche

15. Accident Analysis and Risk Assessment 287
    T. G. Theofanous and Richard Wilson

16. The Waste Disposal Risk 299
    Bernard L. Cohen

Contents

17. Radon Problems 313
    Bernard L. Cohen

18. Risks in Our Society 317
    Bernard L. Cohen

V. Special Nuclear Issues, Past and Present

19. Health Effects of Low-Level Radiation 329
    Bernard L. Cohen

20. Routine Releases of Radioactivity from the
    Nuclear Industry 339
    Bernard L. Cohen

21. Low-Level Radioactivity and Infant Mortality 349
    Karl O. Ott

22. The Myth of Plutonium Toxicity 355
    Bernard L. Cohen

23. Myths about High-Level Radioactive Waste 367
    Bernard L. Cohen

24. The Aging Reactor Myth 369
    A. David Rossin

25. The Police State Myth 371
    Bernard I. Spinrad

26. Insurance and Nuclear Power: The Price–Anderson Act .. 373
    Chris Whipple

27. Solar and Nuclear Power Are Partners 381
    A. David Rossin

Index 383

Introduction and Overview

T. J. Connolly

The reader will find in this collection of essays treatment of a wide variety of questions common to what is known as the nuclear controversy. These include future energy supplies and demands, the economics of nuclear power, and several of the concerns that people have regarding the risks and hazards of nuclear power. Each essay represents the best efforts of its author to address the subject objectively and with perspective, always mindful of the uncertainties inherent in dealing with human activities and future events. Because the authors are exclusively engineers and scientists, however, there are other dimensions to the nuclear controversy—political, sociological, psychological—that are left untreated. This conscious omission does not reflect a belief that these aspects are not important. Indeed they are. For example, the Presidential Commission appointed to evaluate the accident at the Three Mile Island nuclear power plant reported that the principal injury to the surrounding population was one of severe mental stress. We are not inclined to regard this form of injury as an imaginary one or one to be disregarded. We do not possess the expertise to address it in its entirety, but we do believe that the widespread exaggeration of the dangers of nuclear power is one contributory factor; that is a subject that we can and do address in these essays. In recognition of additional political and sociological aspects of the nuclear question, however, we attempt in this introduction to set the scene of the nuclear controversy in the United States as we see it.

First, a bit about ourselves, the authors: We have in common that our scientific and engineering educations and professional experiences have involved us deeply in the field of nuclear energy and, more recently, in energy problems

---

T. J. CONNOLLY • Department of Mechanical Engineering, Stanford University, Stanford, California 94305.

generally. There is, to be sure, a wide variation in the nature of our individual involvements. For some of us, the experience dates to the first nuclear reactors, constructed in the early 1940s. While these essays could only be written out of our collective experience, the motivation goes far beyond professional allegiance. The motivation is a shared deep conviction that the tremendous potential benefits of nuclear energy are being squandered through a host of ill-considered, even superstitious, actions and policies on the part of individuals, local and state governments, and the federal government.

The possession of deep convictions that run counter to state or national policy hardly gives great distinction to any group. However, a part of our conviction is that the information on which nuclear policies are being set is grossly distorted, often deliberately so. Whether it will have any effect or not, we are moved to set the record as straight as we can. At least, we want the record to show that we tried. We do not believe our world view reflects any unusual political ideology, but view ourselves as scattered about the center of the American political scene. Whether we are simply technologists so enamored of the intricacies of our technology and so caught up in its challenges that we do not possess a reasonable perspective on the totality of societal problems, only you, the reader, can judge.

Our involvement with nuclear energy may present a problem, of course. It is common in these days of technological disputes waged in the political arena to discount the testimony of those involved in the field at issue. There is inferred to be an implicit conflict of interest. In our case we would note two relevant personal points on this subject. For most of us, our professional position is sufficiently secure that our future income would be little affected if the United States should order all nuclear power plants shut down tomorrow. In fact, the work involved in implementing such an order would keep many of us busy well beyond retirement. The second point relates to simple fact: the preponderance of the benefits of nuclear energy can only be enjoyed by future generations, but the degree to which they will enjoy them depends strongly on the actions of this generation. Considering our ages, therefore, we do not stand to lose much if this nation chooses to abandon nuclear energy.

In many ways the energy crisis has been more interesting from the standpoint of the societal response that it has evoked than the economic and technological problems that it presents. We try an imperfect analogy. Imagine we are on an ocean liner in the Pacific Ocean sailing serenely for San Francisco. The ship is traveling at 25 mph and is 1000 miles to the west of port. Suddenly someone announces he has just calculated that in 40 hours we will crash into San Francisco. Clearly there is a crisis. An emergency meeting of the passengers is convened. The captain explains that, indeed, sailing at 25 knots we should arrive in San Francisco in 40 hours. He goes on that it is his practice to lower the speed in stages as port is approached. Nevertheless, he admits, under intense examination

Introduction and Overview 3

by a passenger-interest lawyer, that if he should not give the correct orders the ship could go aground. Clearly, that's what will happen. Many solutions to the crisis are proposed. A Congressman aboard immediately wires to Washington a bill to move San Francisco 25 miles to the south. This hardly seems to go to the heart of the matter but in Washington, it's action that counts. Other passengers suggest that the ship be slowed to 5 knots immediately. It is pointed out that the trip would then take 200 hours and there isn't enough food. A vegetarian group has the solution to that. One group begins taking swimming lessons; another Yoga. Several lawsuits are filed against the passenger line, the ship builder, etc.

It won't do to draw this analogy too finely. But perhaps the parallels to the energy crisis in the U.S. are apparent. There is the failure to distinguish symptoms from substance. There are countless actions that carry the facade of a solution but in fact only complicate the situation. There are many self-serving proposals that are advanced as solutions. There is the search for scapegoats. A point is reached where the remedies suggested or adopted become more a problem than the problem itself, if indeed the two can be distinguished.

So let us see if we can get the issues clearly defined. First, the U.S. is, at this writing, importing over 8 million barrels of oil daily, 3 million of these from Arab/OPEC nations. These imports represent over 40% of our petroleum consumption.* This dependence places the nation in an extremely precarious position. It means that the supply of this material, so vital to the economy, is subject to sudden and extended interruption. The cost to the nation of such an event—and it is not a low-probability event—is almost incalculable. Furthermore, the prices that we pay for this oil, prices supported in large measure by our imports, represent a large financial drain on the country and relatively more so on some poorer countries. Until now, we have lacked the unity and the will to face this issue squarely and to deal with it effectively.

An independent but interrelated issue is that the U.S. and the world need new sources of energy to substitute for fossil fuels, particularly natural gas and petroleum. The latter supply about two-thirds of the world's energy at the present time. This is a much longer-term issue. The transition away from petroleum and natural gas can take place over a period of 50 years or more. This is no cause for complacency, however. We need all that time; we should be going about the transition resolutely and with a well-thought-out strategy. To a considerable extent this issue is coupled to the issue of the present-day oil cartel because the less effectively we deal with that problem, the less will be our financial ability to carry on the research and development required to replace oil and gas in a timely manner and with well-developed technologies.

---

* This paper was originally written in 1978 so that here and in a few other places the statistics cited are no longer current.

In keeping the issues straight, nothing is more important than a sense of time and timing. There are many examples, but one case in point will suffice for the moment. In spite of the fact that the world needs substitutes for natural gas and petroleum, it can be confidently predicted that in the next several years there will be reports of oil gluts and natural gas gluts.* These can be expected to occur both worldwide and locally. At other times and other places, shortages can be expected. Whatever may be the interpretations made in newspaper accounts, the gluts will not be a signal that the need for a petroleum substitute was false, nor will shortages indicate that the barrel has finally run dry. These will be the short-run ups and downs in the output of a very complex supply network. They also occur in housing, in hotel rooms, in beef, in almost all commercial goods. The supply of natural gas and petroleum requires many operations: exploration, production, transportation, refining, marketing. Worldwide, these operations are performed by many different parties. It is too much to expect that they will always be operating in unison. It is important that we be careful not to put false interpretations on short-run signals. To be sure, in a system as large as the world's energy, there can be many human tragedies caused by even the smallest wiggles in the supply. Groups of people can be deprived of adequate food and heat, jobs can be lost and businesses ruined. In giving our primary attention to the longer-run problems, we do not mean to say such tragedies are unimportant. They must be addressed. It is only that the response should properly recognize the degree to which a given problem is local and temporary.

To return to the time-frame question, if you are driving a car at 55 mph approaching a curve, then it follows that in the near future and in some coordinated way you must slow down and turn the car. To fail to do so invites disaster. On the other hand, to turn too soon can be just as bad. The replacement of fossil fuels is surely a more complex problem than driving an automobile around a curve. It has many more dimensions. We do know, however, that we are coming to a curve, even if we can't foresee just where it is or how sharp it is. It is only prudent that we prepare for it. We shall have much to say about the kinds of preparations and how long they may take. For now, we shall borrow a phrase from Peter Fortescue, "I believe that our real choice lies not in what has to be done in the end, but only in whether we plan now or are to be forced into action later by the imminent pressures of inexorable events." The challenge, therefore, is to discern what are the reasonable decisions and actions to take now and to have national unity and the will to carry them out.

For the U.S., therefore, the task before us is to set about implementation of a series of measures to reduce our dependence on oil imports from the less

---

* This statement was also made in 1978. In this instance, the passage of time makes the observation particularly pertinent.

Introduction and Overview

stable parts of the world, to hedge against import interruptions, and to lay the groundwork for a sound, balanced energy system that will serve the country well through the transition away from oil and gas. There are many institutional and technological measures that should play a role in the total program. These certainly include demand reduction or energy conservation, the upgrading of many "old" technologies such as coal combustion, the development of new technologies for the synthesis of liquid fuels, and, the subject of these essays, nuclear power. Although nuclear is normally associated with the longer-term energy sources, in the present instance it can also be important in the short run. There are a large number of reactors that have been licensed for construction but have not been completed. These could all easily be completed in the 1980s. Collectively, they would supply the equivalent of 3 million barrels of oil per day, an amount equivalent to what we are now* importing from the Arab/OPEC nations.

Because this work is almost exclusively a plea not to squander the benefits of nuclear energy, it may seem that we are, by default, either opposed or indifferent to other energy forms or technologies. If solar energy or fossil energy were in the deep political trouble that nuclear is today, we hope that we would speak out, although others are obviously better qualified by virtue of their expertise.

One of the most unfortunate developments of recent years has been the tendency to polarize technologies or the proponents of technologies. One reads continually of nuclear advocates and solar advocates, projecting the idea that nuclear energy and solar energy are somehow mutually exclusive. We have many energy resources and technologies. Each has its strengths and its weaknesses. It should be the goal of energy policy to ensure that each has the opportunity to find its efficient role. While it is true that some are good substitutes for others and thus competitive, we should also recognize that more often than not, two energy technologies are more *complementary* than competitive. We hear a great deal these days to the effect that solar energy or geothermal energy can replace nuclear energy; as if there had to be a winner. This is not a football league in which we will have an Energy Super Bowl game some Sunday afternoon with one energy source marching off the field triumphant over all the others. The best and most reliable energy systems are those that have a mix of energy sources and technologies.

So our basic argument is that nuclear has its place, not that it should or could provide all or even most of our energy requirements. The reasons can be put very simply:

- It is the safest and most environmentally benign large-scale energy technology that the world possesses.

---

* This is a 1984 datum.

- Using technology already demonstrated, it makes available an energy resource that is effectively unlimited.
- By virtue of the fact that it today is generating 12% of the nation's electricity, more than was generated in the whole country in the 1940s, the economic viability of the technology is established.

Now, if you are an average American who reads the newspaper, watches television, and/or listens to the radio, you must wonder if we have proofread the above statements very well. Haven't we seen the NBC "documentary" showing the mongoloid children of an atomic plant worker (by false implication, the result of the radiation exposure of the worker), or dead cows being bulldozed into ditches (by false implication, as a result of leaking radioactivity)? Haven't we read newspaper accounts of deadly atomic wastes leaking from tanks in Hanford, Washington? Haven't we read that scientists with impeccable credentials have claimed that hundreds of thousands of people are dying or will die from the low-level radiation caused by nuclear energy activities? Finally, and most important at this moment, haven't we heard of the Three Mile Island "disaster"?

The answer to all such questions and countless more is, yes, we have indeed. We shall try our best to respond to some of them. For now, we state that the above list of qualities of nuclear energy technology is a carefully worded one. We believe the statements to be true and defensible. They raise for us, and we hope they raise for the reader, the question of just why prevailing views should be so directly contrary.

Because it is probably so that nuclear power in the U.S. can only go forward when the issues of Three Mile Island are in some degree resolved, we cannot avoid the matter here. Not that we desire to avoid it, but there are many subissues involved that cannot be adequately treated in the space we can devote to it. One of the most important facts of TMI is that no one, whether employee or member of the public or the working press, was killed or injured. A person from another planet would hardly be able to extract this simple fact from all the news accounts and the relative attention devoted by the media. Then it surely must be so that what we had was an incipient disaster; we were poised on the apex of a slippery slope at the bottom of which lay the state of Pennsylvania in ruins. This picture also is not substantiated by the facts.

It would appear at this point that we are making light of TMI. We certainly don't wish to do so. We do think that facts should receive as much weight as emotions or imaginations. TMI is surely a disaster in an economic sense. There are certainly many lessons to be learned about reactor operation and regulation and emergency preparedness. It is a fact that these matters are receiving a great deal of attention. A matter that is not receiving adequate attention is the issue of the public perception of risk. To be sure, a part of the matter has received attention: What are the risks to people from nuclear power plants and associated operations? What are the nature and the magnitude of these risks to people living near and far from plants? These studies invariably indicate that the risks are small compared to other risks that society has found acceptable: danger from falling aircraft, danger from transport of explosive materials, for example. For whatever reason, however, a perspective has not been conveyed to the public. The risks of nuclear power are distorted by continually presenting them in isolation, as though they were the single blight in a risk-free world. One can mention the fact of 100 or so coal miners killed each year, or the fact that the federal government pays about a billion dollars per year in black lung indemnification to coal miners and their dependents, and countless risk statistics associated with other activities of our society. It has not proven possible to this point to incorporate an evenhanded evaluation of relative risks into public discussions. This situation has proven a bonanza for those demagogues who have been able to isolate attention on the risks of nuclear activities. The result is that many people have been frightened by the TMI incident. We do not take their fright lightly. We do argue that much of their fright is a result of carefully cultivated distortions of the dangers. It is tempting in this situation to recall Franklin Roosevelt's famous statement that we have nothing to fear but fear itself. Somehow, it doesn't quite fit. We need a certain fear or at least awareness of all the risks around us. It is when these fears become obsessions that poor public decisions can be made. In the land of the frightened, the fearmonger is king. In the U.S., there is no lack of aspirants to the throne.

There certainly is more to the antinuclear movement than fearmongers, however. Its various constituencies form a heterogeneous mixture that defies simple cataloging. It is also difficult to know whether the antinuclear movement is a component of some other movement, such as environmentalism, or whether it is a movement in its own right. Llewellyn King, publisher of the Energy Daily, made a good attempt to capture some of the flavor of this movement when he wrote:

> Those who oppose nuclear power are nominally known as environmentalists and sometimes they are joined by so-called consumerists. They are dedicated, articulate, well-educated, middle-class and upper-middle-class Americans, many of whom learned the art of public protest during the Vietnam war, who believe that the industrial–political axis which has nurtured the development of peaceful atomic energy is cynically foisting a dangerous and unnecessary technology on a gullible American public. In their fight against the technology, they have used the tools which come easily to them as a result of their education and social position: litigation, media manipulation and quasi-scientific propaganda. Additionally, they are now penetrating the political structure, as they have done at the California Energy Commission, and are waging a relentless and committed fight at a grass-roots level against which the nuclear industry is almost powerless. . . .
>
> Those of us who have tried to codify and understand the nature of the nuclear opposition have gradually come to the conclusion that we are dealing with what amounts to a new class in American society; one that is unfettered by fear of shortage,

privation or disaster. It is a class whose traumas have been external and national and not personal. It is a class of men and women who, paradoxically, are seeking to hobble the American economic machine when they themselves are the products of its bounty: well-fed, well-housed, and well-educated—a class that has been brought up in a cocoon of personal well-being, in the comfort of a good home, the security of good schools and the luxury of a university education. Their class perception of American society is of a good thing gone wrong; of venal capitalism astride the stallion of technology violating the wholesomeness of America. . . .

Its political solution, therefore, is the decentralized society; its weapon for capitalistic excess is regulation, not nationalization; its means for decentralization are technological and not political. The cutting-edge of this agenda, turning the United States from an industrialized, centralized society into a decentralized, semi-agrarian nation, is to put a tourniquet around centralized energy in particular nuclear power, and to bring about, through the dispersal of energy sources, a dispersal of decision-making and to return power to the people in small, local units. This agenda, though not new, has been given considerable intellectual legitimacy and rhetorical cohesion by Amory Lovins, an engaging renegade technocrat and techno-social philosopher.

Mr. King suggests that a new class with a new ideology is emerging and that the antinuclear movement is its most evident manifestation, perhaps only because the nuclear industry is the target of opportunity. It came along at the right time. The intrusion of ideology creates a dilemma in the nuclear debate. In discussions of health effects of low-level radiation, the disposal of nuclear waste, etc., it is difficult to know whether it is the *real* issue. It may just be the current issue of a group ideologically opposed to nuclear power. In these essays, we are addressing the person whose real concern is nuclear waste or whose real concern is weapons proliferation. We have neither the ability nor the desire to convert the ideologue.

In the course of the nuclear debate, an example of technological decision-making in the political arena, the public can easily be confused as to the position of the majority of experts in the country. There has developed a format for the presentation of controversial technical matters that can easily lead the public to believe the experts are very much divided. This format is used in hearings held by Congressional or other legislative bodies as well as in debates on television programs. A principal objective, of course, is to present both sides of an issue, to have each side presented by an able advocate, and to give equal time to each position. The hearing is thereby said to be fairly balanced. Even if the total expertise in the country may be 10-to-1 or 100-to-1 in favor of one position, the hearing can leave the listener or viewer with the impression that experts are split right down the middle. Herein lies one of the secrets of success of the antinuclear movement. In countless hearings, they have been able to give their view equal time. The number of their experts are few but the same individuals show up repeatedly in hearing after hearing across the country. It reminds one of the ruse of a military general in antiquity who marched his few troops repeatedly past an opening in a mountain pass to create the illusion of great numbers. We suggest that, were the true experts in the country to be objectively defined and polled, it would prove to be the case that in most instances the pronuclear side is favored by the overwhelming majority. And yet the public sees a 50–50 split. A further irony of this situation is the implicit assumption that because these individuals may be employed by or sponsored by a self-proclaimed public-interest organization they are objective and disinterested. We doubt that many of them could pass any objective test of objectivity.

Certainly, one of the reasons for the position of the U.S. in the world is its possession of technological expertise. If it is now going to be possible for people to "neutralize" that expertise by the simple act of forming a public-interest organization, then it will be interesting to see what the new basis for the position of the U.S. in the world will be. We readily concede that just as war is too important to be left to the generals, and law to the lawyers, technology is too important to be left to the technologists. That hardly makes it reasonable to ignore them as entirely as they are being ignored in the nuclear controversy.

Yet another social phenomenon that the antinuclear movement has found possible to exploit is the attitude, "not in my backyard" or NIMBY. Its operation is simple. Most of us, for example, would not like to live next door to a sewage treatment plant. This fact would hardly justify a finding that sewage treatment plants lack public acceptance. On the other hand, the fact that nuclear facilities have been voted down in a large number of local elections has been given precisely such an interpretation. The sewage treatment plant analogy has its weaknesses, however. It is generally accepted that a community is responsible for its own sewage and a treatment plant has to be located within or near the community. Nuclear power, on the other hand, is a highly centralized activity. A single facility can serve many communities or even the whole nation. Therefore, it is understandable that many single communities should feel that there is no need for them to assume the burden of a nuclear plant of whatever type. Actually, for the nuclear facility involved, no "burden" not associated with the siting of conventional (and often welcome) industrial activity is involved. It is here, however, that the antinuclear groups have done a very effective job of fearmongering. The result is that the local authorities or the local voters often believe that the prudent course is to refuse the facility.

There are countless such cases but a recent one at this writing is the banning of nuclear shipments in the port of Miami. It is argued that whether the shipments are safe or not, the adverse publicity could jeopardize the port's cruise passenger business. This action was promoted by an antinuclear group called the Conchshell Alliance whose announced goal is to choke off a nuclear plant in the vicinity. It is a prime example of the antinuclear legal tacticians at work exploiting the NIMBY attitude. This kind of decision is very easy for the local authorities to make because they pay none of the costs. No one is going to shut off electricity to the port. The costs are very real, however. The piper will come back to be paid. The trouble is that the piper takes his time and is rather indiscriminate about whom he charges.

The people of the country as a whole can be deprived of the benefits of nuclear power by literally thousands of these local NIMBY actions. If it is the case that the concept of a common good is lost or the national good is sacrificed to parochial interests, then nuclear energy has a bleak future indeed. Far more serious, of course, is that the same may hold true for the nation.

The question of whether the waste from nuclear power operations can be disposed of in a manner safe to this and future generations is the question that appears to give most people most concern. The risks are treated in detail in the essay by Bernard Cohen. Here, we shall just briefly touch on the national drama of trying to come to grips with the problem. If the overall issue was not so important, the drama could only qualify as low comedy.

Nuclear wastes have their origin in the physical fact that the products of the energy-releasing fission reaction are a collection of highly radioactive materials, called radioisotopes. A 1-million-kilowatt nuclear electric station produces about 1 ton of these radioisotopes each year. If these isotopes are concentrated, they are referred to as high-level waste. In the course of many operations, some of these radioisotopes become diluted with other materials; the collection is then known as low-level waste (or non-high-level waste), which has less radioactivity but is more voluminous than high-level waste.

It may serve to get things into perspective to point out that radioactive materials are a *natural* phenomenon. They occur in our bodies, in our drinking water, and in the air we breathe. It has been thus since before the first life appeared on this planet. These natural radiation levels are often high compared to standards set by regulatory agencies such as the Environmental Protection Agency or the Nuclear Regulatory Commission. They are *very* high compared to new standards that antinuclear groups are urging these agencies to adopt. It may give some feeling of security to realize that if the EPA or NRC had been around at the formation of the earth, it wouldn't have happened.

Since we have not yet stored or disposed of radioactive wastes in the quantities and for the length of time required, how is it that we are so sure it can be done? There are several good reasons for certainty. One is that, as we've said, nature is itself a vast storehouse of radioactive materials (uranium, thorium, radium, etc.). Quite frankly, Mother Nature has not been all that tidy in tucking them away. In Canada, for example, they explore for commercial uranium by putting radiation-measuring instruments into an airplane and flying over the countryside at a safe speed and height. The instruments show uranium deposits very clearly. It is not at all difficult for nuclear waste disposal technologists to get nuclear waste radioisotopes further out of the human environment than that.

A second reason for believing that nuclear waste can be disposed of is that the major part of its original intensity dies out in a few hundred years. This may seem like a long time, but we have geologic formations that have been in place for millions of years. You have also heard that many of these radioisotopes will last for hundreds of thousands of years and that is true. But here we are dealing with quantities that are comparable to those found in nature and we have a solid basis for claiming we can do better than nature has in retaining them. Remember that radioisotopes with lives of thousands, millions, and billions of years are all found in nature.

The fact that waste from nuclear power operations was a matter to be taken seriously was recognized from the earliest days. Furthermore, because there was concern on the part of Congress that the workers in the field might be underestimating the job, the Joint Committee on Atomic Energy asked the National Science Foundation to review the nuclear waste problem. The NSF issued Publication 519, "Disposal of Radioactive Wastes on Land," in 1957. We refer to this document to stress the point that the nuclear waste problem has not just recently been discovered. Since Publication 519, the nuclear waste problem has been studied by many groups made up of experts—chemists, geologists, physicists, etc.—from outside of the nuclear industry. There are now in the record studies by the American Physical Society, the National Academy of Sciences, the U.S. Geological Survey, The Federal Energy Resources Council, and the Comptroller General of the U.S. General Accounting Office, not to mention a large number of studies in other countries. The consensus that develops in these many diverse studies is amazing and, one would think, convincing. Representative conclusions are: "We believe that nuclear wastes can be disposed of permanently in geological formations in such a way that there is very little prospect of material escaping into the environment" (Ford/MITRE Study, 1977). "The central conclusion that emerges from this report is that institutions can be developed which will provide reasonable assurance of safe management of radioactive waste in the U.S. and elsewhere in the world" (MIT Energy Laboratory Report MIT-E1-76-011, 1976). "The safety analysis shows that radiation doses for large population groups attributable to final storage will be virtually insignificant in that the long term effects on health will be negligible" (Karn-Bransle-Sakerhat, Stockholm, 1977). These studies are all a matter of public record. If we are quoting out of context, anyone is free to so demonstrate.

So, if that's the situation, why aren't we getting on with the task of disposing of nuclear waste? That's where politics enters in a big way. The federal agencies, primarily the AEC, ERDA, and DOE, have mounted a series of projects. Each in turn has encountered political opposition and the project has been halted or modified. With each wave of advance and fall back, it has become more obvious in Washington that here is a political hot potato. It is therefore receiving the Washington Hot Potato Treatment, i.e., throw the problem to a committee, or task force, or whatever you call such things these days. The hotter the potato, the larger the committee must be so that more people can fumble it around. The objective is not to get on with a nuclear waste disposal program but to create the impression that you are until you can gracefully pass the whole mess on to the next administration. The very size and number of committees understandably create in the public mind that here we have a problem that truly has no technical solution. The technical reason why we can tolerate these long delays is that nuclear waste occupies such a small volume that we can go along for years with interim storage. This is expensive and unsatisfactory but it can be done and is the political path of least resistance.

At the heart of the problem is the fact that nuclear waste disposal represents a case of NIMBY writ large. The nation will only require a few locations. (That, of course, is one of the great advantages of nuclear energy; its wastes are small in volume.) So, at any given location, the citizenry can reasonably ask, "Why here?" There are reasonable answers to this question. One answer would presumably relate to the geological formation underlying the region. Another would concern the impact that the entire waste disposal facility and operation would have on the region. There is no reason why it could not be the equal in all respects to other, sought-after, industrial activities. However, that is not the way the discussions are going to go. The antinuclear establishment has focused on nuclear waste operations as the Achilles heel of nuclear power. They will be able to thwart attempts to find and develop suitable waste disposal sites by exploiting the NIMBY attitude. It is, of course, a dream issue for local politicians. "I shall not stand idly by while you make our town the nuclear garbage pit of the nation and the world," is the stuff of great speeches. Perhaps this view is unduly pessimistic but we don't think so. Nuclear waste disposal is a political problem. Its solution will have to be a political one. At the present time there does not appear to be the political will to work for such a solution. The prospect is for an indefinite stalemate, with the antinuclear establishment achieving their objective of making the nuclear waste problem appear to be insoluble.

The question of nuclear waste leads rather naturally to another interesting aspect of the political phenomena surrounding nuclear power, namely, the military aspect. We say it leads naturally because the disposal of wastes arising from military nuclear activities does not differ in any substantive way from the disposal of waste from civilian nuclear power plants. One of the strangest features of the debate on nuclear power has been the absence of the military activities from discussions. Perhaps it is because neither side to the debate feels that it has anything to gain by merging the issues. Perhaps it is because of the secrecy required. Whatever is the reason, the stark fact remains that we have a nuclear military and have had one for a long time. This means that a large number of the Navy's capital ships move on nuclear power. They sail in and out of American ports practically unnoticed except by the crews and their families. Only when one separates the military activities, is it correct to speak of plutonium as a commodity of the future. For years the military has been producing it, processing and transporting it around the nation and around the world in quantities large by any standard. Prior to 1963, a rather large quantity (about 5 tons) was released to the atmosphere during weapons testing. Even NASA managed to dump a very large quantity into the Indian Ocean on one of the Apollo missions. NASA also left even larger quantities of plutonium on the moon to supply power to various experiments. If it is quite as dangerous a material as it has been painted, wouldn't it seem reasonable to expect that we would have some concrete evidence in the form of accidents, injuries, etc. by now? In a more general sense, wouldn't it be more reasonable to carry on the nuclear debate in the context of *all* of our nuclear energy activities rather than isolating civilian nuclear energy as if it were a unique activity?

As a case in point, let us focus for a minute on the plutonium that the U.S. and other nations have as weapons. It amounts to hundreds of tons, quantities large enough so that they will exceed commercial plutonium quantities for many years whatever the future of nuclear power may be. What do we finally do with this plutonium? If it is fired in anger, then the concerns for the risks of nuclear power are indeed misdirected. Let's hope for the best, that we will live to see the weapons dismantled. What then? Either we must dispose of the plutonium as waste or, much more beneficial, convert it into useful energy in power reactors. This option is open only if we have power reactors.

In addition to the question of plutonium is the question of the high-level waste that has built up from our past and ongoing military nuclear activities. These do not differ significantly from the wastes produced by nuclear power plants. In the words of a report of the General Accounting Office to Congress in 1977, "Even if all activities which generate radioactive waste were stopped today, the United States still would be faced with a major radioactive waste disposal problem." Most of that problem, at this time, comes from military residues. The ultimate irony of the present situation has been pointed out by Dr. Pollard of Oak Ridge: we are in danger of rejecting the peaceful atom with all of its benefits while retaining the military atom with all of its risks.

We attempt in these essays to address some of the legitimate concerns and fears that people have about nuclear energy. Much more difficult to evaluate are the political and social currents that are inexorably mixed into the scene. If what we are witnessing is the emergence in the U.S. of a movement of the people toward a simple life, one less demanding on the resources of the earth, then we applaud it and certainly would not wish to impede it in any way. If this hypothetical slackening in material demands and, therefore, energy demands would permit forgoing nuclear power and if, among the alternatives, that is the one chosen by democratic decision-making, then we could hardly object. We do insist that such decisions should not be based on the distortions that are so prevalent in the debate at the present time.

However, the evidence of any popular movement toward voluntary reduced consumption is very weak. And yet many actions and inactions that are occurring today will significantly reduce the future supply of energy, electric and nonelectric. The outcome can be that supplies in the 1980s or 1990s fall short of people's expectations. In the modest, short-run shortages we have experienced to date, we have seen how strong the reactions can be. Recent history indicates that the nation is unlikely to make the best decisions once events are running out of control. Remedial actions tend to be directed to symptoms and not to causes. Moreover, to the extent that technology can be a part of the solutions, most technological options will have closed if we wait until the symptoms of the problem are a daily reality. It is not possible for us to make the dangers of energy shortages as gripping and frightening as antinuclear spokesmen can make radiation and nuclear power plants. We can only try to get the totality of the problems into perspective.

I Energy and Society

The first part of this book is about energy in general. It is introduced by "Energy Futures: A World Study," B. I. Spinrad's summary and impressions of a major study of world energy futures, by the International Institute for Applied Systems Analysis, which was published in 1981 under the title *Energy in a Finite World*. The next contribution is V. P. Kenney and J. W. Lucey's "Energy Demand–Energy Supplies," in which various energy supply systems—fossil fuels, renewable sources, direct use of sunlight and nuclear power—are analyzed with regard to their general properties, their risks, and their capabilities for meeting the demands of an industrial society.

These two papers set the stage for informed thinking about our energy problems. The next three papers in Part I analyze the consequences of attempting to follow one of the more popular notions (in academic circles) of how to deal with them. This is the facile and simplistic concept of relying on conservation alone to solve our energy problems. An extreme example of this thinking is Amory Lovin's philosophy of abandoning "hard" energy and using only "soft" energy. I. A. Forbes and J. C. Turnage undertake a realistic examination of the hard vs. soft controversy in "Exclusive Paths and Difficult Choices." Then, R. E. Uhrig, in "An Energy-Deficient Society," examines the coercive tactics used by opponents of energy supply growth and the consequences of shortages that these tactics are likely to bring about. Finally, A. D. Rossin places the possibility of energy shortage in the proper context, that of a risk, in his paper "Energy Shortages: The Downside Risks."

The argument of Part I is that energy is a valuable commodity in the world today. It is essential for the aspirations of poor people throughout the world—including the U.S.—for a better life through a more productive society. Even assuming that conservation is practiced to the full extent that is economically rational, we are likely to need more of it. A diverse energy supply is more resilient to disruptions and responsive to economic opportunities than any system that restricts particular modes arbitrarily. In the intermediate and long term, nuclear energy must be an important part of that diverse supply. Measures taken arbitrarily to restrict its use represent serious risks. They include the risks of environmental catastrophe from too much coal burning or wood use, the risks of a society impoverished as a result of enforcement of a "soft" energy ideology, or the downside risks of a society regimented to deal with energy shortages. These are much greater risks than any that are associated with nuclear energy.

B.I.S.

1 Energy Futures

A World Study

Bernard I. Spinrad

1. INTRODUCTION

Between 1973 and 1979, a major energy project was conducted at the International Institute for Applied Systems Analysis (IIASA) in Laxenburg, Austria. IIASA is a sort of international think tank, set up on the initiative of countries of the Western and Communist blocs to consider nonpolitical problems that transcend their rivalry. The energy problem certainly fits into this category.

What IIASA calls "Phase I" of the study has been completed and was written up as a thick book, whose title is *Energy in a Finite World: A Global Systems Analysis* (Ballinger, 1981). As befits IIASA's mission, the study must cover energy as a world problem, and as one that takes a long view of the future. Phase I essentially scopes the problem. Its stated aim is to provide a framework for thinking about future world energy supplies and demands. However, Phase I has actually done much more than that. It has accumulated data on resources, explored possible demand scenarios, examined world problems that arise from the energy situation, studied energy system supply opportunities and capabilities, and has not refrained from making broad observations as to how all these aspects fit together.

This author participated as a senior researcher in the final stages of Phase I. What is reported is his interpretation of the results achieved so far.

BERNARD I. SPINRAD • Department of Nuclear Engineering, Iowa State University, Ames, Iowa 50011.

2. THE PROBLEM

Today, the world has a simple and well-defined energy problem. From 1948 to 1973 an industrial structure and a way of life and work had been built on the basis of cheap and abundant oil. This oil was priced low, relative to its value—until 1973. The high prices oil has commanded since then reflect more realistically the value of this depletable resource; but buying oil has put the economies of many of the world's nations under severe stress.

Since 1973, many types of adjustments have been tried: substituting other energy resources for oil; substituting other resources for energy; and doing with less energy and accepting as a consequence lower material standards of living. These adjustments have been only partially successful in reducing oil demand and resulting economic stresses. There are many local and parochial reasons for this, but all of them have a common basis: it takes time to build a new world that is optimized on the basis of expensive rather than cheap oil.

Looming ahead of us is another related problem. The development of the world has resulted in a large increase in population and a separate increase in the material and experiential expectations of people. Energy demand is positively correlated with both these increases. In the years immediately ahead that still means mostly oil demand. But this demand has now become so large that it will not be long before production capacity will be strained. A number of studies have concluded that the expectations for demand will begin to outstrip supply capabilities by the turn of the century—the year 2000. What then?

Beyond that, the real energy problems begin. The core of our technology is based on fossil fuel, and it is essentially irreplaceable for certain major applications. In particular, liquid fuels are vital both for transportation and for dispersed applications of energy needed on an occasional or sporadic basis. Reduced carbon in any form is the basic "ore" for the important petrochemical industries. We must expect a transition from "clean" oil and gas of today to "dirtier" hydrocarbons such as oil shale, coal, and primary oil and gas extracted from difficult locations at increasing ecological cost; and even these hydrocarbons must be considered depletable at the scale of future world use. In the long run, the world will have to rely upon renewable, durable, and virtually infinite energy supplies. The problem, then, is how to manage these two transitions: from clean fossil fuels to dirty ones, and from a fossil fuel base to essentially a nuclear and/or solar base.

All this is what we might call the "general energy problem." But we cannot look at the energy problem in a void. Energy is a vital input to the world and especially to the regions that have not reached an advanced state of industrial development. To support their growing populations in dignity and comfort, these regions must build up material infrastructures and productive capabilities to a more advanced level. Energy conservation can not be allowed to inhibit this sort of meaningful development. We have to look at energy supply and demand in particular, not merely at an aggregate level, but at a level that takes into account the heterogeneity of energy supplies, demands, and aspirations.

3. SOME FACTS AND FIGURES

The unit in which we describe our rate of energy use is a unit of power, the terawatt (TW). This is a standard unit of international science, 10^{12} (1 trillion) watts. Our energy unit is the energy equivalent to 1 TW of power supplied or consumed over a year, and is the terawatt-year (TWa). One TWa is 8.77×10^{12} kilowatt-hours. To indicate average annual energy use, which is again a unit of power, we use the terawatt-year per year (TWa/a). One TWa/a is formally equivalent to a terawatt, but distinguished by being an average value, as stated.

The terawatt is a very large unit, but appropriate for describing energy use rates on a world scale. For example, the world today is using energy at a rate of 8 TWa/a. The U.S. is using energy at a rate of 2.5 TWa/a, of which a little less than 1 TWa/a goes into the production of electricity. The size of the terawatt unit takes some thinking to get used to: if we rate electrical power plants at 1000 MW (megawatts) of electricity, a size that is in the conventional range for large nuclear, fossil, and hydro units, then it takes 1000 such plants to generate at a 1 TW rate. Correspondingly, it would take 1,000,000 plants at the 1 MW level to do the same job.

Energy use in the world is unevenly allocated, particularly between rich and poor regions. Developing countries, with 70% of the world's population, use only 16% of all manufactured energy. This works out to a ratio of about 0.08 between the energy used by an average citizen of a developing country and that used by an average citizen of an advanced country. Yet this is better than it has been in the past. Since 1950, there has been an eightfold increase in energy usage in developing countries and only a threefold increase in advanced countries. A large part of this difference is attributable to differential population growth, but the average citizen of a developing country is still *somewhat* better off now, even on a relative basis.

We may scope the range of future energy demands in the following way: suppose that we define one-fourth of the per capita energy consumption of the advanced world today as a threshold value for a dignified life. Then, to bring developing regions to this level, our energy use would have to rise immediately to almost 11 TWa/a.

Now, suppose that the world's population were to double in the next half-century. This projection, which is due to Keyfitz [1], has been used at IIASA. Following population and fertility trends, more of the growth would be in developing regions than in advanced regions, as seen in Table I. Maintaining the

TABLE I
Population Projection

	Population (millions)		
Year	Advanced regions	Developing regions	World
1975	1115	2831	3946
2000	1400	4682	6082
2030	1562	6414	7976
Population ratio, 2030/1975			
Advanced regions	1.40		
Developing regions	2.27		
World	2.02		

present per capita energy use in advanced regions, and again a per capita use of one-quarter that amount in developing regions, would then lead to a requirement of about 18 TWa/a by 2030.

A more sophisticated systems study of possible future energy demands has been made at IIASA by considering trends toward urbanization and urban energy use [2]. Taking the population projection already exhibited, a largely urbanized world by 2030, reasonable population densities in urban areas and an energy demand density (energy per unit urban area) characteristic of today's conurbations, we arrive at a requirement of about 80 TWa/a for 2030. This is not necessarily a luxury world, but it defines an upper limit for our aspirations.

Another way of defining a target would be to ask what the energy demand might be if all the world's population had, by 2030, energy on the average of today's rate in the advanced regions. This would come to about 45 TWa/a.

These rough considerations are *not* projections, but they serve to describe two sets of range parameters. We may describe the range of study of energy futures by 2030 as falling somewhere between 16 and 80 TWa/a—2 to 10 times today's world consumption. Within this range, we can use the normative descriptions of "high" for about 40 TWa/a and up and "low" for about 20 TWa/a or less.

4. THE SPACE–TIME FRAME OF THE IIASA STUDY

The IIASA Energy Systems Program deals with aspects of global energy supply and demand between now and 2030. During this period the world energy system will enter into a major transition. The time until the year 2000 may be characterized as a "muddling through" period, during which the world will face increasing shortages of clean conventional fuel. Somewhere in the distant future, society will have to meet most of its needs from such inexhaustible resources as solar energy and nuclear power through breeding or fusion. From this perspective, one must consider a transition period, beginning around the year 2000, and see how it might be accommodated smoothly. In planning this, we must consider realistically the long lead-times that are always associated with new economic and social conditions, and new technologies. The year 2030 serves as a limit to the forward projection. It allows a 50-year time horizon, which is about two times the life of a typical major energy installation and more than one human generation. Beyond that point, we can only look for major indicators of what might happen.

While the IIASA study formally extrapolates to 2030, some aspects of the energy problem must be considered for even longer periods, and where necessary this was done. Specifically, the study goes beyond 2030 in considering the proper use of the coal resource, the limitations on fossil fuel use imposed by the CO_2 problem, and the implications of deriving nuclear fuel for burner reactors from very-low-grade uranium ores.

In order to cope with heterogeneity in a manageable way, the world has been divided into seven regions. These are listed in Table II. The regions have been identified, not necessarily on geographical proximity, to reflect similarities in economic infrastructure, life-styles, and energy resources. The population projections used for these regions are shown in Table III, which is a detailed breakdown of the numbers already seen in Table I.

TABLE II
The Seven Geographical World Regions in the Energy Program Scenarios

Region	
I	North America (U.S. and Canada)
II	Soviet Union and eastern Europe
III	Western Europe, Japan, Australia, New Zealand, southern Africa, and Israel
IV	Latin America
V	Africa (except northern and southern Africa), South and Southeast Asia
VI	Middle East and northern Africa (Egypt, Algeria, Libya)
VII	China and centrally planned Asia

TABLE III
Population Projection by Region

Region	Population projection (millions)		
	1975	2000	2030
I	236	284	315
II	363	436	480
III	516	680	767
IV	319	575	797
V	1422	2528	3550
VI	133	249	353
VII	912	1330	1714
	3901	6082	7976

5. CONSTRAINTS

The world imposes constraints on energy use. These are not absolute limits but problems that arise if we try to do too much too quickly. Constraints can arise from every quarter, and they can also be observable even when we don't know the causes.

Among the most easily identifiable constraints are those that are categorized under climatic impact, environmental impact, risks to life and health, and limitations of resources, capital, and personnel. IIASA studied these either singly or, in some cases, in an aggregated way.

With regard to climatic impact of energy systems, a general consensus exists among scholars in this area. Even a very large increase in world energy production and consumption is still a small fraction of global energy flux from solar and geothermal energy inputs and outputs. By itself, it cannot be a significant factor. What can be significant are the catalytic effects that energy use can have in altering the natural flows. Such catalytic effects could arise from concentrating our energy use in the wrong places; from interfering with the flows of air or water that distribute energy around the globe; or from interfering directly with the phenomena that balance global energy inputs and outputs. Of these, the first (overconcentration) does not seem a real problem so long as world industry and population do not all cluster in a few small subregions; the second, interference with atmospheric and oceanic circulation, sets upper limits on the amount of wind power or oceanic power [as with ocean thermal energy conversion (OTEC)] that can be harnessed. The third, interference with the global energy balance, is the most serious and ominous. It is particularly associated with the carbon dioxide problem: the increase in atmospheric CO_2 concentration that is positively correlated with the world consumption of fossil fuels for energy. Atmospheric CO_2 interferes with the radiation of the globe's "waste heat" to space, and too large a concentration would result in significant global warming. The outcome could range from shifts in weather and agricultural patterns among latitudinal bands on the globe to, ultimately, large-scale melting of polar ice-caps and continental flooding. It is considered prudent to limit our consumption of fossil fuels to not much more than it is at present, until we can develop more quantitative information about the problem. Consumption at more than twice the present rate would definitely not be prudent.

All energy systems present risks to people—risks of sickness and of death. On a statistical basis, coal is the riskiest resource in both of these categories. Nevertheless, the risks are low enough so that they can be tolerated if proper attention is paid to them. That is, good design and operation of the mines, transportation systems, and power plants that are associated with using coal can keep these risks down to a level that is very small compared with "natural" or "societal" risks we routinely tolerate. If we shift our attention to catastrophic risks, then large-dam hydroelectricity and nuclear power are the riskiest, hydro being somewhat more so. But the worst catastrophes imaginable—the sudden failure of a dam above major human settlements, or the release and dispersal of most of the radioactive fission products in a large power reactor—are so improbable that they hardly contribute at all to total human risk. Moreover, these catastrophes, as well as other failures of energy systems (such as explosions of stored natural gas, major seepages from nuclear waste deposits, etc.) do not lead to serious impacts from a historical perspective.

Thus, there is no energy technology that is "too risky to use" in the medicobiological sense. Perceived risk, which for example is very high in the case of nuclear fission power, is a social phenomenon and one that can be manipulated. It is therefore important to political decisions, but not a phenomenon we can rely on. We can only assume that the average effects of misperception will be small: that over time perception and reality will roughly coincide. Finally, social risks, such as the risks of diversion of nuclear fuel to make crude atomic bombs, or the risks of sabotage of large dams or gas-storage depots, have specifically *not* figured in IIASA's considerations.

With regard to environmental impacts, a number of the major ones were examined within a descriptive system called WELMM. The acronym stands for water, energy, land, materials, and manpower and describes these inputs to different energy systems. WELMM diagrams, such as those exhibited in Fig. 1, permit intercomparison of different systems on a relative basis. Needless to say, the WELMM pattern also correlates well with the capital requirements of various energy systems.

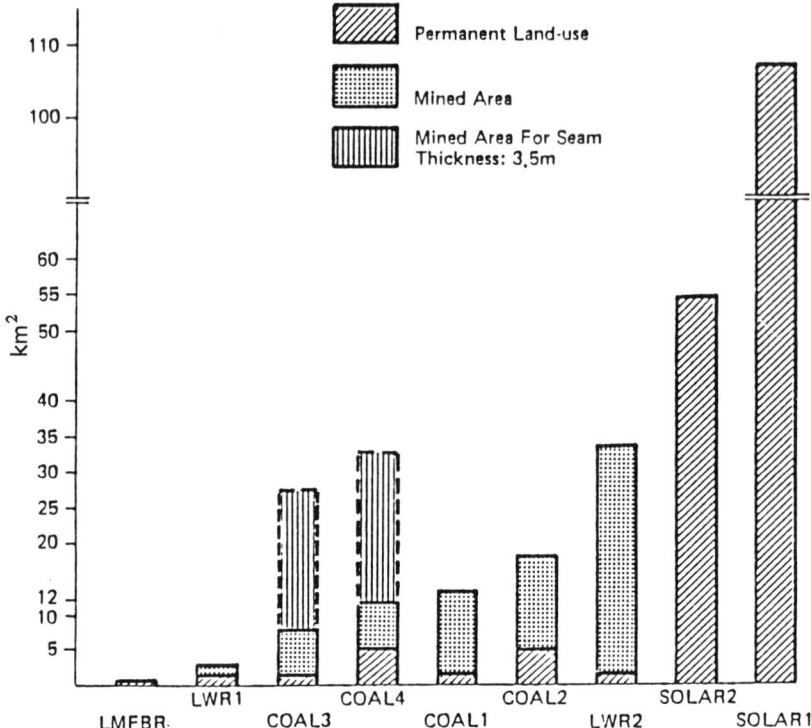

FIGURE 1. Cumulative land requirements for energy chains producing 6.1 TWh of electricity. Coal 1 and Coal 2: U.S. western underground coal, 1.5 m seam thickness, transported by rail (1) or slurry pipeline (2) over 900 km; Coal 3 and Coal 4: U.S. western mined coal, 9.2 m seam thickness, rail transported (3) or pipeline transported (4) over 900 km (the effect of decreased seam thickness is also shown); LWR 1 and LWR 2: Light water reactor, fueled with uranium from conventional present-day ores (1) or from Chattanooga shale (2); LMFBR: Phenix-type fast-breeder reactor, fueled with uranium from Chattanooga shale; Solar 1 and Solar 2: solar thermal energy converters (Solar 1 receives 1500 kwh/m²-year of solar irradiation, operating 1500 hr/year with 6 hr thermal storage, and Solar 2 receives 3000 kwh/m²-year of irradiance, operating 2700 hr/year).

IIASA investigated in considerable depth one further institutional constraint: how long it takes new technologies to penetrate the market [3]. In the case of energy technologies, the characteristic time seems to be 20 to 25 years; that is, once the technology has been introduced, its market share tends over the long run to grow not faster than 4–5% per year. Figure 2 illustrates the past history of fuels and demonstrates how rigidly this has been followed.

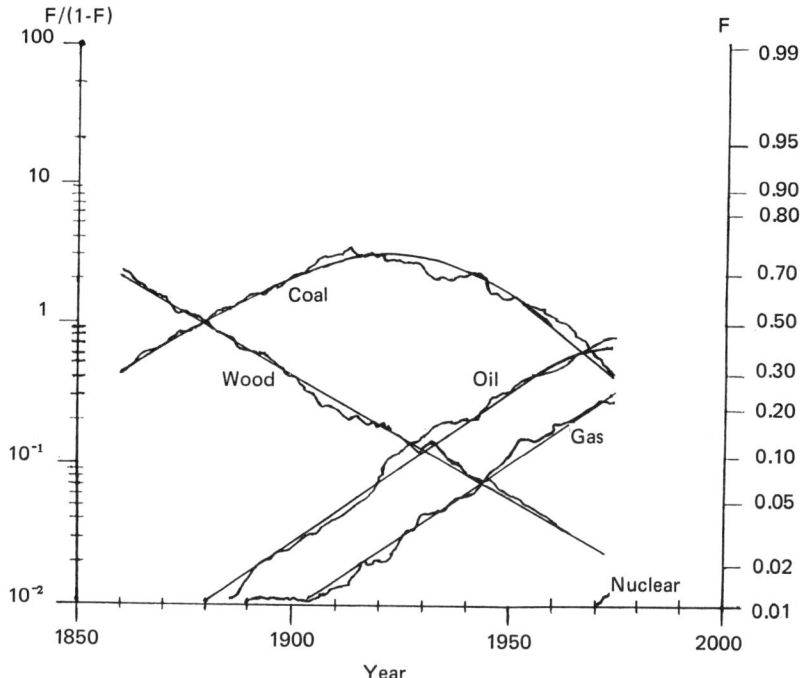

FIGURE 2. The global market share of energy sources. Uneven curves are actual data; smooth curves are fits to a logistic substitution model. F is the market share. The plotting ordinate is F/(1-F) on a logarithmic scale. On that scale, periods of both market penetration and market decline are well represented by straight lines.

6. PRIMARY, SECONDARY, AND FINAL ENERGY

Another aspect of the picture is the necessity to differentiate between energy at various stages of its conversion and use. Figure 3 helps to explain this point. *Primary energy* is the energy recovered from nature: water flowing over a dam, coal freshly mined, oil, gas, uranium. Only rarely can primary energy be used to supply *final energy*—energy that is actually used to supply *energy services*. One of the few forms of primary energy that can actually be used as final energy is natural gas, which is why it is a fuel of preference whenever it is available.

For the most part, primary energy is converted to *secondary energy*. This is defined as an energy form that can be used over a broad spectrum of applications. Examples are electricity, gasoline, natural gas; at lesser convenience

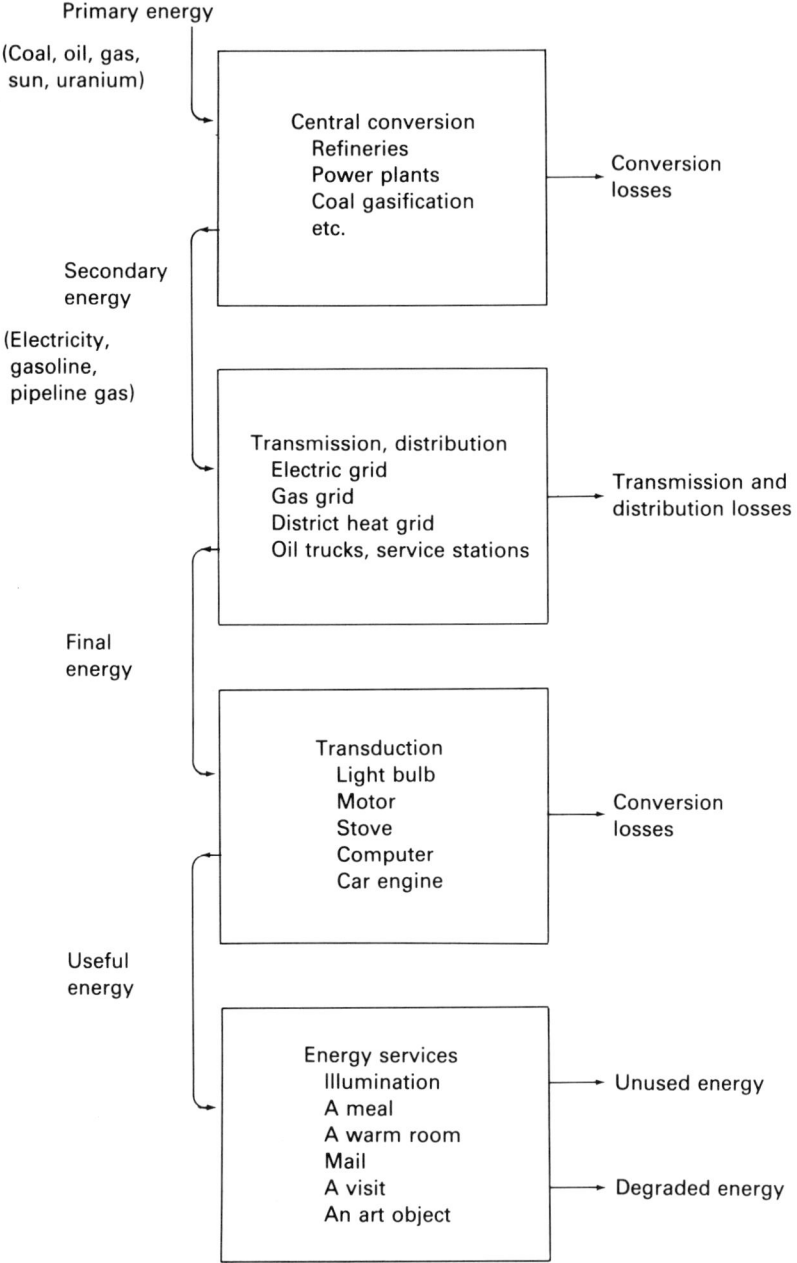

FIGURE 3. Stages in energy utilization.

(which is why they are declining in their market shares) one could also consider charcoal, sorted and graded coal, and cut and split firewood as secondary energy forms. The point is that, in order to apply energy without making undue demands on the consumer, it must be converted into a form that is readily transported and distributed, and that can be used in a variety of devices. The trend has been toward grids, for obvious reasons, specifically electricity, gas, and district heating grids. For convenience of storage, portability, and easy transportability, the trend has been to liquids, of which gasoline and diesel oil are the best examples.

Primary energy is converted to secondary energy in a number of different ways. Central power plants produce electricity and sometimes district heat. Refineries convert petroleum, which is *not* an easy fuel to use at the end point, to more convenient liquids—gasoline, jet fuel, diesel oil, naphtha. When gas is not available, coal conversion plants can make it. Sometimes the conversion plant is the end product of a system, as with nuclear fission energy (for which chemical conversion, isotopic enrichment, and fuel fabrication all precede the power plant).

7. FOSSIL FUELS

We still have a lot of our traditional fuels, coal, oil, and natural gas, and we will be using them, particularly those that are cheap and clean. Then, in the next 50 years we see a transition. Many parts of the globe will have to go to more expensive and dirtier oil and coal products—products for which the effort of extraction and refining becomes very much greater, and for which more WELMM inputs are needed. We will have to use dirtier oil: secondary and then tertiary recovery from existing oil fields; oil from far-offshore wells and from polar fields; heavy oils from Venezuela; and "minable" rather than "drillable" fossil fuels: shale oil, tar-sand oil, and, particularly, coal. By 2030, coal could probably be a principal source of liquid fuels: gasoline, kerosene, and methanol. Coal will also be a principal feedstock to the petrochemical industry after the oil is gone.

From a perspective of a few centuries, we will ultimately have to recover carbon from carbonate rocks or atmospheric CO_2 (as from biomatter) if we continue to exploit the convenience of reduced carbon as a fuel. On these terms, fossil carbon is a resource to be husbanded. Both for this reason, and to avoid the most severe impacts of the CO_2 problem, the energy and hydrogen that are used to make liquid fuels and feedstocks should be derived from nonfossil sources. Nuclear and/or solar power are the best bets for now. Both are effectively "infinite" and neither of them has an intractable emissions problem.

8. NUCLEAR POWER

Central-station nuclear power is in an excellent position to supplement fossil fuel over the next half-century. Light-water- and heavy-water-cooled burner reactors are commercial now and are the cheapest source of electricity in many parts of the world. More advanced systems, producing heat at higher temperature, both for chemical processing and for more efficient conversion of heat to electricity, are ready for commercialization. These are high-temperature, gas-cooled reactors and liquid-metal-cooled fast breeder reactors.

Breeders are particularly important because they require only small amounts of uranium and thorium for fueling. Burners require 100 times more uranium and would use up the good uranium resources very rapidly. We would then be forced into mining so much rock containing small amounts of uranium that we call the scheme "yellow coal." But with breeders, the "good" uranium will last a very long time, and even if we had to use some "yellow coal" several centuries from now, the amount needed would be very small.

IIASA explored several scenarios of reactor deployment that lead to emplacement of 10 TW of nuclear electric capability (17 TWa/a of thermal energy) by 2030. This is an upper limit of what is considered possible from maximum market penetration of nuclear power by a vigorous, accepted world industry. These scenarios require an "investment" of 10 to 15 million tonnes of uranium before a resource plateau is reached; after that, no mining for about 1000 years, and trivial mining after that, could supply all the uranium and thorium needed, virtually indefinitely. The uranium investment is about three times as much as existing estimates of world resources; but it is only about half of what can be estimated when we take into account the fact that most of the world has not been well explored for uranium.

Figure 4 illustrates one such supply scenario. The mix of reactors includes pure burner reactors for such uses as shipping and remote or undersea settlements; light-water reactors for dispersed domestic utility and industrial use; high-temperature reactors for use in chemical and metallurgical industries (including coal hydrogenation), and as efficient electric power sources; and breeders that produce enough nuclear fuel to keep the system operating, while also producing electricity. The system would have some part of its electrical output—10 to 30%, perhaps—dedicated to producing hydrogen by electrolysis.

This scenario is possible. Whether it is achieved rests on public acceptance of nuclear power, and this means that questions of reactor safety and nuclear waste disposal must be resolved. In a technical sense, they are already resolved or resolvable. Indeed, the recent Three Mile Island accident, with its almost trivial consequences and its positive implications for future reactor design and operation, shows primarily a gulf between reality and perception. Let us be clear

1 • Energy Futures

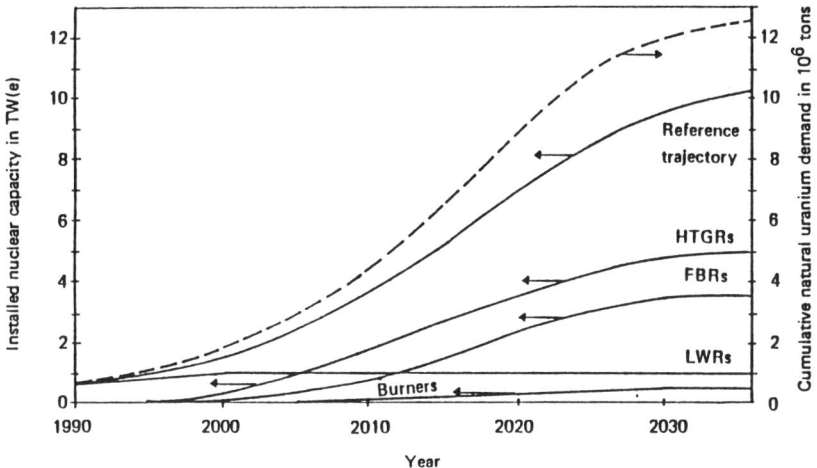

FIGURE 4. A multipurpose nuclear reactor strategy. HTGR, high-temperature reactors for central-station power or process heat; FBR, fast breeder reactors for baseload heat, some process heat; LWR, light-water reactors for isolated power grids or grid network balancing; burners, special-purpose reactors for ship propulsion, space, arctic, and undersea power, etc. (Capacity figures are nominal electrical design ratings.)

as to what TMI means: it means that accidents will happen, but that they will not lead to catastrophes and that we learn from them.

Nevertheless, there will be active nuclear opponents for quite a while. We can only guess at motivations and at why this opposition has been effective. IIASA likens the situation to an iceberg: there is much more underneath that we can't see. A suspicion is that a large part of the hidden iceberg is the fear of nuclear weapons and nuclear war. If this is correct, then these nuclear scenarios are only realizable in a stable, peaceful, and optimistic world. Above all, this world must have a sense of the future—a desire that life be improved, and a faith that this can be done by the cooperative effort of humankind.

A note about fusion: it is clearly a "late starter" in the energy race, and it seems incapable of significant market penetration in the next half-century. If it becomes both technically and economically feasible, so much the better. It is worth a lot of effort. But its significance is as a possible future alternative and not as something we can plan for.

9. CENTRAL-STATION SOLAR

Central-station solar power can provide electricity and hydrogen, just as nuclear power can. It uses much more land and materials, but it is otherwise free of risks, it is very well accepted, and it is truly inexhaustible. Again comparing it with nuclear power, it has larger WELMM impacts and its capital costs are larger. This will doubtless affect the pace at which it can be installed, and its potential could only be realized one or two decades later than nuclear power's. By 2030 it would only be beginning to become a supply option at the terawatt level. However, it is possible that it could by that time be penetrating the market in a major way.

Continuing the comparison with nuclear power, in a competitive situation relative economics will be decisive. (We must observe that money is a surrogate for the ability to make investments of human and physical capital, and must not, therefore, be ignored.) For the fission breeder, cost targets exist, starting from considerable experience with pilot systems. Targets also exist for central solar stations; they are not based as much on experience, but the systems tend to be simpler and the targets therefore are more readily derivable from general industrial experience. If present targets are met for solar power, it will be competitive with burner reactors; if present targets are met for breeder reactors, they will provide still cheaper power. So, we can only speculate which will ultimately capture the larger market.

It is both likely, and prudent for planning purposes, to assume that ultimately the world use *both* central solar and fission power. In visualizing a synergy between these two types of systems, it is then proper to ask where each works best. The ultimate advantage of nuclear power is that it is relatively site-independent. It can be built where the need exists, and would be used for general-industry and utility purposes. Solar energy performs best when its large land requirement is not otherwise needed, and when it is placed where the sunshine is reliable. Essentially, this means desert siting. Large desert installations could be particularly attractive as sources of pipeline hydrogen (from electrolysis) and might take over this function.

Finally, it is amusing to note that the oil-rich Region VI (North Africa and Middle East) is the most favorable part of the world for solar energy exploitation. In a solar energy future, we might continue to see this region as a world energy supplier. The capital acquired by selling oil and gas could very well end up reinvested in solar farms to sustain that region's economy indefinitely.

10. SMALL-SCALE SOLAR AND OTHER RENEWABLES

IIASA is more optimistic about large-scale solar power installations than about small-scale ones. In the latter case, considerable investment burdens are placed on the ultimate consumer, who cannot profit from technical economies of scale nor invest in as sophisticated a manner. Moreover, as the world improves its efficiency of energy consumption in end-use, the market for small-scale solar

1 • Energy Futures

devices shrinks. For example, conservation-minded building design reduces domestic energy demand for hot water and comfort heat in temperate climates, and the demand is in any case small in tropical and equatorial climates. Most of the solar energy used would be in "passive" construction techniques, which is a demand reduction rather than a supply increment.

A possible exception exists for household photovoltaic systems, which could provide power for air conditioning in warm climates. Their development is, however, speculative. The problem is not with photovoltaic cells but with their interconnection and with the power conditioning to provide inexpensive, maintenance-free power at the point of application. They also must be compatible with utility electricity which is needed for periods of maintenance or unusual cloudiness or simply to supply peak power for which household provision is uneconomic.

Therefore, IIASA looked at other renewable energy sources. The significant ones are hydroelectricity, wind, and biomass. There are others, but they are either low in technical potential at a terawatt level or require unknown technological breakthroughs. In almost every case, most of the potential is for centralized rather than dispersed energy collection and distribution.

Table IV summarizes the situation. The world has a very considerable technical potential to generate or capture energy from renewable sources: 20 TW is the total demand of a "low" energy scenario in 2030. Only a fraction of this is likely to be realized, however. For example, a decision to harvest 7.5 TW of

TABLE IV
Estimated Potential of Renewable Energy Supply

Source	Potential (TW)		Constraint
	Technical	Realizable	
Forest and fuel farms	7.5	2.5	Ecological, climatological
Solar panels, soil panels, heat pumps	5.0	1.0	Economic, technological
Hydropower	2.9	1.5	Ecological, social
Wind	3.0	1.0	Economic
OTEC	1.0	0–1	Ecological, climatological, technological
Geothermal	<0.4	0.2	Economic
Organic wastes	0.1	0.1	Balanced
Glacier power	0.1	0	Technological
Tidal	0.04	0	Computational
	20	6.3–7.3	

biomass has major implications for the world ecology; 2.5 TW, which is about the present level of harvesting crops and trees for both food and fiber, is about all we could responsibly plan on for energy. We have already commented on the low level of demand for small-scale solar devices. With hydropower, the problem is that the last 50% of the available resource is remote from consumers, would therefore be very expensive in transmission-line investment and losses, and is likely in any case to be held back in order to preserve wilderness ecologies. With wind, the problem is again matching the resource to the customer; most of the best wind sites are, as with the last bit of hydro, not near customers. When wind-power is examined in that context, it loses its appeal as compared with central solar installations. All in all, *realizable* renewable energy potential is reduced to somewhat less than 8 TW. Even this is, of course, a respectable number, about as much total energy as the world now uses.

Of the realizable energy supplies, about 3 TW should be accessible for direct use at or near the point of collection. Most of this would be biomass from community-sized tree plantations. An appreciable charcoal industry could develop on that basis. Additional "soft" energy would come from community systems generating gas from agricultural residues, from wind-power for irrigation in suitable agricultural settings, and from solar heating in prosperous suburbs. Except in the latter case, the soft use of renewable energy could be a transitional phase as regions develop their industrial sectors; the uses described essentially recapitulate the experience that advanced countries passed through, a century ago.

11. SUPPLY SUMMARY

Semiquantitative estimates of long-term energy supply capabilities are shown in Table V. We have omitted minor sources that make the "renewables" total add up to more than the sum of wood and hydro. We have lots of resources that are either vast, effectively infinite, or truly infinite. We cannot, however, be confident that these resources can be *produced* at a rate commensurate with our needs. Taking into account that we probably could not get more than 17 TWa/a of primary nuclear energy by 2030, production by that year might not be capable of supplying more than 45 TWa/a. When we examine constraints, we might not be able to produce even that much; for example, 8 TW of oil and gas and 10 TW of coal would give us CO_2 emissions at 3–4 times the present rate. So 45 TWa/a, a "high," but not ridiculous aspiration, might not be achievable, and we would have to wait until we could build up a *lot* of hard solar power and even more nuclear.

TABLE V
Summary of Resources, Production Potentials, and Constraints

	Production (TWa/a)	Resource (TWa)	Constraints
Renewables			
Wood	2.5		Economy–environment
Hydro	1–1.5		Economy–environment
Total	6–(14)		Economy–nature
Oil and gas	8–12(?)	1,000	Economy–environment–resources
Coal	10–14(??)	2,000	Society–environment–economy
Nuclear			
Burner	12 for 2020	300	Resources–public acceptance
Breeders	17 by 2030	300,000	Buildup rates–resources–public acceptance
Fusion	2–3 by 2030	300,000	Technology–buildup rates
Solar			
Soft	1–2		Economy–land–infrastructure
Hard	2–3 by 2030		Buildup rates–materials

12. GLOBAL ENERGY SCENARIOS

Another aspect of the IIASA study has been to construct scenarios: projections of the way energy demand and supply might evolve over time. These are not, of course, predictions. Rather, they are attempts to see how some of the variables that might affect the problem interact with each other. Among these variables are population, economic growth, other economic variables such as investment capital, the market penetration process, and social constraints.

The scenarios have used the single population projection that was already noted and variable economic growth targets. They have also taken, as inputs, trends in technological efficiency, and rates of technology deployment were constrained below subjective maxima. Within these limits, energy was to be supplied according to secondary energy demand, at minimum investment and cost.

The process employed a set of models [4], at the heart of which were MEDEE, in which energy demands were prescribed, and MESSAGE, in which supplies were assigned. The models were computerized, but they were not cross-linked by machine. Instead, all inputs and outputs were examined to make sure that credible results were appearing at all steps. Finally, interpretive data on

TABLE VI
Average Growth Rates of GDP (1975–2030) and GDP/Cap

Region	Growth rate of GDP (% per year)		GDP/cap in 10^3 $ 1975		
	High	Low	1975	High 2030	Low 2030
I	2.9	1.7	7.05	25.16	13.24
II	3.9	3.0	2.56	15.95	9.82
III	2.9	1.9	4.26	15.25	8.68
IV	4.4	3.5	1.07	4.48	2.80
V	4.3	3.3	0.24	0.98	0.56
VI	5.1	3.6	1.43	8.27	3.71
VII	4.2	2.6	0.35	1.80	0.79

conservation trends, investment needs, interregional trade analysis, and economic price and elasticity determinations, were also examined to see whether the outputs formed a pattern that was consistent with the inputs. If not, inputs were varied and the system was rerun. It was hoped that linking the pattern-recognition capabilities of humans to the data-processing capabilities of machines would use people and machines optimally. The model set, in addition, assures that the results are internally consistent.

The models were run on a region-by-region basis. The main linkage among regions is international trade, balanced by reassigning energy supplies in each region until energy trade—chiefly oil—was cleared.

The analysis tried to bracket world energy demand by 2030 between a "high" value around 40 TWa/a and a "low" value around 20 TWa/a. The economic figures corresponding to a high-growth and a low-growth world, as so defined, are given in Table VI. It may be seen that the economic productivities of Regions IV and VI, judged by GDP/cap, reach "advanced" levels by 2030 in both scenarios, but that Regions V and VII remain underdeveloped, particularly in the "low" scenario. The generally more rapid economic growth rates of the developing regions are counteracted by their population growth, again particularly in the "low" scenario.

13. NATURE OF ENERGY DEMAND

To appreciate the global scenarios of primary energy consumption, one needs to look into the nature of energy demand in different world regions. Primary energy is only a means and not an end in itself; it only gives the resource consumption for meeting the ultimate objective of energy services. The real measure of value is useful and final energy, which is much less than primary energy because of conversion and transmission losses. Even the ratio of secondary to primary energy will change as the secondary form is varied: compare direct burning of coal to conversion to gas and to conversion to electricity.

Table VII shows the sectoral shares of *final* energy demand. This has implications for the form of secondary energy: transportation is postulated to use almost entirely liquids whereas buildings are strong consumers of electricity.

The transportation activities in the developing regions have a relatively high share of the final energy demand throughout. This is mostly freight transport, accompanying growth in industrial output, but it also includes increased personal travel, which is far below the saturation mark. On the other hand, the final energy share of buildings is much higher in the developed regions, mainly on account of their large space- and water-heating requirements. The low share of buildings in the developing regions is due both to their considerable dependence on the renewable (presently noncommercial) fuels, which are estimated to meet about 45% of the useful energy requirements of the household/service sector, and to modest space heat requirements and some saturation effects.

Table VIII summarizes the projected final energy demand and the shares of electricity and liquid fuels in three regional groups chosen to highlight regional differences. The electricity share in final energy demand increases in all regions whereas the demand of liquid fuels shows distinctly different evolution in the three groups. The share of liquid fuels in final energy decreases in Regions I and III. These are regions that are heavy consumers of imported liquids, and reflect policies designed to restrict oil use to such premium applications as

TABLE VII
Percent Shares of Sectors in Final Energy Demand[a,b]

Region	1975			High 2030			Low 2030		
	T	I	B	T	I	B	T	I	B
I	29	40	31	28	52	20	26	50	24
II	18	59	23	19	64	17	19	63	18
III	20	51	29	25	52	23	23	49	28
IV	41	47	12	44	46	10	44	43	13
V	30	59	11	29	62	9	32	55	13
VI	39	47	14	37	52	11	36	50	14

[a] Abbreviations: T, transport; I, industry; B, buildings.
[b] From *Energy in a Finite World: A Global Systems Analysis,* Table 16-8, p. 487, International Institute for Applied Systems Analysis, Ballinger, Cambridge (1981).

TABLE VIII
Shares of Electricity and Liquid Fuel in Final Energy[a,b]

		1975	High 2030	Low 2030
I + III	Final energy (TWa)	3.5	8.0	5.6
	% electricity	12	21	21
	% liquids	56	46	46
II	Final energy (TWa)	1.3	3.7	2.6
	% electricity	10	23	20
	% liquids	34	32	30
IV–VII	Final energy (TWa)	1.0	10.6	6.0
	% electricity	6	13	13
	% liquids	48	50	53
World	Final energy (TWa)	5.7	22.8	14.6
	% electricity	11	17	17
	% liquids	50	45	46
Share of motor fuel and feedstocks in liquid fuel demand (%)		64	92	88

[a] Figures are subject to revision for Regions II and VI.
[b] Projections for Regions I to VI are based on MEDEE whereas those for Region VII are based on SIMCRED.

transportation and feedstocks. In other regions, the share increases: rapidly in Region II and more slowly elsewhere. Globally, demand for liquid fuels continues to increase at about the same rate as the total final energy demand, even though 84 to 90% of its use is restricted to premium applications in 2030, as against 58% in 1975. Thus, liquid fuels will continue to dominate the world energy market at least for the next several decades.

14. PRIMARY DEMAND

As already noted, there is a worldwide trend to mediate between primary energy supplies and final energy demand through the use of secondary energy carriers. The reason, convenience of final energy use, cannot be circumvented or denied. However, the price of this convenience is the dissipation of primary energy as waste or losses.

Countervailing this trend is the process of technological improvement, which increases the efficiency of conversion steps and of applications of energy. And finally, there are structural differences between the coupling of energy and economic growth. For a given absolute growth of GDP, advanced regions, emphasizing service activities, need less energy; developing regions, emphasizing a buildup of material infrastructure, use more. This last point is illustrated in Fig. 5, in which this "Final-Energy-GDP" coefficient is plotted against time.

The breakdown of primary energy demand, by regions, is given in Table IX. The data of Table IX are preliminary but show a demand of 37.1 TWa/a in the "high" scenario and 23.9 TWa/a in the "low."

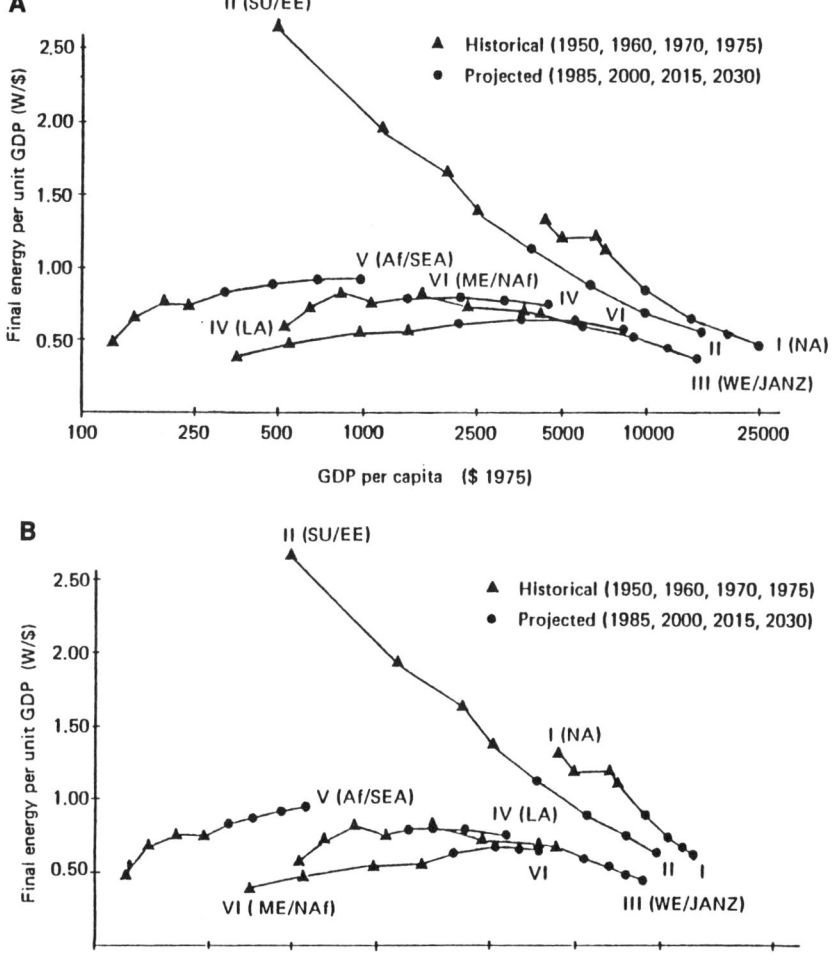

FIGURE 5. Time series of energy intensiveness, by region. (A) high scenario; (B) low scenario. (See Figure 7 for region identifications.)

TABLE IX
Primary Energy Demand by Regions, in TWa/a

Region	1975	Low 2030	High 2030
I	2.66	4.37	6.02
II	1.91	6.50	9.02
III	2.26	4.54	7.14
IV	0.34	2.31	3.68
V	0.33	2.66	4.65
VI	0.13	1.25	2.11
VII	0.46	2.29	4.46
	8.09	23.92	37.08

15. SUPPLYING THE DEMAND

Table X gives the model-constructed breakdown of supply allocations to meet the regional demands. This allocation is listed by type rather than by region. Figure 6 presents the same data as market shares. Of importance to note are:

- The large share of nuclear (for the electrical supply sector)
- The growing share of synthetic fuels, from coal, in the liquid supply sector

TABLE X
Two Supply Scenarios, Global Primary Energy, 1975–2030 (TW)

Primary source	1975	High 2030	Low 2030
Oil	3.61	6.96	4.85
Gas	1.51	5.45	3.00
Coal	2.26	12.28	7.92
Nuclear 1	0.50	3.31	1.40
Nuclear 2	0	6.46	4.53
Hydro	0.12	1.40	1.40
Solar	0	0.41	0.29
Other	0.14	0.81	0.51
	8.14	37.08	23.90

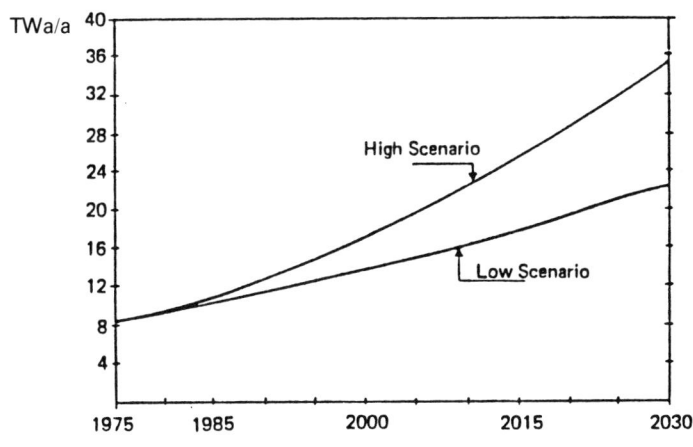
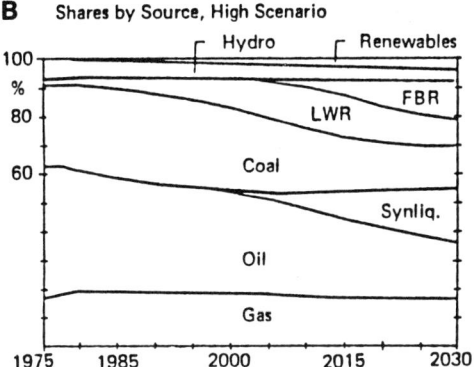
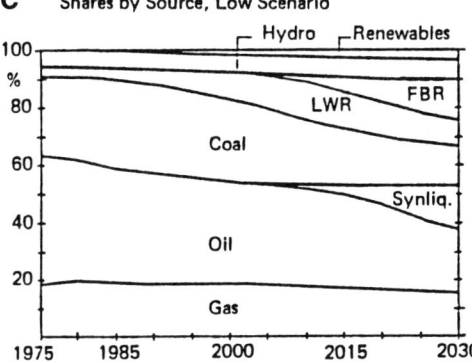

FIGURE 6. Global primary energy (1975–2030). LWR, light-water reactors; FBR, fast breeder reactors; Synliq., synthetic liquids.

TABLE XI
Production Costs of Liquid Fuel and Electricity

	Fuel cost	Final product cost ($/kWa)
Liquid fuel		
Crude of refinery	12 $/boe[a]	74
Coal liquefaction	20–25 $/boe[a]	119–147
	25–50 $/tce[b]	109–154
Electricity generation		
Coal-fired	25–50 $/tce[b]	151–235
Oil/gas-fired	12–25 $/boe[a]	214–431
Nuclear	80–130 $/kgU[c]	138–150
Hydro	0	70–112
Solar	0	335

[a] boe = barrels of oil
[b] tce = tons of coal equivalent
[c] kgU = kilograms of uranium

Both of these trends are paced by developments in advanced regions, where maximum efforts to move toward self-sufficiency in primary energy supply were assumed.

These allocations were arrived at by cost minimization. Within each region, the lowest cost supply option, for either liquids or electricity, was chosen to the extent that the region has a particular resource. The costs chosen (in 1975 U.S. dollars) are given in Table XI. Scarcities of resources in various regions were made up by a continued oil trade. This is presented in Fig. 7. As can be seen, even in 2030 oil from Region VI will be the world's swing fuel. There would be more of it available, and therefore more of it traded, in the "low" as compared with the "high" scenario. Note that it is possible, and has been assumed, that Region I (U.S. and Canada) no longer will need imported oil by that time.

16. NUCLEAR AND SOLAR POWER

Returning to Table VII, we can now note an inconsistency with our climatic constraints. The rule of thumb is to assess CO_2 emission rates by taking energy supply rates, multiplying coal by 1, oil by 1/2, and natural gas by 1/3. On this basis, the emission rate of CO_2 by 2030 would be 2.5 the 1975 value for the "low" scenario and 3.8 times for the "high" scenario. Unless new results are more reassuring, 2.5 times the present rate is likely to be imprudent, and 3.8 times most probably *is* imprudent. We should not count on sustaining these rates for long.

The resources picture is more reassuring. We may look at the scenarios and ask what fossil fuel has been consumed and what is left for future use in 2030.

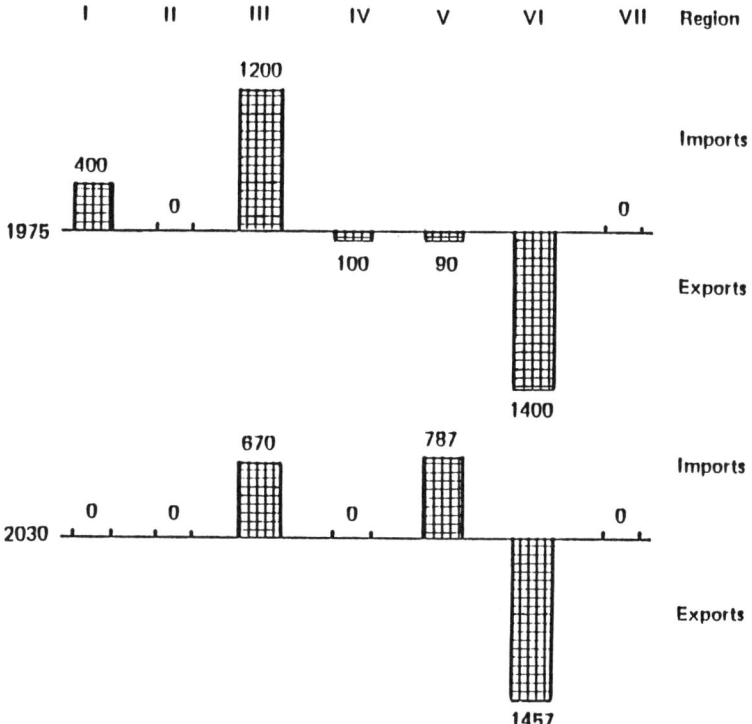

FIGURE 7. Oil trading between regions (high scenario). Regions: I, U.S. and Canada; II, Soviet Union and Eastern Europe; III, Western Europe, Japan, Australia, New Zealand, South Africa, Israel, etc. (advanced private-enterprise economies outside North America); IV, South and Central America and the Caribbean countries; V, Central Africa, South and Southeast Asia (non-Communist developing countries in the Old World); VI, Middle East and North Africa; VII, China, Mongolia, North Korea, and Indochina.

As shown in Table XII, even in the "high" scenario the world will have consumed only 69% of the conventional oil, 40% of the conventional gas, and 72% of the category I (cheaper and more accessible) coal. The unconventional gas will not be touched, and only 2% of category II coal and 1% of unconventional oil will have been consumed. However, coal conversion will be a big industry. Thus, of about 3000 TWa of total fossil fuel resources, more than 2000 TWa will still be available for use after 2030. The transition to dirty fuels will therefore just be beginning then, and real depletion of fossil fuels will occur only later in the century.

Both because of higher levels of demand and increasing lead times to develop new sources of supply, the resilience of the energy supply system will be limited.

TABLE XII
Cumulative Uses of Fossil Fuels, 1975–2030, High Scenario

	Total resource available (TWa)	Total consumed TWa	%
Oil			
Conventional (categories I + II)	473	325	69
Unconventional (category III)	365	4	1
Natural gas			
Conventional (categories I + II)	400	161	40
Unconventional	120	0	
Coal			
Category I	562	402	72
Category II	1024	24	2

Diversification of supply alternatives, particularly by invoking more nuclear and solar power, can cover part of the risk of future fuel embargoes and/or societal and environmental pressures against large-scale exploitation of the "dirty" fossil fuel resources. Could we have more nuclear and solar power? What are the problems?

With nuclear power the answer is straightforward. The "high" scenario only contemplates 12 TWa/a of nuclear power by 2030, while the supply studies suggests that 17 TWa/a would be achievable if breeder reactors were given a green light. Under these circumstances, a further buildup of nuclear power could continue after 2030. Thus, a technical capability exists. Yet, from today's perspective, it is unlikely that even 12 TWa/a could be deployed. There is too much social opposition.

One must try to understand this opposition, but it is difficult to find a logical basis. Neither Brown's Ferry nor Three Mile Island were catastrophes, or even near misses: very simply, hardly anyone was hurt, and the chances of anyone being hurt remained extraordinarily low throughout the accidents. The worldwide fear must come from other sources. Similarly, repeated conceptualizations of the risks that might arise from failures of nuclear waste repositories, even of failures we consider virtually incredible, show that these risks are small. There is fear of low-level radiation, sponsored by a coterie of inventors of bizarre scientific concepts, yet these risks are also small. There remains the possibility that all these fears stem from displaced fears of nuclear war. If this is so, we can only suggest that a progressive and benign technology is hostage to worldwide desire to control war and the causes of war, and ask the world of politics to take heed of this desire.

For solar power, there are, of course, technical difficulties, particularly WELMM requirements and capital cost. It may never be cheaper than nuclear power. Yet, it is a doubly important insurance policy for the world. As a companion to nuclear power, it can help to relieve a variety of burdens caused by overexploitation of fossil fuels; if necessary, it could be called upon to take a larger role, particularly after 2030. But unless we start seriously to build a solar industry very soon, even at costs that run against conventional economics, it may be impossible to deploy it on a large scale by 2030. The world system always slows things down. Behind such a capital and materials intensive technology as solar power are needed a host of supply industries, capable of expansion as needed, and these essentially do not exist.

17. CONCLUSIONS AND OBSERVATIONS

There are a number of questions that have been identified as important to any energy study:

- How much energy do we really need?
- Do we have enough resources?
- Should they be used now or preserved?
- Do we have the appropriate matching technologies?
- If so, how much do they cost, and how soon can they be deployed?
- Are there side effects of new, insufficiently tested technologies? If so, are they serious? At what scale?
- What about side effects of known technologies, particularly when their scale of deployment is expanded?
- Who can, and who should, develop new alternatives? How should they be deployed?

The questions are the same, whether they are posed in Saudi Arabia, Bangladesh, Japan, Poland, Argentina, Canada, or China. The answers will vary, of course, with the perceptions of the countries and regions and according to personal ideologies and values. These seem to reduce to value issues concerning four points:

- Carrying capacity
- Interaction
- Equity
- Legacy

Carrying capacity refers to global, regional, national, or local capability to support human society. It has the extensive factor of population and the intensive factor of individual activity, so that a small, mobile, and rich population can saturate the capacity of an area as much as a larger, more stationary, and poorer population can. Limits on carrying capacity are set by resources and ability to make productive use of resources, but also by effluents and ability to control them. Advancing human knowledge tends to increase carrying capacity by improving production and control capabilities.

Interaction refers to the degree to which goods, people, and ideas flow from one place to another. A highly interactive world is characterized by an emphasis on trade, on shifting personal relationships, and on rapid dissemination of, and response to, ideas. A weakly interactive world emphasizes self-sufficiency, stable relationships, and traditional thought. Technology favors interaction, but at any given time a saturation point may be reached, such that overspecialization results from trade, shifting relationships lead to personal confusion, and the kaleidoscope of ideas leads to fad and fashion.

Equity refers, of course, to distributional values. An equitable society is one in which, by and large, everybody has access, to an equal degree, to the same opportunities and amenities. It is also a society in which duties and responsibilities are shared equally. An inequitable one assigns both amenities and responsibilities in different shares for different people. Perfect equity of this sort is virtually unachievable, because the abilities of people to carry out responsibilities are variable and because tastes in amenity are likewise variable. Therefore, compensation for greater responsibility is usually made through more access to amenity; and so long as the measures of responsibility and amenity are considered to be "appropriate," this much inequity is considered equitable.

Finally, legacy refers to a time sense. A society that consumes its resources without making something out of them is wasteful and has a low sense of legacy. A society that pollutes its environment is also low in legacy: the pollution is a burden on the future rather than an inheritance. A society of builders and planters has a high sense of legacy. Knowledge is probably the highest legacy of all; a society that learns and communicates its learning is doing the maximum for its posterity. A society that simply preserves is neutral. Technology is neutral to legacy, in the sense that it can be directed toward consumption and pollution or toward construction and growth of carrying capacity.

How do different societies, or societal visions, rate on these scales?

Traditional, low-technology societies are low in carrying capacity and interaction, and neutral in legacy. They are highly variable in equity, ranging from free bands of hunters or farmers to highly hierarchical tribes and fiefdoms.

Today's advanced, high-technology societies are high in carrying capacity and so high in interaction that saturation effects are beginning to appear. They are moderate in equity; a trend toward increased equity is correlated with higher carrying capacity, but there is no demonstrable cause–effect relation, and there are examples of egregious inequity. They are variable in legacy: construction and advancement of knowledge go hand in hand with resource consumption and pollution.

One view of the world considers that the saturation of interactions, the imperfect equity, and the variable legacy of today's advanced societies are inevitable and structural. D. Meadows and the Club of Rome, and separately A. Lovins, place the fault at attempts to improve carrying capacity by advanced technology. They argue that this can only be done by jeopardizing legacy through pollution and resource consumption, by oversaturating interaction, and by moving toward inequity. In consequence, they argue that world problems can only be solved by deliberately reducing the role of advanced technology and accepting the reductions of carrying capacity that go with it. In particular, they see centralized "hard" energy systems as employing prototypically "bad" technology: technology that degrades both equity and legacy. Their world of "soft" energy systems is one that has an admittedly lower carrying capacity than today's advanced societies, and one that is less interactive. It is claimed that such a world would be optimal in interaction and high in equity and legacy.

The IIASA Energy Project has taken a different view of the world problems. A world of lower carrying capacity is considered unacceptable, because there are already many people, there will be more, and their material aspirations are high. Further, it is impossible to correlate inequity with high technology. If any correlation exists, it may be the other way; but equity is essentially a social problem at every level of carrying capacity and technology and not one that has a technical cause or a technical fix. So, while one must favor an equitable world, one must be humble enough to admit that it is metatechnological. Finally, the soft society, as a less productive one, has reduced potential for helping the development of poor regions. One is thus led toward futures that have high carrying capacity and a high sense of legacy. This means using technology, but differentiating between good (constructive, nonpolluting) and bad (consumptive, polluting). As to saturation of interactions, one can hope that a highly productive world will give people the time and knowledge to work problems out.

These considerations have led to emphasizing the role of resource *investments* in the evolving world. In the energy field, this means building an infrastructure that can be used by future generations to reduce or eliminate the continued shoveling in of resources and labor; an energy utility, in the best sense of the word (i.e., something maximally useable), that is abundant and nonpolluting.

There are a variety of systems that exemplify such investments in energy facilities:

- Breeder reactors (and their associated reprocessing plants and waste disposal facilities)

- Central solar plants
- Coal conversion plants
- Suitably selected hydroelectric facilities
- Geothermal power plants
- Wind energy farms

In addition, there are a host of research ideas that are very worth pursuing and/or developing further. Among them are:

- Fusion
- Photovoltaic power sources
- Storage batteries, fuel cells, flywheel systems, and other schemes for storing high-quality energy
- Electrolysis and thermolysis of water

Research, development, economic rationalization, and commercialization are all part of the investment process. It is expensive, but it will be more expensive in the long run not to have the energy when it is needed, or to be forced further along the road of consuming resources and polluting the world. A society that takes the notion of legacy seriously simply cannot discount the future at 10% per annum. And a responsible society cannot discard a capability because it is ideologically defined as "hard" or "soft"; the list of desirables includes systems of all sizes and types to be used as appropriate instruments of society, rather than as tools to shape it.

In summary, the IIASA studies indicate that we have plenty of energy resources. If we follow the path of least resistance, we can shoot for a "low"-energy world or "high"-energy world. Neither is very desirable. The low-energy world doesn't develop the economies of poor regions very well—it just drags the current picture further along. The high-energy world does marginally better on economic development, but it pulls us down the road of depleting resources and increasing emissions. When we look at the solutions proposed in "soft" energy worlds, we find them not very pertinent to existing human aspirations. We are led to the notion of energy investments, examples of which have just been listed. They are achievable, with perhaps some sacrifice of consumption now. It would be worth the price.

ACKNOWLEDGMENTS. The energy project at IIASA is under the general direction of Professor Wolf Häfele, and his contributions to it have been so generous and comprehensive that any description of the project, this paper included, must acknowledge, with gratitude, his involvement. In addition, this paper contains many ideas derived from fruitful discussion with Drs. Wolfgang Sassin and Arshad M. Khan. The paper itself has been adapted from: W. Häfele, A. M.

Khan, and H. H. Rogner, Geographic diversity in energy supply and demand, First Arab Energy Conference, Abu Dhabi, March 4–8, 1979; and W. Häfele, A. M. Khan, and B. I. Spinrad, Nuclear power in the developing world, Forum on Third World Energy Strategies and the Role of Industrialized Countries, Royal Institution of Great Britain, London, June 20–22, 1979.

REFERENCES

1. N. Keyfitz, Population of the World and Its Regions 1975–2050, Report WP-77-7, International Institute for Applied Systems Analysis, Laxenburg, Austria (1977).
2. W. Häfele and W. Sassins. *Annual Reviews of Energy, 1977*, pp. 1–30 (see in particular pp. 17–20), M. Hollander, ed., Annual Reviews Inc., Palo Alto (1977).
3. C. Marchetti, N. Nakicenovic, V. Peterka, and F. Fleck, The Dynamics of Energy Systems and the Logistic Substitution Model, Reports AR-78-1A, -1B, -1C, prepared for the Stiftung Volkswagenwerk, International Institute for Applied Systems Analysis, Laxenburg, Austria (1978).
4. P. Basile, The IIASA Set of Energy Models: Its Design and Application, Report RR-80-31, International Institute for Applied Systems Analysis, Laxenburg, Austria (1977).

2 Energy Demand–Energy Supplies

V. P. Kenney and J. W. Lucey

Just a few years after the U.S. celebrated its first centennial it passed another milestone. In about 1885, coal replaced wood as the nation's primary energy source. Wood, properly managed, is a renewable resource. Coal is not. Nor are the other energy sources developed and put to use in the last 100 years. Coal, natural gas, petroleum, uranium, thorium, and deuterium are all resources that can be used but once.

Coal has since been replaced by oil as the country's major energy source. But oil is a very limited resource, soon to be replaced by another. Whether that replacement is uranium, deuterium, the sun, or revitalized coal is still to be determined.

The growth in energy consumption since 1850 has averaged about 3 to 3.5% per year [1] (Fig. 1). This growth has not been smooth and steady because energy consumption is closely related to the health of the national economy. Energy consumption dropped during the economic depression in the early 1920s, during the Great Depression of the early 1930s, and during the recession tied to the Arab oil embargo in the 1970s. On the other hand, energy consumption increased very rapidly during the two World Wars as industry worked round the clock to sustain the war effort.

There are two components to the growth in energy consumption. One is the simple fact that there are more people each year, and they each need energy to support their standard of living. The second component is the fact that our standard of living is constantly changing, yearly requiring more energy per person.

A few simple examples demonstrate how per capita energy consumption

---

V. P. KENNEY • Department of Physics, University of Notre Dame, Notre Dame, Indiana 46556.   J. W. LUCEY • Department of Aerospace and Mechanical Engineering, University of Notre Dame, Notre Dame, Indiana 46556.

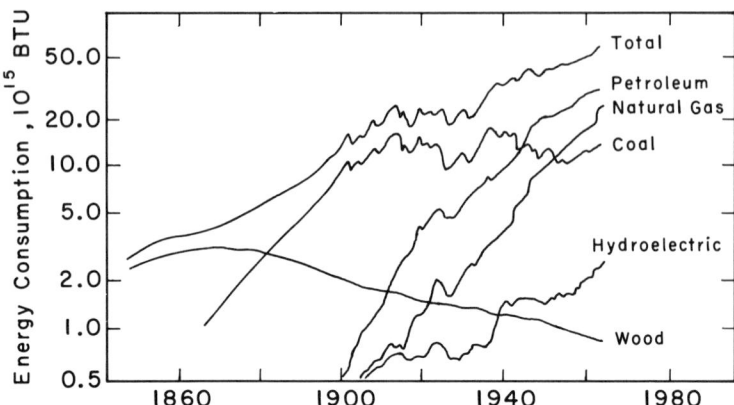

FIGURE 1. U.S. energy consumption since 1850. Note increases in consumption in 1918 and 1941–1945 during wartime and decreases during recessions of 1914 and 1929.

has increased. There are some of us who can remember a time when a family's evening entertainment was provided by the radio. In the late 1940s and early 1950s radio was replaced by television, and now the original black and white televisions have been replaced by color models. Each of these entertainment centers, radio, black and white television, color television, requires more energy to operate than the one that preceded it.

The same progression appears in the kitchen, from ice box (which *used* ice to cool rather than making ice), to electric refrigerators, to refrigerator freezers, to the "frost-free" refrigerator freezer units available today.

Just as the items in the home have come to require more energy, industrial processes have also become more energy-intensive. The shift from labor-intensive to energy-intensive processes is apparent in agriculture where the transition from family to corporate farming has produced greater yields with fewer workers.

1. ENERGY DEMAND

Energy demand has historically been stimulated by the ready availability of cheap energy. Steadily falling constant-dollar prices of energy of the U.S. over the past century have been matched by steadily increasing energy consumption, right up until the sudden fourfold increase in OPEC oil prices in 1973–74. During the preceding decades 1950–1970 the real cost of energy consumed in the U.S. decreased by 28%, and our energy consumption doubled [2]. Postwar prosperity is particularly evident in the living standard of the American worker. It is not unusual for a Detroit factory worker to own a comparatively spacious single-family suburban home, complete with lawn, garden, appliances, and two-car garage—a living standard reserved for the managerial class in western European society, where energy costs have been higher. Meanwhile, the "have-nots" among our people, blacks and Chicanos especially, present an enormous "pent-up" demand for still greater per capita consumption of energy. These are the people who are reminded each day by television that they do not yet share the prosperity of their more fortunate, or more solidly entrenched, blue-collar brethren.

The correlation between gross national product and per capita energy consumption for various nations is shown in Fig. 2 [3]. The U.S., with 6% of the world's population, accounts for 32% of the world's energy consumption, and 31% of its GNP. In the two-decade period 1950–1970, during which domestic energy consumption grew at an average annual rate of 3.5%, the GNP showed a very similar growth, averaging 3.6% per year. Power consumption and GNP increase with time (since 1900) are compared in Fig. 3 [3]. The upward trend in energy consumption is interrupted by periods of economic depression and joblessness, most notably the Great Depression of the 1930s, the periods of

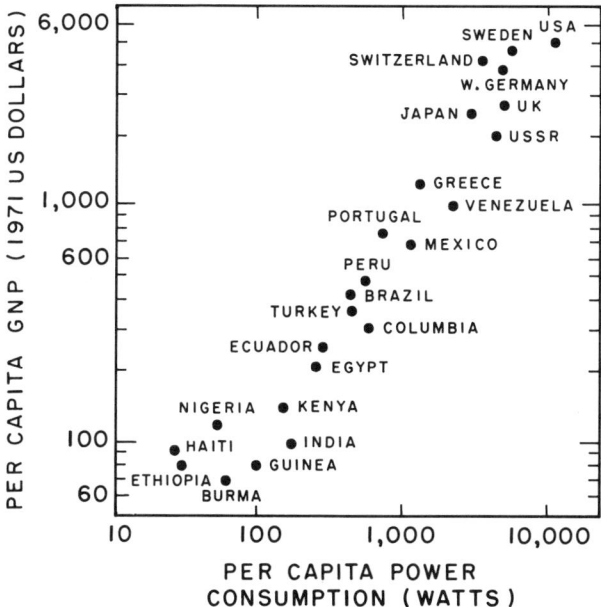

FIGURE 2. Per capita GNP and power consumption from fossil fuels, nuclear fuels, and hydropower in 1971.

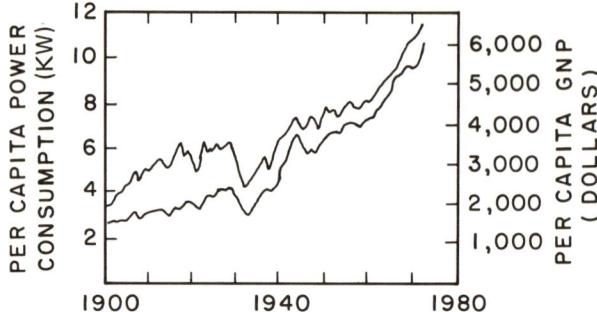

FIGURE 3. Per capita average power consumption (top curve) and per capita gross national product (1973 dollars; lower curve) in the U.S.

economic adjustment following both World Wars, and minor recessions since. Decreases in per capita energy consumption [4] are matched by drops in per capita GNP. The correlation between energy consumption and employment is striking, as factory layoffs during power crises of the 1976–77 and 1977–78 winters remind us.

Even if per capita energy consumption were somehow to be held constant, the *total* energy consumed in the U.S. would continue to grow via population increase. The average rate of population growth in the U.S. since 1900 has been 1.4% per year, so that the population doubled in approximately 50 years. From 1960 to 1974, however, the population growth rate decreased significantly, to 1.1% per year, and the popularity of "abortion on demand" may further slow population growth in future years. With falling birthrates, immigration becomes the dominant factor in population growth of the U.S. The pattern of immigration has changed since the 1950s, from central–southern European predominance to immigration largely from Mexico and Latin America, where per capita energy consumption is 10–20 times lower than in the U.S. and where the standards of living and employment are correspondingly low. Quotas for legal immigration have 5- to 10-year waiting lists, and the Immigration Service estimates the number of illegal immigrants presently residing in the U.S. at 6–8 million [5].

Can the patterns of energy consumption established over a century, or longer, be reversed within the time span of a few years? Should they? If they are, what social costs may we expect? Ethical values overlap the technological, economic, and political. According to classical "free market" doctrine, the imbalance between demand and supply exists only because of government interference in the pricing of energy, and this artificial interference should be modified or eliminated. The thesis of unrestricted capitalism is widely opposed on the grounds that demand pricing of energy places a disproportionate burden on the poor, for whom energy costs are a larger fraction of their family budgets. Orthodox Marxist ideology, on the other hand, sees the energy issue as one more chapter in the class struggle between the "oppressed" and their "oppressors." Marxist economists [6] see decentralization of energy production and conversion as the solution. Those who follow this line have been criticized as being more concerned with the elimination of industry and utility dominance of the economy than with solutions to the energy crisis, or the desires of the people directly affected. Benjamin Hooks [7], former executive director of the NAACP, has made it clear that his organization favors increasing energy supply to meet the growth in demand, and that blacks oppose measures that would deny them the opportunity to continue to improve their standard of living. This point of view is echoed on a global scale by Third World nations pressing for increased industrialization, distrusting what they see as efforts to freeze the status quo for those who "have" as well as those who "have not."

2. ENERGY CONSERVATION

Many observers have commented on the fact that approximately half of all energy consumed by the American economy is "wasted." An electrical generating plant converts to useful electrical energy only 30–40% of the heat energy of the fuel consumed. For an automobile engine the corresponding energy efficiency is only about 25%. Lost energy is, of course, not destroyed in these processes but is degraded into heat that no longer has entropy sufficiently low for general use. It has been proposed [8, 9] that energy wastage be reduced by more energy-efficient factories, homes, transportation, and commercial establishments, by more careful matching of energy sources with their design usage, and by recovery of "lost" energy generated at primary sources for distribution to such secondary activities as can use this energy in its degraded form.

"Cogeneration" of energy for primary use as electrical power or process heat, and secondary use for district heating of homes and offices, is quite feasible with existing technologies. In New York City, for example, Consolidated Edison generated electricity and supplied steam heat to office buildings in lower Manhattan for many years. The procedure fell into disfavor because it was capital-intensive, adding to the cost of both electricity and heat, and was a casualty of rising construction costs and decreasing energy costs in the 1950s. With rising energy costs, cogeneration could again become attractive, particularly if government incentives offset the high installation costs of the required systems. The same comment applies generally to the energy conservation option—installing additional insulation in private homes, new and more energy-efficient machinery, recovery of "waste" heat, or whatever—the high initial installation costs of capital-intensive energy recovery systems must be paid by someone, passed on to the consumer or spread over the entire population through governmental subsidy.

Beyond this, there are three factors that must be kept in mind with regard to conservation measures to reduce energy "waste."

1. A certain energy loss is unavoidable in any energy usage process, as the second law of thermodynamics dictates that no such process be 100% efficient. For an absolutely perfect heat engine, operating with intake temperature T_{in} and exhaust temperature T_{out}, the theoretical energy efficiency is just

$$\varepsilon = 1 - (T_{out}/T_{in})$$

If one were to conceive of a "perfect" steam engine, for example, whose intake and exhaust temperatures were 600° and 300°K (absolute temperatures), respectively, the energy efficiency could not exceed

$$\varepsilon = 1 - (300/600) = 1/2 \text{ or } 50\%$$

even ideally. The second law of thermodynamics requires that even a "perfect" heat engine, operating between these temperatures, must "waste" half the heat supplied. When we take into account the fact that "real" engines are necessarily imperfect (real bearings have nonzero friction), it is less surprising that auto engines, for example, lose or "waste" 75% of the energy supplied. It has been suggested [10] that the practicality of our goals would be clearer if we defined them in terms of "second law efficiencies"—normalized in each case to the maximum efficiency allowed by the laws of thermodynamics.

2. Energy efficiency is not an absolute. A certain amount of energy loss can be regarded as one of the legitimate costs, social or otherwise, of an advanced society. A conventional 120-kV power line loses, for example, 7.5% of the electrical energy it carries for each 100 miles of its length as heat dissipated in the electrical resistance of the wire cable. To supply New York City with energy generated at Niagara Falls, 400 miles away, with conventional 120-kV power lines would mean that 30% of the generated energy would be lost between Niagara Falls and New York City. It is not practicable to move either New York City or Niagara Falls to conserve energy. Power companies are constructing higher-voltage (more energy-efficient) power lines, up to 760 kV at present, with higher voltages predicted in the future. If we object to these on environmental and safety grounds, as many do, we accept a certain amount of energy loss as a necessary cost of operating our society, rather than as "waste," in the usual sense of the word. As another example, mandatory increases in home insulation standards would significantly increase the already-high capital costs of home ownership. We might therefore accept a certain amount of energy loss as a reasonable

compromise in our society, in order to keep home ownership within the grasp of our "blue-collar" population [11].

3. Energy conservation, while an important component of any energy strategy, is not a long-term solution. Stringent measures to enforce energy conservation can produce a significant "notch," or temporary leveling, of the long-term energy increase trend, but after the energy conservation economies have been taken, the demand curve will again rise. A two-car family can contribute to energy conservation in the transportation sector, for example, by getting rid of one car. Further energy conservation could be achieved by getting rid of their second car, and using public transportation. They can contribute no further to energy economy by getting rid of any more cars, however. Similarly, the homeowner who doubles the thickness of his wall insulation can reduce energy loss significantly, but redoubling the insulation thickness will not produce a proportional energy savings, since overall energy loss does not decrease linearly with insulation thickness. The net effect of energy conservation measures is to postpone the day of energy demand/supply reckoning. It appears that some acceptable way of increasing energy supply must be found.

3. ENERGY SUPPLY: HYDROPOWER

For the sake of subsequent discussion we define an "acceptable" energy supply as one that is technically capable of replacing a significant percentage of our present dependence on petroleum fuel (oil and gas) by the year 2000, and do so with the least detrimental impact on health, safety, and the environment. We first examine hydropower—electrical energy, primarily, produced by water-driven turbogenerators at locations where the water has sufficient height or velocity to make energy conversion feasible.

One might expect that hydropower would be the ideal energy resource: it is renewable (water will continue to flow out of the mountains whether the flow is dammed or not, and the water itself retains its usefulness for agriculture, commerce, and drinking supplies), it is remarkably clean (no radioactivity, no smoke, no pollution of air or water), it provides useful by-products (flood control, recreation, navigation, economic development), and would seem to involve minimal environmental degradation. Pumped-water storage, in which excess energy supply is converted to gravitationally stored energy during the night, to be recovered as electrical energy during peak daytime demand hours, is an important hydropower subset. Pumped-water storage does not add to our energy supply, but allows that supply to match demand more efficiently.

There are obvious objections that can be raised in the case of hydropower. The most important has to do with safety. Dams, like most structures, can collapse, and when they do the damage and loss of life can be enormous. The worst dam disaster in recent years was the 1962 failure of a dam structure in Belluno, Italy, in which 2300 persons drowned. One can readily conjure up a "worst case scenario," however unlikely the probability, of all the dams affected by the San Andreas fault collapsing simultaneously, with a loss of life (assuming, for the worst case, no evacuation) of the order of a half-million persons. The objections most frequently raised, however, have to do with other consequences of hydropower. Agriculturalists are concerned with the inundation of valuable agricultural land or, in the case of arid regions, excessive evaporation of valuable water from the lake formed behind the dam. Others are concerned with the silting of slow-moving rivers, as in the case of Egypt's Aswan dam, and the adverse economic effects on fishing and navigation in the river and estuaries.

The most significant objections raised, however, have come from environmental preservation interests. The Sierra Club and, later, the Friends of the Earth gained national prominence through their (largely successful) litigation against further development of the Colorado River and the Pacific Northwest. A consortium of preservationist organizations prevented development of an "unsightly" pumped-water storage facility in the Hudson River Valley. Arguments have stressed preserving the status quo wherever possible—wilderness and "wild waters" in the case of Western rivers, the serenity of the Hudson Valley experience in the East. There is also concern over adverse effects of power-dam construction on the natural habitats of birds, wildlife, and marine life in the waters and areas affected. Construction of the TVA dam in Tellico, Tennessee, was delayed following discovery that the "snail darter," one of many species of the small darter fish family, but one that was on the "endangered species" list, would be imperilled by construction of the dam.

It is difficult [12] to weigh the esthetic objections of conservation/preservation interests on the same scale as the demand for jobs and a better living standard for people who lack both. It is certainly necessary that someone speak up on behalf of wilderness and wildlife for our future generations. Some observers see, however, an antipeople bias ("people are slobs") in much of the conservationist argument, and it seems clear that the conservationist/ preservationist ranks are drawn largely from the well-to-do middle class who "have theirs" and may have a vested interest in maintaining the status quo. There are few black and Chicano "back-packers." It has been suggested [13] that blockage of the Hudson River pumped-storage facility at Storm King Mountain originated with wealthy, land-owning interests, living out the predictions of Thorstein Veblen's "Theory of the Leisure Class."

For discussion of hydropower as a solution to the energy crisis, all this is somewhat immaterial: hydropower presently provides less than 5% of our national energy needs, and even if maximum utilization were achieved, over all opposition, this could at most increase threefold. Hydropower therefore fails our test that energy sources be capable of providing a significant fraction of the energy supply that will be needed as our oil and gas resources dwindle. It is instructive, nonetheless, to note that there are probably no energy sources that can be tapped in the future which will be completely benign environmentally, or which will involve no potential health or safety problems for the population at large.

4. ENERGY SUPPLY: WIND, TIDES, OCEAN, BIOMASS, AND GEOTHERMAL ENERGY

In a time of approaching crisis, it is not surprising that panaceas should appear all over the energy landscape. Many of these give legitimate promise of becoming important, if limited, energy sources at some future time. Most suffer from having been oversold by their advocates, and their full environmental implications will probably not become evident until such time as their utilization is imminent.

Harnessing the energy of the tides is a typical example. A 340-MW power station has operated on the Rance estuary, in Brittany, for a number of years. It was completed in 1966 at a cost of $100,000,000. Peak tides at the entrance at the estuary are in excess of 40 feet. A proposal for a tidal power project at Passamaquoddy Bay, on the border between the U.S. and Canada, has been studied for a number of years but has until now been considered uneconomical. It is possible that two or three tidal power stations in the 300- to 1000-MW range will be constructed in the U.S. by the year 2000, and they will have considerable impact on their immediate localities. Tidal stations will nonetheless supply less than 0.1% of the total U.S. energy demand, and they do not constitute a major "supply" solution to the energy problem. Environmental concern for the ecology of saltmarsh and estuarine wet lands will create serious opposition to tidal power development when such proposals are given serious national consideration. Proposals for extracting energy from temperature differences in the ocean depths and from wave power have earnest advocates, but would be extremely expensive and subject to failure from storms, barnacles, etc. The appropriate technology for large-scale utilization by the year 2000 is, in any case, not in sight.

The earth's crust contains much radioactive material. There is, as a consequence, a great deal of "geothermal" heat energy trapped beneath the surface of the earth. In such locations as the Geysers area of northern California, and in Italy and New Zealand, subterranean water flows past large volumes of hot rock, and power-generating plants have been established to tap the energy of the resulting hot water and steam. It would be possible in additional locations to inject surface water deep into the earth to release geothermal energy artificially.

Suitable geothermal locations are generally areas of significant earthquake activity, however, and injection of water along fault planes is known to stimulate seismic action. Subsidence, and pollution of groundwater supplies by undesirable minerals (arsenic, mercury, sulfur and its compounds) brought to the surface are major environmental problems [14]. Geothermal energy will, like tidal power, be of significance in certain localities, but does not have the potential of making a major supply contribution to national energy needs.

Extracting energy from the wind is a more attractive supply alternative. Average wind velocities capable of delivering significant amounts of energy to modern wind turbines (such as those designed by NASA) are found throughout the U.S., but primarily across the Great Plains and in coastal areas. Typical wind turbines with blade diameters 60–70 m (two-thirds the length of a football field) in an ideal location (energy fluxes of the order of 50 W/m^2, such as are found on the flat coastland of Denmark) have been shown [15] to produce an average energy output of 0.5 MW each. To produce the 1000-MW equivalent of a modern power plant, one might consider a network of 2000 such windtowers, preferably interconnected to send power to where the winds were temporarily inadequate. The cost of each of the 2000 installations is estimated [15] at $700,000, and it would be necessary to provide energy storage, at additional cost, to cover inevitable periods of calm.

Catching energy "on the fly" and storing it is not easy. An average American family of four uses 1000 kW-hr daily as their per capita share of U.S. energy consumption: it would require a thousand 12-V auto batteries to store that much energy, and one loses ("wastes") 25–30% of this energy in each storage cycle. Improved batteries are under development, pumped water storage is an attractive alternative, but energy storage will always be expensive and carry inherent efficiency losses of the stored energy.

Introduction of "decentralized" wind-power generators, as opposed to wind generators in a power grid, is advocated by some. Although these could certainly provide the personal power requirements of "rugged individualists," they could not provide the uninterrupted power required for commercial and industrial utilization.

Despite the adequacy of wind velocity on the average, the inherent *variability* of wind power from day to day and season to season, the need for energy storage (which is technically difficult), and the high construction costs involved will probably prevent widescale development of wind-power. One imagines there will be esthetic objections to 200-foot-high wind towers, as there are to powerline towers, particularly to deployment of wind towers in massive arrays. Farmers will object to the rights-of-way necessary to ensure maintenance access for the wind towers, and to interference with aircraft carrying out aerial spraying and seeding activities. Wind towers are considered a safety hazard in areas prone to hurricanes and tornadoes, a major limitation to their installation in coastal areas and other regions of high population density. Although of considerable local importance in some areas, wind-power conversion is not likely to make a significant energy contribution on the national scene before the year 2000.

"Biomass conversion," including the use of forest materials, agricultural residues, and other nonwood plant fibers for fuel, diversion of food grains to alcohol production, and development of generators to recover methane fuel from biological wastes, has been widely advocated. Significant progress in all these areas may be expected. The use of alcohol as a vehicle fuel may be particularly important. To the extent that biomass conversion to fuel energy competes with food production from the same acreage, however, food needs of the nation and the world may provide an "ethical imperative" limiting such utilization. Fuel use of nonedible forest and plant products has a further serious limitation in that the fuel content per unit volume tends to be small, placing inordinate demands on transportation, which is itself an energy problem area. Though it will be of considerable importance in some localities, we would not expect biomass conversion to make a major contribution to the national energy problem in the near future.

5. ENERGY SUPPLY: SOLAR ENERGY

The direct use of solar energy to supply electrical power and heat energy for residential, commercial, and industrial use is the most interesting of the technological alternatives we have to consider. The sun radiates energy at the rate of 4×10^{26} W. 1.4 kW/m^2 reaches the outer layers of the earth's atmosphere, and one-half to two-thirds of this reaches the earth's surface. At noon on a clear day on the 21st of June in Phoenix, Arizona, the solar insolation (average solar power striking a flat, horizontal surface at ground level) is approximately 1.0 kW/m^2, a maximum for the continental U.S. Solar energy decreases markedly from this value depending on the time of day, season of the year, latitude north or south of Phoenix, degree of cloud cover, and clarity of the air. The average solar insolation in Indiana, for example, varies from a maximum of 0.263 kW/m^2 (June) to a minimum of 0.064 kW/m^2 (December).

Solar energy is therefore a somewhat diffuse and highly variable source. For electricity generation, one therefore needs high conversion efficiency and energy storage or backup systems to carry over nighttime hours and cloudy days. Solar cells, which convert incident sunlight directly into electricity, are available with effective energy efficiencies ranging from 2–3% (CdS/Cu$_2$S and GaAs) to 10–13% (single crystals of silicon) [16], efficiencies that are substantially lower than for any other form of energy conversion. In operation, efficiencies are further reduced by the covering needed to protect solar cells from weathering and from the inevitable deposit of dust, dirt, and pollen on the cover surfaces.

Solar cells are still very expensive ($100–180/foot^2 for large flat-plate arrays), and electricity produced from them is 50 times more expensive than electricity generated from conventional power plants. With improved technology and mass production, efficiency may be expected to rise slightly while costs decrease by at most a factor of 5–10. Direct conversion of solar energy at ground level will always be very expensive. Orbiting satellites carrying double arrays, each 2.5 × 2.5 miles in linear dimensions, have been proposed, beaming energy back to earth by microwaves, but this appears to be an even more expensive option. Solar cell-generated electricity has the advantage of being a self-contained, discrete power source and will find application in providing power at remote locations. It seems unlikely that it will ever provide the nation with a significant fraction of its electricity, and certainly not before the year 2000.

Focusing the sun's heat onto a boiler mounted atop an 800-foot tower, surrounded by an array of reflecting mirrors that track the sun from morning to night, maintaining focus on the boiler, is being considered as an alternative form of generating electricity. The boiler would produce steam, as in a conventional power plant. A demonstration plant designed to produce 100 MW (one-tenth of the size of a typical conventional power plant) would require 20,000 mirrors or "heliostats," each 20 × 20 feet in area, each separately steered by motors either clockwork- or computer-controlled to keep the sun's rays focused at the top of the tower [17]. Energy storage for only 6 hr is provided (costs of storage adequate to provide reliable power on cloudy days and also meet nighttime base-load requirements are prohibitive), and the power plant is designed for peak-load applications on a major electrical power grid.

A better understanding of the costs of constructing, operating, and maintaining this complex array should be available when construction of such a demonstration project is completed. It should be noted that direct utilization of solar energy for power generation is, for all practical purposes, limited to locations in the arid areas of the Southwest where solar insolation and atmospheric clarity are adequate for the purpose. Transmitting power out of such areas over great distances is undesirable because of the power-line losses ("waste") discussed previously. The alternative would be a massive relocation of industry and working populations to the arid Southwest. There has been little consideration as yet of the environmental consequences of such industrialization and population influx on the fragile, water-limited ecology of these regions.

The amount of electricity that could be generated by either the solar-cell or the heliostat boiler technique is, in any event, not large, possibly 0.5% of the total national energy consumption by the year 2000 [18]. The comments of W. G. Pollard [19] of the Institute for Energy Analysis, who has studied solar energy options extensively, would seem to apply:

> It is unfortunate that so many people continue to entertain high hopes for satisfying all our needs for electricity through direct or indirect means of generating it from the

sun. In remote locations, where cost is not a factor, a small amount may be produced with wind or solar cells and battery storage, and the potential for small self-contained total energy systems is significant. But for any appreciable contribution to future national requirements for central-station electricity, neither direct nor indirect solar energy is really suitable. There is practically no chance of realizing such a contribution regardless of how vigorously it is promoted and funded by Congress in response to public aspiration.

We note finally that use of solar energy to provide hot water and space heating for homes and commercial establishments is well within the scope of present-day technology. That it is not more popular has to do with the high initial costs of installing solar energy conversion and storage systems ($4000–$7000 for the average house for parts and materials, with plumbing and construction labor extra) and the need for a standby "conventional" heating system for supplementary use in very cold weather or during prolonged cloudy weather. (The least expensive "standby" service from the viewpoint of installed equipment cost is through electrical heating. Its use poses an ethical problem, however, in that the additional installed generating capacity necessary to meet peak demand when all solar energy users switch to electricity should presumably be paid for through higher rates to the peak-demand consumers, rather than spread over the larger, and presumably less affluent, population at large.)

Regardless of cost, solar heat is not likely to make a significant contribution to national energy needs. Solar input is greatest in the "Sunbelt" where space-heating demands are already small, so little savings will be achieved there. In the North-Central area and in the Northeast the winter sunshine is weak (which is, of course, why winters are cold there) and cloud cover extensive, and there is not enough solar energy input to make a significant contribution toward needs. Future solar-heating developments will be most effective in the northern Great Plains and Mountain States, where weather is cold but sunshine is reasonably abundant. The fraction of the total U.S. population living in these areas is, however, small.

6. ENERGY SUPPLY: COAL

Coal is the most abundant energy resource in the U.S. Some 20% of the world's reserves of coal are in this country, and coal constitutes 90% of our total energy reserves: Coal is to the U.S. what oil is to Saudi Arabia. The National Energy Plan [2] envisaged increasing coal production by 66% over the next 10 years. But even if our coal utilization should triple, coal reserves [20] in the U.S. would still last for 1000 years. There is no question but that coal is an *adequate* energy resource in the face of diminishing petroleum fuel supplied.

Prior to the OPEC oil embargo, our national consumption of coal had been decreasing, in part because of cheap availability of other fuels and in part because of enactment of tough new air-quality standards. Air pollution control equipment (limestone "scrubbers" and electrostatic precipitators) has been developed to alleviate pollution problems. In 1970, a study by the National Academy of Engineering concluded [21] that "contrary to widely held belief, commercially proven technology for control of sulfur oxides from combustion processes does not exist." We may anticipate continued improvement in pollution-control technology through the year 2000 while continuing to doubt that such equipment will ever be completely effective in eliminating pollutants from coal-burning factories and power plants, particularly the submicrometer particulates, which are known carcinogens.

It is instructive to consider briefly the chemistry of coal combustion. Under ideal circumstances, the carbon of the coal combines with the oxygen of the air

$$C + O_2 \rightarrow CO_2$$

and the only reaction product is carbon dioxide. Some authors [22] have raised the possibility that CO_2 from fuel combustion could alter the global climate by enhancing the "greenhouse effect" through which the earth stores its heat. Recent studies [23] suggest that the CO_2 so released is so small a fraction of the CO_2 naturally released from the biosphere that this possibility is unlikely.

Under less than ideal circumstances, combustion of carbon is incomplete, and the relevant reaction is

$$2C + O_2 \rightarrow 2CO$$

and carbon monoxide is formed. Carbon monoxide is a toxic gas when present in significant concentrations, and tall smokestacks are used at coal-burning power plants to disperse CO and other pollutants over a wider area. (In the 1960–1970 decade, the average height of smokestacks in the U.S. increased from 240 feet to 600 feet.)

A problem arises in that one does not combine carbon with pure oxygen in the combustion process but with air, which is mostly nitrogen. At the elevated temperatures of coal combustion, nitrogen and oxygen combine readily,

$$N + O_2 \rightarrow NO_2$$

to form nitrogen dioxide, a toxic gas. Similarly, the coal we burn is not pure carbon but contains numerous other minerals such as sulfur, mercury, etc. These also oxidize and are released into the atmosphere as toxic gases, as in the case of sulfur

$$S + O_2 \rightarrow SO_2$$

which forms the toxic chemical sulfur dioxide.

But the chemistry is still not complete. So-called "synergistic" reactions take place between the chemicals released and the chemical constituents of the air and new poisonous materials are produced:

$$2SO_2 + O_2 \rightarrow 2SO_3$$
$$2SO_2 + 2NO_2 \rightarrow 2NO + 2SO_3$$
$$2NO + O_2 \rightarrow 2NO_2$$
$$2NO + 2SO_2 + O_2 \rightarrow 2SO_3 + 2NO$$
$$SO_3 + H_2O \rightarrow H_2SO_4$$

releasing toxic nitrogen oxides (NO_x), sulfur oxides (SO_x), and sulfuric acid (H_2SO_4) into the atmosphere. Coal soot, very finely divided particles of unburned carbon, a known carcinogen [24], is also carried out with the flue gas, and significant amounts of radioactivity—considerably more radioactivity release than is tolerated in nuclear power operation—are also released [25].

Given the amount and variety of toxic substances released in the combustion of fuel coal, it is not surprising that polluted air is a major cause of death and sickness in the U.S. and elsewhere. Considerable publicity has been given to acute incidents, as in Donora, Pennsylvania (1948), and Greater London (1952, 1956) when temperature inversions in the atmosphere trapped pollutants and impeded their dilution, with large increases over normal death rates resulting (3900 in the case of the 1952 London "episode"). Chronic illness and death from air pollution go comparatively unremarked. Yet cancer incidence is known to be correlated with air pollution for 16 separate categories of cancer [26]. Studies in Great Britain and the U.S. show that mortality from lung cancer among nonsmokers is three times higher in urban populations than in rural populations [27]. In 21 areas around Buffalo, New York, it has been found that mortality rates from respiratory disease (bronchitis, asthma, emphysema) were twice as high in the areas of worst air pollution (>135 μg of suspended particulates per m^3 per day) than in the areas of least air pollution (<80 μg per m^3 per day) [28]. Incidence of cardiovascular disease is twice as high in areas of polluted air compared to areas of clean air [29].

Some communities prefer to burn coal for power generation at locations remote from their own populations. Coal burned at the "Four Corners" power station at the intersection of Utah, Colorado, Arizona, and New Mexico produces electric power that is transmitted to user communities in distant states. Aside from the inefficiency of the power losses in transmitting electricity over long distances, an ethical problem arises. Indeed, the same ethical problem arises wherever those who desire to preserve the esthetic values of their own communities nevertheless place major demands on energy generated at locations elsewhere. When it comes to the black lung disease of coal miners, the degradations of Appalachia, and the despoliation of land stripmined for coal all across the nation, we all share in the guilt.

We conclude that coal mining and combustion are technologically effective techniques with ample recoverable resources, to provide all the energy we are apt to need by the year 2000 and well beyond. The price we pay in human health and safety, not to mention degradation of the environment and esthetic sensibility, may simply be too high. New types of coal "reactors" involving fluidized bed combustion, for example, may ease the pollution problem, at least, by the end of the century. Were coal not such an established technology, and the hazards accepted by so many of us, a moratorium on new coal-fuel power plant construction at least until such time as new coal combustion concepts were put into effect would certainly be at the top of the energy agenda.

7. ENERGY SUPPLY: OIL AND NATURAL GAS

Just as coal replaced wood in the 1880s, oil and natural gas replaced coal as the major energy source in the U.S. in the 1940s. As oil and gas generally occur in the same geologic formations and their benefits and drawbacks are similar, it is useful to consider them together.

The motivation for changing to oil and gas is easy to understand. They are easier to transport from well to user (by pipeline rather than train), they are easier to burn and produce far less waste. Fluid fuels have proven to be the most convenient to use for transportation—auto, rail, sea and air.

Unfortunately, the known reserves of these convenient fuels are quite small compared to coal. Assuming a total worldwide available oil of 1350×10^9 barrels, reserves will be effectively depleted in about 2020. An upper estimate of 2100×10^9 barrels of available oil does not increase the time of depletion significantly; only to about 2030.

To understand this seemingly anomalous set of estimates, we must consider the pattern of consumption of any finite resource (see Fig. 4) [30]. When exploitation of a resource begins, use grows slowly as consumers become familiar with the resource and learn how to use it. Use accelerates as technology to utilize the resource becomes widespread. As discovery and recovery of new reserves becomes more difficult, costs increase and use declines until the resource is exhausted. A doubling of the available reserves doubles the area under a production curve such as that given in Fig. 4 and increases slightly but does not double the time required to use the middle 80% of the resource.

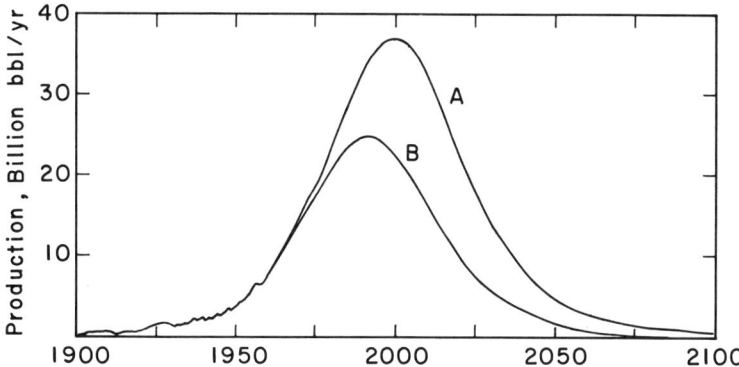

FIGURE 4. Cycle of world oil production based on two estimates of recoverable oil. Curve A assumes a total production of 2.1×10^{12} barrels and curve B, 1.35×10^{12} barrels.

Figure 4 shows world production of oil reaching its maximum in the late 1990s. In the U.S. production reached its peak in the early 1970s and has been declining since.

The statistics for natural gas are similar to those for oil, although it is more difficult to estimate gas reserves than it is oil reserves (just as it is more difficult to estimate oil reserves than coal reserves), due to the way they occur in nature.

There are vast reserves of oil shale in the U.S. that if fully utilized, contain more oil than the Middle East oil fields. The problems involved in tapping this resource, however, are many. Oil shale contains a substance called kerogen, which, when heated, becomes a liquid very similar to crude oil. Expected yields vary from 50 to less than 10 gallons per ton of shale. In other words you must mine a lot of rock to get a tank of gasoline. A consequence of this is that, based on today's technology, synthetic crude oil from shale is not economically competitive with natural crude, even at the arbitrary prices imposed by OPEC.

Another problem with oil shale is that the retorting (heating) process that releases the synthetic crude also increases the volume of the residual shale. Disposing of this residue is both an operational and an environmental problem.

In the midst of World War II, cut off from Middle East petroleum, Germany fueled its war machine with synthetic oil produced from coal. The technology to produce a synthetic gas or synthetic crude from coal still exists and is being vigorously developed by a number of organizations. The reasons these technologies are not commercially utilized at the present time are, again, economic. The cost of oil or gas produced from coal substantially exceeds the cost of the natural product. As the price of natural gas and crude oil increases, however, "syn gas" and "syn crude" become more economically attractive.

The most likely coal to be used in any synthetic fuel process will be from the vast near-surface (strip-minable) coal deposits in the Great Plains—North and South Dakota, Montana, and Wyoming. All processes for converting coal to a liquid or gas require significant amounts of water. But water is a scarce, valuable resource in these regions, and available supplies are already allocated to agricultural and domestic use. It is not clear at this time who would be required to give up their water rights to facilitate "synthetic" fuel production.

8. ENERGY SUPPLY: NUCLEAR ENERGY

Finally, we consider the most recently developed energy source, the atomic nucleus. Nuclear power is the source of more controversy than any other at the present time. Objections to its exploitation are both ideological and technical.

If "small is beautiful," then surely "big is bad." Nuclear power plants are today's best example of the economy of scale. Plants coming on line are sized in the 1100- to 1200-MW (electric) range, a size unheard of 20 years ago. They are capital-intensive (current cost estimates range around $2000 per kilowatt of installed capacity). They make economic sense only as a part of a large interconnected utility distribution system. Those who are opposed to the centralization of capital, particularly in the private sector, are perforce opposed to nuclear power. These factors should be considered by the body politic when deciding the nation's energy future (or lack thereof).

In this Section, however, it is the technology of nuclear power that concerns us. These concerns are not totally divorced from the ideological concerns mentioned above, for ideological opponents of nuclear energy often clothe their political opposition in technical arguments.

The prime technical concerns are safety, both of the operating plant and of other components of the nuclear fuel cycle, and, to a lesser degree, the relative cost and availability of nuclear power for some reasonable period in the future. Our purpose in this Section is simply to highlight these topics and leave detailed consideration of them to authors in subsequent sections.

8.1. Operational Safety

We first address the question of operational safety, which can be further divided into normal and abnormal (accident, sabotage) operation.

In normal operation it is difficult to imagine a more benign source of energy. Where fossil-fueled plants release tons of particulate matter, sulfur dioxide, and nitrogen oxides to the environment per year, the nuclear plant releases none. The radioactivity released to the environment in normal operation of a nuclear plant is less than that from a comparable fossil-fueled plant [31]. (Coal deposits

2 • Energy Demand–Energy Supplies

normally contain such radioisotopes as radium, thorium, and uranium.) Even the most dedicated opponent of nuclear power will agree that, like the little girl with the curl on her forehead, "when she was good, she was very, very good."

It is the balance of the nursery rhyme that opponents stress, "when she was bad, she was horrid." How likely are nuclear plants to be "bad?" If bad, how "horrid?"

The best available answer to the question is given in the "Rasmussen Report" [32]. The report is the result of a 2-year study directed by Professor Norman C. Rasmussen of MIT, funded by the Atomic Energy Commission and continued by the Nuclear Regulatory Commission after its formation in January 1975. The study involved 70 man-years of effort and an expenditure of about $4,000,000.

Risk assessment and accident analysis are treated later in this volume by Lellouche and by Theofanous and Wilson. We can summarize the hazard level of nuclear reactors in normal operation by noting that the only naturally occurring event that presents a hazard to life comparable to an energy economy with 100 nuclear reactors of the 1000-MW size is a meteor fall. The probability of death due to a falling meteor is not zero. The large crater in the Southwest, the vestige of an event that occurred 30,000 to 50,000 years ago [33], is impressive and a good indication of the consequences of such an event, should it occur. Fear of being struck by a meteor would be regarded by most as a psychological disorder comparable to paranoia.

8.2. Radioactive Waste

If the operation of a nuclear reactor is not itself dangerous, say its opponents, then surely its wastes are. They represent "an intolerable legacy for generations yet unborn" if opponents of nuclear power are to be believed.

The volume of high-level radioactive waste to be generated depends on whether or not we elect to reprocess spent nuclear fuel. The present policy of the U.S. government is not to reprocess commercial reactor fuel, thereby retaining plutonium generated in normal operation of the reactor in the spent fuel, inaccessible to anyone inclined to use it in a crude nuclear weapon. The government has announced a policy of accepting spent fuel from electric utilities for storage and ultimate disposition.

If, on the other hand, reprocessing were permitted, unburned uranium and plutonium would be recovered and recycled, reducing demand on uranium ore reserves, presumably reducing the cost of the electricity generated. The extracted high-level waste would eventually be converted to solid form, probably a glass. Each 1000-MW reactor would produce about 30 cubic feet of this high-level waste per year. All of the high-level waste projected to be produced between 1975 and 2000 would fill a football field to a height of 12 feet. An interesting comparison is the volume of solid waste (ash) from a coal-fired plant. Thirty-three coal cars of ash per day must be removed from a 1000-MW coal-fired plant.

The main difference between coal ash and nuclear reactor wastes, other than their relative volume, is, of course, their degree of radioactivity. There are two basic contributions to the radioactivity level of spent fuel. Most serious are the fission products, the isotopes produced when the ^{235}U (or ^{239}Pu) nucleus fissions, releasing energy. The laws of nature require that the most intensely radioactive isotopes are also the most short-lived, and vice versa. The longest half-life (the time required for one-half the atoms originally present to undergo decay) among the fission products is about 30 years. ^{137}Cs and ^{90}Sr are typical long-lived fission products. In 10 to 20 half-lives the fission products will have decayed to levels that are innocuous. The other contribution to the radioactivity of the wastes is from the "transuranics," elements above uranium on the periodic chart that result from the chain of nonfission neutron absorptions in uranium and its subsequent radioactive decay. The isotope that seems to cause the most concern is ^{239}Pu with its 24,000-year half-life. The toxicity of plutonium is discussed later by Cohen.

What we have not discussed is its economic value. As a fuel ^{239}Pu has an anticipated value about that of gold. It is this economic value of plutonium that provides the motivation for reprocessing. But no chemical process man may devise is 100% efficient, so some plutonium will remain with the fission-product waste. The economic incentives are such, however, that only some small fraction of a percent of the plutonium will be lost in this fashion.

When the 500- to 600-year period required for fission products to decay to negligible levels has passed, the radioactivity of the waste will be due entirely to plutonium and other transuranic elements. But the level of radioactivity then present will be no greater than that of naturally occurring radioactive ores such as uranium.

Five hundred years is a long time. If we wish to ensure that the radioactivity from nuclear reactor wastes is not released to the environment during that period, it should be sequestered in a geologically inert environment. Bedded salt deposits appear to be a reasonable storage location. Many suitable sites have been identified and a demonstration facility is currently under development. The technology for reprocessing and ultimate storage has been available for years. Its commercial development is delayed, not by technological problems, but by political decisions (or the lack of them).

8.3. Sabotage and Diversion

The Rasmussen study did not deal with the possibility of deliberate sabotage of an operating nuclear reactor. The reasons are simple. It is difficult to estimate the probability that a large, dedicated group would attempt the invasion of a nuclear plant. If a group did make such an attempt, normal security coupled with local law enforcement, and the design of the plant both delay the attacking group and serve to minimize the potential damage they might do. A terrorist group that wishes to kill a large portion of the populace has easier and more effective means available, e.g., contamination of a municipal water system with commercially available pesticides.

Diversion of nuclear material, principally plutonium, and subsequent construction of a "basement bomb" is another potential risk that nuclear opponents have publicized. The facts of the matter are that plutonium is well guarded in its movements from place to place and that construction of even a crude nuclear device is a difficult and dangerous task. A terrorist group or nonnuclear nation that wants a weapon would be much better advised to steal one ready-made from a U.S. military facility. In the more than 30 years the U.S. has had a nuclear weapons capability there has been no such theft.

8.4. Practicality of Nuclear Energy

We turn now to the availability of nuclear fuel and the economics of its utilization. Even if nuclear reactors were eternally harmless, if fuel is not available, or if the cost of nuclear power is greater than power from other sources, there is little incentive for its use.

The economics of nuclear and alternative energy sources in the contemporary marketplace are discussed in a later section by A. D. Rossin. For electric power plants now operating, there appears to be a clear financial advantage for nuclear fuel over any of the fossil fuels.

Today's economics seem to favor nuclear power over coal. What about tomorrow's? The availability of uranium depends on the number of operating power plants that require it, the known deposits of uranium ore, and the cost of exploiting these deposits. Depending on the assumptions made for each of these factors, it is possible to develop very optimistic or very pessimistic estimates of uranium availability. Both have been published.

A recent Ford Foundation study [34] takes the optimistic view that "assuming the incentive of higher prices, more intensive and extensive exploration should produce enough uranium, both in the United States and abroad, to supply the domestic and foreign requirements of light water reactors (LWRs) well into the next century." The study group's estimate has been criticized as being too high or too low, depending on the critic. What is clear is that there will be ample uranium at reasonable cost at least to the year 2000, and probably beyond.

Development of the breeder reactor would extend these uranium reserves substantially. (Development was curtailed by the Carter administration and is endorsed by the Reagan Administration but slowed by a Congressional decision to stop funding for the Clinch River breeder demonstration project.) In the breeder the isotope ^{238}U, which does not fission, is converted to ^{239}Pu, which is an excellent fuel material. An alternate breeding cycle involving the conversion of ^{232}Th to fissile ^{233}U is also being investigated. The possibility of large-scale commerce in ^{239}Pu was given as the reason for the Carter administration's reluctance to pursue this option. The escalating costs of the Clinch River Breeder, due in part to project delay, contributed to the 1983 decision of Congress to halt funding for the project. It should be noted, however, that there is enough ^{238}U already in storage aboveground as residue of the uranium enrichment plants to fuel a breeder reactor economy for several hundred years.

The interesting argument that a "Carter-type" nonproliferation policy would have exactly the opposite effect from that intended is made by Rose and Lester [35]. They contend that by restricting the availability of nuclear power, the U.S. increases the demand for the limited amount of crude oil in international commerce (U.S. imports currently account for 25% of international traffic in oil). This increased demand could then cause nations to develop nuclear power on their own, rather than utilizing U.S. technology under international supervision. Due to the presumed destabilization of international relations, due in large measure to competition for the limited sources of energy available, these nations continue their development past nuclear power to the production and deployment of their own nuclear weapons.

9. SUMMARY

If the U.S. wishes to continue its economic growth, it must do so by increasing its capacity to generate and use energy efficiently. Conservation, while it may postpone increases in energy demand, cannot be the total answer to our energy needs.

Some sources of energy that we have discussed, notably hydropower, are nearly fully developed, with little prospect of expanding their capacity significantly to meet expanding demand. Others, such as natural gas and oil, though widely used in the past, have become uneconomic as reserves become depleted. Potential sources such as solar, tidal, wind, nuclear breeders, nuclear fusion, and biomass cannot be developed rapidly enough to make a significant contribution by the turn of the century (even later in some cases).

The energy sources that have a well-developed technology and are economically viable to meet expanded energy demand in the near term are coal and uranium. Both, as do all the other sources mentioned earlier, can have significant, potentially adverse, effects on the environment and on public health. Failure to meet expanding energy demands will also have an adverse effect on the economy and, consequently, on the environment and public health.

The question of the nation's (and the world's) energy future is a difficult and complex one. It does not have a simple solution, nor a cheap one.

REFERENCES AND FOOTNOTES

1. H. C. Hottel and J. B. Howard, An agenda for energy, *Technol. Rev. 74*, No. 3 (1972).
2. Executive Office of the President, The National Energy Plan, p. 2, Superintendent of Public Documents, U.S. Government Printing Office, Washington, D.C. (1977).
3. R. H. Romer, *Energy: An Introduction to Physics*, Appendix K, Freeman, San Francisco (1976). Note that energy is the product of power (kilowatts) and time over which the power is produced or consumed.
4. Energy consumption by sector is: residential, 19%; commercial, 15.4%; industrial, 39.6%; transportation, 26%.
5. M. Wilson, A. P. news release in *South Bend Tribune*, Feb. 19, 1977.
6. H. L. Parsons, *Marx and Engles on Ecology*, Greenwood Press, Westport (1977).
7. *Time Magazine* (Jan. 21, 1978), p. 33, quotes Mr. Hooks: "If Black unemployment is to be reduced, the nation's economy must grow rapidly. Since that requires more energy, oil and gas prices should be allowed to rise so that companies would have more incentive to step up exploration and output."
8. Executive Office of the President, The National Energy Plan, p. 35, Superintendent of Public Documents, U.S. Government Printing Office, Washington, D.C. (1977).
9. B. Commoner, *The Closing Circle*, pp. 138–175, Bantam Books, New York (1972); *The Poverty of Power*, pp. 6–29, Bantam Books, New York (1977).
10. W. Carnahan, B. M. Casper, K. W. Ford, A. Prosperetti, G. I. Rochlin, A. H. Rosenfeld, M. H. Ross, J. E. Rothberg, G. M. Seidel and R. H. Socolow, Efficient Use of Energy: A Physics Perspective, *Phys. Today, Aug. 1975*, 23–29.
11. *Time Magazine* (Sept. 12, 1977), p. 54. Note that it is even more expensive to "retrofit" old homes with insulation than to provide it for new homes.
12. See, however, an attempt at cost-benefit analysis for the damming of Hell's Canyon, in C. J. Cicchetti, A Primer for Environmental Preservation: The Economics of Wild Rivers and Other Natural Wonders, pp. 9–14, MSS Modular Publications, New York (1974).
13. W. Tucker (Environmentalism and the leisure class, *Harper's*, Dec. 1977, pp. 49–56) quotes Thorstein Veblen: "The leisure class is in great measure sheltered from the stress of those economic exigencies which prevail in any modern, highly organized industrial community. The exigencies of the struggle for the means of life are less exacting for this class than for any other; and as a consequence of this privileged position we should expect to find it one of the least responsive of the classes of society to the demands which the situation makes for a further growth of institutions and a readjustment to an altered industrial situation. The leisure class is the conservative class."
14. R. H. Wentorf, Jr., Formation and Consumption Rates of Major Energy Sources, Power Systems Laboratory Report 77CRD005 (March 1977), p. 5.
15. R. Axtmann, Environmental impact of a geothermal power plant, *Science 187*, 795–803 (1975).
16. H. Kelly, Photovoltaic power systems: A tour through the alternatives, *Science 199*, 634–643 (1978).
17. A. F. Heldebrandt and L. L. Vant-Hull, Power with heliostats, *Science 197*, 1139–1146 (1977).
18. M. Wolf, Solar energy utilization by physical methods, *Science 184*, 382–386 (1974).
19. W. G. Pollard, The long-range prospects for solar energy, *Am. Sci. 64*, 424–429 (1976).

20. J. M. Fowler, *Energy and Environment*, p. 295, McGraw–Hill, New York (1975).
21. National Academy of Engineering–National Research Council, Abatement of Sulfur Oxide Emissions from Stationary Combustion Sources, National Technical Information Service, Springfield, Va. (1970).
22. C. F. Baes, H. E. Goeller, J. S. Olson, and R. M. Rotty, Carbon dioxide and climate: The uncontrolled experiment, *Am. Sci.* 65, 310–320 (1977).
23. M. Stuiver, Atmospheric carbon dioxide and carbon reservoir changes, *Science 199*, 253–258 (1978); G. M. Woodwell, R. H. Whittaker, W. A. Reimers, C. E. Likems, C. C. Delwiche, and D. D. Botkim, The biota and the world carbon budget, *Science 199*, 141–146 (1978).
24. B. Commoner *Poverty of Power*, p. 69, Bantam Books, (1977) recalls the classic studies of cancer of the scrotum among London chimney sweeps, in 1775, later traced to certain chemicals in coal tar and soot.
25. R. Wilson and W. J. Jones, *Energy, Ecology, and the Environment*, p. 203, Academic Press, New York (1974).
26. M. L. Levin, W. Haenszel, B. E. Carroll, P. R. Gerhardt, V. H. Handy, and S. C. Ingraham, II, Cancer incidence in urban and rural areas of New York State, *J. Natl. Cancer Inst.*, 24, 1243 (1960).
27. P. Buell and J. E. Dunn, Jr., Relative impact of smoking and air pollution on lung cancer, *Arch. Environ. Health, 15*, 291 (1967).
28. W. Winkelstein, Jr., S. Kantor, E. W. Davis, C. S. Maneri and W. E. Mosher, The relationship of air pollution and economic status to total mortality and selected respiratory system mortality in men, *Arch. Environ. Health, 14*, 162 (1967).
29. L. D. Zeidberg, R. J. M. Horton, and E. Landau, The Nashville air pollution study, *Arch. Environ. Health, 15*, 225 (1967).
30. M. King Hubbert, The energy resources of the earth, *Sci. Am. 224*, 60–70 (1971). Also available as *Energy and Power*, pp. 31–40, Freeman, San Francisco (1971).
31. J. I. McBrade, R. F. Moore, J. P. Witherspoon, and R. E. Blance, Radiological impact of airborn effluents of coal and nuclear plants, *Science 202*, 1045–1050 (1978).
32. WASH-1400, Reactor Safety Study, An Assessment of Accident Risks: U.S. Commercial Nuclear Power Plants, U.S. Nuclear Regulatory Commission (Oct. 1975).
33. E. J. Olson, Close encounters of the zeroth kind, *Field Mus. Nat. Hist. Bull. 49*, 6–13 (1978).
34. Nuclear Power Issues and Choices, Report of the Nuclear Energy Policy Study Group, Ballinger Publishing, Cambridge, Mass. (1977).
35. D. J. Rose and R. K. Lester, Nuclear power, nuclear weapons and international stability, *Sci. Am. 238*, 45–57 (1978)

3 Exclusive Paths and Difficult Choices

An Analysis of Hard, Soft, and Moderate Energy Paths

Ian A. Forbes and Joe C. Turnage

This paper has been prepared as a contribution to the growing debate concerning the evolution of viable alternative energy futures. We bring to it certain biases, the preeminent one being our agreement with the World Council of Churches on the ethical requirement that alternative futures on a global basis be "just and sustainable." In this contribution to the dialogue and with this ethical bias we argue basically for a process by which one may objectively analyze the desirability of future energy goals. To us this process includes an equitable assessment of risks, a formulation strategy that is realistic, and an explicit consideration of the uncertainties of the current data base and those that always exist when one looks to the future [1–3].

1. INTRODUCTION

1.1. The Nature of the Problem. (The Need for Difficult Choices)

There are certain aspects of the energy problem that we feel should be identified in an introduction to viable energy strategy.

First, the energy problem is international in scope. Global society in the future will, of necessity, be more interdependent; we see this today clearly in

---

IAN A. FORBES • Energy Research Group, Inc., Waltham, Massachusetts 02154.   JOE C. TURNAGE • Delian Corporation, Del Mar, California 92014.   An earlier version of this work was supported in part by funding from the Oak Ridge National Laboratory, U.S. Department of Energy. The authors themselves, however, bear total responsibility for the present paper.

the interdependencies of energy supply and demand. The industrial countries of the First World look to their neighbors in the Third World not only as a marketplace for goods, services, and knowledge but also as a source of important raw materials. The developing countries of the Third World look to the industrialized nations for the capital, education, and technology to satisfy the rising expectations of their peoples.

Second, the nature of the energy problem can be understood best if it is recognized that the energy crisis is in part a political crisis. It is a crisis in which difficult decisions often get postponed and reconsidered, in which there is no visible constituency supporting any of the key decisions, and in which the time scales of the problems encountered are much longer than those of political accountability.

Third, the energy problem is a problem of how to implement a process of option-creation. To us energy policy is not today's decision of energy supply mix fixed for the next 25 or 50 years, but rather today's decisions regarding the creation of options that future generations may elect to select or reject.

Fourth, the nature of the energy problem is that it is dynamic. The implication is that solutions have to be dynamic. A "snapshot" approach to energy supply and demand today or in the year 2000 or a trajectory guessed at today is not a good model upon which to base energy policy. Typically, when one sets about to deterministically orchestrate tomorrow's energy future, there is neglect of second-order effects, there is neglect of political feedback considerations, and there is seldom the kind of concern for feasible transition that is essential to the creation of what we call viable options for the future.

1.2. Energy as a Proxy Variable. (Real Constraints and Hidden Agendas)

A major part of the energy debate concerns itself with what energy means to the various protagonists. We choose Dixie Lee Ray's definition of energy [4]:

> Energy is the *sine qua non* of a modern society's ability to do the things it wants to do. Such goals as maintaining the standard of living for a growing population, national security, improved quality of life, increased affluence, or increased assistance to less developed societies can only be obtained with increasingly large amounts of energy. While lower energy costs allow society more freedom of action in seeking its goals, the availability of energy is the first requirement for having any freedom of action at all.

Energy therefore acts as a necessary condition for the creation of certain aspects of that thing economists refer to as social welfare. Obviously, energy consumption need not correlate with happiness. Alternatively, energy is not a proxy variable for environmental damage or for feelings of alienation and loss of control.

In the U.S. our use of energy has been growing. Energy consumption per employed worker increased by about two-thirds between 1948 and 1970. It is relevant to ask, as Starr has done [5]—what did the nation get in return? The answer:

> A 66% increase in energy consumption per employed worker supported a 90% increase in real family income. The percentage of personal income required to furnish necessities such as food and clothing declined, whereas purchases in discretionary areas such as medical care, education and recreation increased substantially. The rise in mean family income moved a fifth of the population out of the poverty bracket to a level above the threshold for acceptable nutrition and housing.
>
> These economic changes triggered a profound social impact as well. One example was the massive increase of women to the work force. This entry was no doubt hastened by new labor saving devices for the home and the increased use of office and factory machinery that could be operated by brains rather than brawn. Had the percentage of women in the work force been frozen at the 1948 level, . . . 7.8 million fewer women would have held paying jobs in 1970.

Starr's point is that in terms of productivity per worker, energy is fuel for increased productivity and opportunity. Energy is the food of George Bernard Shaw's mechanical slave. As Shaw put it:

> The fact is, is that civilization requires slaves. The Greeks were quite right there. Unless there are slaves to do the ugly, horrible, uninteresting work, culture and contemplation becomes almost impossible. Human slavery is wrong, insecure, and demoralizing. On mechanical slavery, on the slavery of the machine the future of the world depends.

All of this is not to imply that over the 1948–1970 period, we in the U.S. could not have done the job more efficiently. Perhaps too much energy was used to do the things we wanted done. The issue is not materialism viewed as energy gluttony. The issue is whether the U.S. can survive the transition from inexpensive energy to expensive energy and respond at the same time to the challenge of creating options for a just and sustainable world future.

1.3. Rationale for Making Difficult Choices. (Faulty Paths and Utopian Solutions)

The environment within which one considers energy policy is one that often compels difficult choice under uncertainty. Keep in mind that the preferable choice is not what energy future shall be dictated but rather what options of choice should be invested in today. Our requirement is to develop the data, methodology, and vision that can then be applied toward managing the transitions in energy supply and demand that our future will require.

As an example of a lack of this kind of assessment, one need only turn to recent advocacy [6–8] of nonnuclear, "soft" technologies. The "soft" technology proponent typically proposes the use of the following "hard" technology sparingly or not at all: large coal-fired plants, coal conversion plants, urban-sited LNG terminals, Arctic petroleum extraction, most "unconventional" hydrocarbons, "exotic" large-scale solar technology. Such a view would abandon nuclear fission. Under the assumption that the demand growth will turn around and become negative, the most vocal of the proponents of the soft energy path, Amory Lovins, sees 37% of the energy supply in the year 2000 from solar heating and cooling, biomass conversion to fuels, wind, solar-process heat, and small-scale storage. In contrast, a report of a 2-year effort by the Workshop on Alternative Energy Strategies (which included international representatives from industry, finance, and academia) predicted the following global prospects [9]: (1) energy demand will continue to grow; (2) energy sources need to be developed *now;* (3) nuclear fission is important; so is conservation and efficiency improvement; (4) coal is important; (5) intercontinental gas transport systems will be required; (6) renewable sources of energy are not likely to provide significant input before the year 2000.

If Lovins is wrong, the Workshop on Alternative Energy Strategies spells out the consequences:

> Failure to recognize the importance and validity of these findings and to take appropriate and timely action will almost certainly result in a world different from the one on which these projections have been based. Failure to act could lead to substantially higher energy prices as the supply/demand imbalance becomes more apparent—with depressant effects on the economies of the world and the consequent frustration of the aspirations of the less developed countries. The major political and social difficulties that might arise could cause energy to become a focus for confrontation and conflict.
>
> In addition, the longer the world delays in facing this issue, the more serious the outcome will be. Even with prompt action the margin between success and failure in the 1985–2000 period is slim. Time has become one of the most precious of our resources. Recognizing the importance of time and the need to respond can help us through the period of transition that lies ahead.

2. UNDERSTANDING THE PROBLEM

2.1. Information from the Past

Past trends in supply and demand have been remarkably steady and the transitions in resource uses and technologies quite consistent. There is a tendency in the "invention" of the "energy crisis" to believe that we have discovered much by way of new alternatives in the last several years and that energy systems can be molded in short order in the hands of a few capable strategists. History suggests that neither point is correct. (For an incisive overview, see [13].)

On the demand side, possibly the only really "new" discovery of today is an examination of growth and a questioning of consumption. While there is a tendency to be preoccupied with conservation to the extent that some feel no need to worry about energy production, it is evident that the details of demand are not as well understood as the details of supply. This is evident in debates over energy and jobs, life-style and conservation versus growth. To the extent that these debates can move toward a careful researching of the basic roots of energy demand, this trend would be a healthy one. But we must accept that present relationships of economy and growth do exist and that they have a momentum of public acceptance and advocacy.

Similarly, on the supply side, there is an inertia associated with the introduction of new energy sources to the marketplace. Marchetti [15] has analyzed historical trends and transitions in technologies and energy sources. The fractional contributions of the principal energy sources show amazingly steady trends over long periods of time even through events like the Depression and the Suez crisis. Marchetti projects future energy supply fractions with year 2000 values close to current estimates on oil and nuclear, but much higher on natural gas and much lower on coal. To turn these latter two trends around will be no minor feat in policy and strategy, as gas shortages during the winter of 1977–78, and problems in increasing coal production might suggest.

Thus, "depending on the nature of the product or process substitution, the total time taken for basic research, development and commercialization, lies typically in the range from a minimum of 50–60 years to a maximum of 100–150 years. Wars, economic booms and disasters, government and statutory controls and regulations appear to have remarkably little effect on these substitution times" [49]. There is a limit to the rate at which new sources can build into a system that requires affordability. In the meantime, the needs of a whole generation must be met with current technology.

2.2. Patterns of Energy Use

As has frequently been pointed out, the use of energy in industrialized societies is restricted to a relatively small number of end-use applications, largely covered by the broad categories of low- and high-temperature heat, mechanical work, and feedstocks. Energy consumption also ranges greatly from country to country, generally rising with increasing standard of living but varying significantly even among countries with similar per capita gross national product.

Often, when shown the kind of data on energy consumption depicted in Fig. 1, we in the U.S. are exhorted to mend our ways and get down to the kind of levels of energy efficiency exercised by, say, West Germany or Sweden. The fact is, however, that energy use represents an aggregate of many decisions, personal, corporate, and governmental, which reflect not only price and availability but also productive efficiency, geography, climate, and a host of other

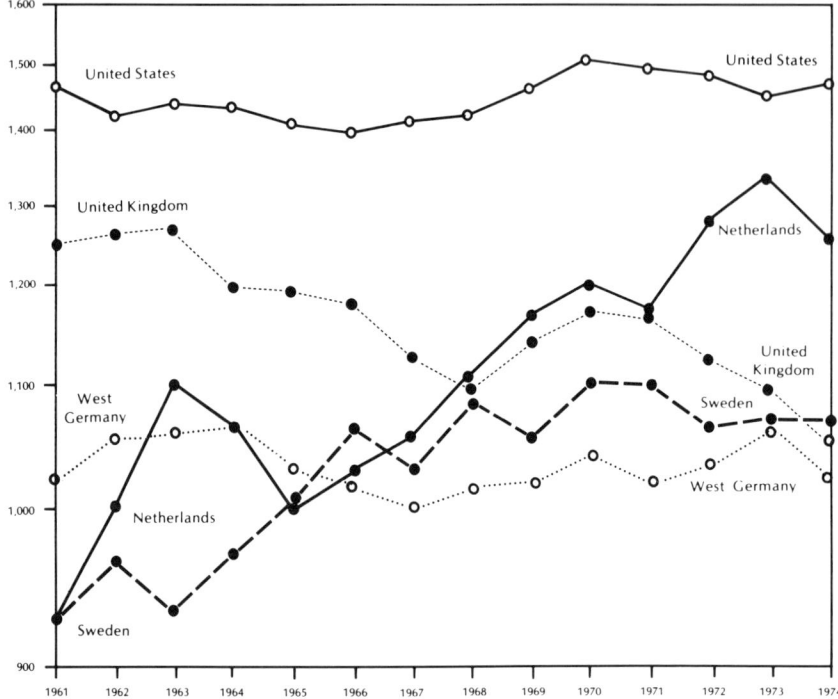

FIGURE 1. Energy/output ratios for five selected countries. *CDP is constant 1972 dollars converted at purchasing-power-party exchange rates.

variables—some of which are controllable by society, some of which are not. A report by Darmstadter et al. of Resources for the Future [18] explores how the ratio of energy consumption to gross domestic product varies in five industrialized nations. The report reveals that about one-half of the difference between the energy/GDP ratios of those countries and that of the U.S. is attributable to differences in the composition of goods and services relative to the U.S. The other half (roughly) is a product of two factors—the energy intensity per task in each country and an "interaction" effect attributable to both the kinds of tasks and the energy intensity of each task. Marchetti has derived an empirical relation for per capita GNP [19] that suggests a similar conclusion.

As an example, the report of Darmstadter et al. gives the breakdown of energy/GNP ratios of nine countries for 1972. West Germany's ratio was 70% of that of the U.S. in 1972; Sweden's was 72%. In the case of West Germany, 43% of this difference is associated with transportation; in the case of Sweden,

50% of the difference is associated with transportation. The use of energy for transportation is, in turn, a product of multiple decisions about transportation modes (e.g., public vs. personal, large cars vs. small cars). These decisions reflect consumer preferences, cost, geography, economic efficiency, and a host of other parameters.

A similar point may be made regarding energy use within the U.S. Table I shows the ratio of energy consumption to personal income by region. In this case, one may observe that New England has an energy/personal income ratio about one-third that of the Texas-dominated West South Central region. Before concluding that the people of Texas are energy gluttons, one needs to understand that most of the difference in energy/personal income ratios comes from the fact that energy-intensive industries are to be found preferentially in the West South Central region rather than in New England. Here, the difference seems to be mostly in the kinds of tasks people do rather than in the energy intensity per similar task. Even in transportation, one can speculate that the differences in energy use result more from differences in geography than from thrifty decisions regarding miles per gallon or mass transit. Likewise, it would make little sense to suggest that the New England and Middle Atlantic states, which have energy/personal income ratios similar to Sweden's, have no need to conserve. There is very substantial room for efficiency improvements and conservation of energy in every region of the U.S. and in every sector—independent of regional or national comparisons—and with the compelling logic that it is frequently the least expensive alternative.

All this leads us to a few simple conclusions: First, relationships between energy use and economic welfare (not necessarily to be confused with individual happiness) are complex products of multiple decisions. Second, in understanding these decisions, one should be as concerned with understanding the kinds of tasks to be performed as with the energy intensity per task. Third, since energy use is tied both to the things people do and to how they do them, fundamental changes in relationships are not likely to be sudden. Finally, there is at least one additional point to be made regarding uses of energy. The vast majority of the world today is poverty-stricken and experiencing a population explosion. There are roughly 4 billion people in the world today and current projections are that this number will climb to between 6 and 7 billion by the year 2000. The creation of options for a just and sustainable future requires massive institutional efforts, and, we suspect, substantial amounts of energy.

Unfortunately, as practically everyone is saying these days, we are attempting to fuel this global environment with depleting liquid hydrocarbons. As Patton points out, "while world petroleum supplies are adequate to maintain the industrial democracies in prosperity for several decades, they are inadequate to offer half of the world's people who live in the poor countries hope for attaining great improvement in living standards." Without aggressive action, this inade-

TABLE I
U.S. per Capita Energy Consumption
and Personal Income by Region for 1972

Region	Per capita energy consumption (10^6 BTU)				Personal income per capita ($)	Energy/income ratio (10^3 BTU per dollar)
	Residential, commercial	Industrial	Transport	Total		
New England	141.3	42.6	68.0	251.9	4754	52.99 (0.72)
Middle Atlantic	121.0	79.6	68.7	269.3	5005	53.81 (0.73)
South Atlantic	104.2	88.1	84.2	276.5	4131	66.93 (0.91)
East North Central	134.6	136.5	74.6	345.7	4699	73.57 (1.00)
West North Central	140.9	88.5	97.6	327.0	4281	76.38 (1.04)
East South Central	105.0	160.7	90.7	356.4	3448	103.36 (1.40)
West South Central	115.7	330.5	113.4	559.4	3849	145.34 (1.97)
Mountain	150.5	137.0	110.5	398.0	4158	95.72 (1.30)
Pacific	109.2	84.5	92.7	286.4	4880	58.69 (0.80)
Total U.S.	121.9	123.1	85.1	330.1	4478	73.72 (1.00)

3 • Exclusive Paths and Difficult Choices 81

quacy is likely to grow rather than diminish. First, technology is in general controlled by the First World; 98% of new patents (excluding the Communist Second World) originate in the First World. Second, geographical factors as well as the technology to produce cereal grains combine to favor the industrialized First World as a provider of food for the world's peoples and while Third World countries are experiencing continual population growth, the First World population is leveling off. Finally, to those who would hold out an exclusively "soft" energy path for the developing world, there is the stark reality that "most people in the developing world live in an environment that does not even remotely resemble a renewable paradise with intensive and regular sunshine, plentiful and non-fluctuating stream flows, sprightly winds and abundance of resilient and fast-producing biomass" [20]. The fact is that there is very little diversity of renewable energy resources (wind, hydropower, biomass, etc.), some are not truly renewable, and others are already committed for indispensable uses that may not involve energy production (e.g., crop by-products).

All this implies that whatever global stability may be possible in the future will emanate in large measure from the recognition of global interdependencies and the creation of global institutions capable of transferring meaningful technological options to the Third World where decisions on appropriate implementation must be made. The depleting liquid hydrocarbon base that supports the First World's energy consumption and the political instability of a world in which most of the people are "have-nots" combine to produce one of the future's most difficult challenges.

3. ENERGY FUTURES: REQUIREMENTS AND CONSTRAINTS

3.1. Growth and Conservation

We will not attempt in this paper to replicate the many excellent detailed estimates of conservation potential and structural analyses of resulting primary energy demand reduction—such as those of Hirst [21], Schipper [22], Von Hippel and Williams [23], the Institute for Energy Analysis [16], Ford Foundation's Energy Policy Project [24], etc. Rather, we will attempt some extrapolative quantification of our own.

Figure 2 shows an estimate of the increase in *technical* efficiency* of the U.S. energy system over the past century, drawn from the work of Putnam [13],

---

* The deficiencies of First Law efficiencies and advantages of Second Law efficiencies have been noted by several authors [2, 4]. We would suggest that it is the First Law efficiency that is of principal concern and that the value of Second Law efficiencies is in their use for optimization of First Law efficiencies.

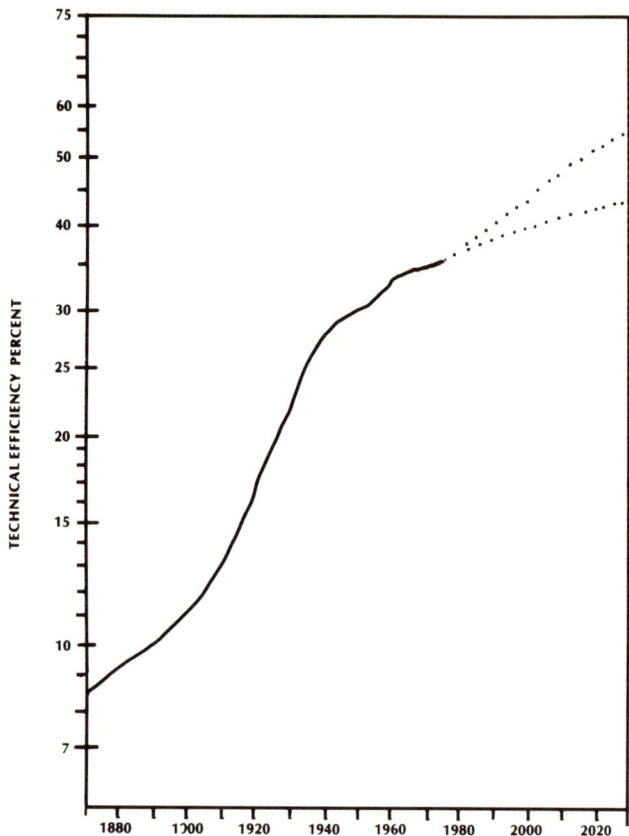

FIGURE 2. Technical efficiency of the U.S. energy system.

Cook [26], and Battelle [27]. Primary U.S. energy input can be multiplied by this average technical efficiency to yield the energy *output* delivered to end-use functions. (As Fig. 3 shows, energy output per capita has increased more consistently than energy input.)

It is this output energy that we would expect to be more closely related to the economic and social benefits received, and this appears to be the case. Figure 4 shows the classic energy/GNP ratio as well as the energy output/GNP ratio. In contrast to the classic input ratio, the energy output/GNP ratio increased during the continuing industrialization of the first third of the century, then flattened out to a fairly constant value during the second third. (A ratio of energy output to gross private domestic product appears even more constant in the

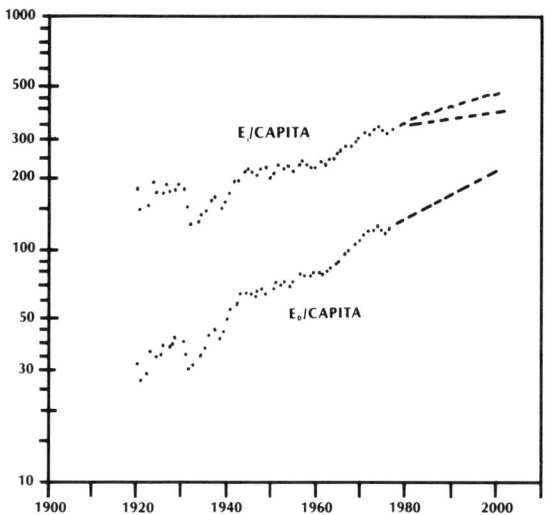

FIGURE 3. Per capita use of input and output energy in the U.S.

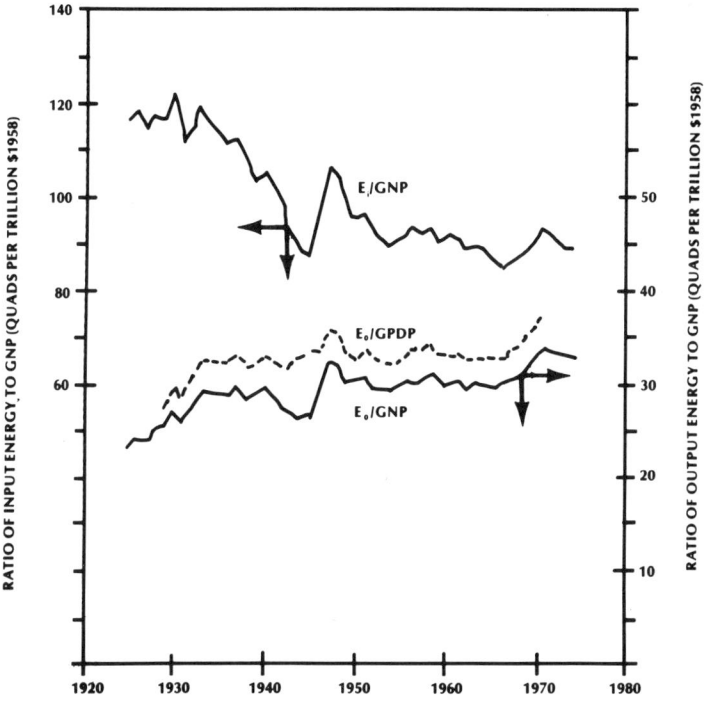

FIGURE 4. Historical energy/GNP ratios. (GPDP) Gross private domestic product.

1933–1966 period.) Then, in the late 1960s, the energy consumed per dollar of GNP took a sudden and dramatic jump.

If the technical efficiency of the U.S. energy system were increased from the current 36 or 37% to around 42% by 2000, and the energy output/GNP ratio brought back only to its historic (1933–1966) level by improvements in end-use applications, this would be sufficient to double GNP with slightly less than a 60% increase in primary energy supply (118 quads in 2000). The *potential* for conservation is considerably greater (ranging to 25% increase and less for GNP doubling), although there is far from unanimous agreement over what can be expected in terms of actual response and implementation. But we suggest that this level at least is necessary for a viable energy future, as discussed further below.

Accomplishing this will require that national policies are adequately addressed to achieving significant improvements in technical efficiencies and end-use requirements. This means careful attention to matching energy supplies to end-use requirements and the continuing development of efficient and appropriate energy conversion technology.

3.2. Capital Constraints on Energy Futures

Investment in energy-related facilities averaged somewhat less than 2% of GNP in the U.S. in the mid-1970s. (Energy purchases have accounted for about 5% of GNP.) With increased costs, the availability of investment capital looms large as a major constraint to the achievable range of energy futures. Total business fixed investment averaged 10.25% of GNP from 1966 to 1975 (declining to 9.35% in 1975), and new plant and equipment expenditures averaged 7.86% of GNP in the same period (declining to 7.26% in 1975). If a major struggle for available capital between the energy sector and other sectors is to be avoided, either energy investments would need to be held to around 2 to $2\frac{1}{2}$% of GNP, or the savings rate increased to make more capital available. Since the U.S. in recent years has had one of the lowest records of capital investment among the major industrialized nations, it is not unreasonable to assume that increased investment rates could be achieved through appropriate economic policies [28]. Even so, capital expenditures for energy-related facilities and equipment above about 3% of GNP would appear to be difficult to maintain in a balanced economy.

At an average annual growth rate in real GNP of 3.15% between 1975 and 2000, cumulative GNP would total 57 trillion constant 1975 dollars. Energy-sector capital expenditures between 2 and 3% of cumulative GNP would thus amount to between 1140 and 1710 billion dollars from 1975 to 2000.

Table II and Fig. 5 show estimates by the Institute for Energy Analysis [29] using the Bechtel Energy Supply Planning Model, by the Control Analysis Corporation [30] using the ETA model, and by the Ford Energy Policy Project

TABLE II
Estimates of Cumulative Energy Facilities
Investment Requirements, 1975–2000
(Billions of 1975 Dollars)

Estimate	2000 demand (quads)	Cumulative investment to 2000 (billions)
IEA, low	101.4	854
IEA, high	125.9	1110
ETA, high	145.4	~1550
EPP, TF	124	1468[a]
EPP, HG	185	2435[a]

[a] Adjusted from 1970 dollars by a factor of 1.392.

[23], of the cumulative capital investment in energy-related facilities required for a wide range of supply projections. Roughly speaking, very high year 2000 supply scenarios of over 150 quads would require in excess of 3% of cumulative GNP, while only moderate to low scenarios below 120 to 125 quads can stay within a level of about 2% of GNP.

In this sense, then, the high growth or "hard path" is not only unnecessary

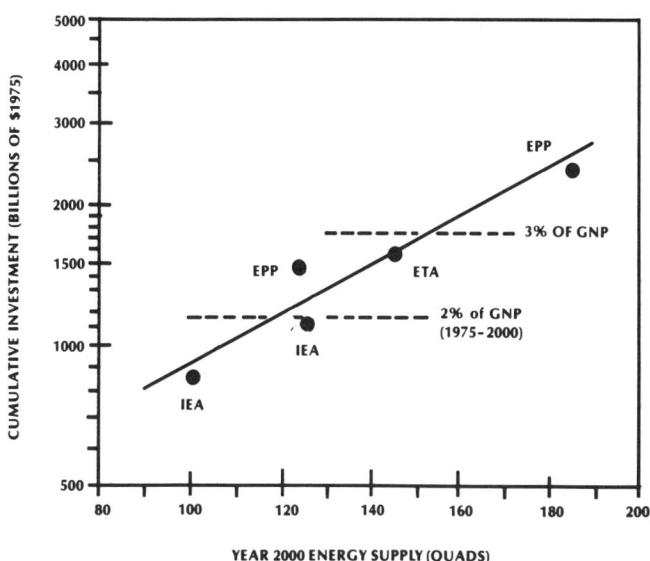

FIGURE 5. Estimates of cumulative 1975–2000 energy facilities investment.

because of the potential for conservation, but most likely intractable in terms of capital availability.

3.3. Estimates of Future Demand and Supply

Table III presents a comparison of several projections of U.S. energy demand in 2000 [16, 23, 24, 31–33]. (We have purposely omitted projections that we consider to be unrealistically high.) Projected energy growth in the residential and transportation sectors ranges from little or no growth to about 2.2% per year—reflecting the potential for conservation. Projected growth in the commercial sector ranges from −0.5 to 4%, averaging about 2% per year. Growth in the industrial sector is highest, ranging from about 1.5 to over 4.5% and averaging about 3% per year—reflecting the importance of industrial output to stable economic growth. Projected 1975–2000 growth in the industrial sector is substantially similar to the 1950–1973 average of 3.5%. Growth in the other

TABLE III
Comparison of Estimates of U.S. Energy Demand in the Year 2000 (Quads)

Estimate		Residential	Commercial	Transport	Industrial	Total
IEA	Low	18.3	10.9	22.2	50.3	101.4
	High	24.3	15.4	28.1	58.1	125.9
Von Hippel	Low	18.3	8.2	21.7	41.0	89.2
and Williams	High	26.9	11.6	27.8	46.2	112.5
EPP	ZEG	17.0	18.8	17.2	47.0	100.0
	TF	19.3	16.9	24.7	63.1	124.0
ERDA-48	COAT (V)	30.4		27.0	73.7	131.1
	IE (III)	40.1		32.7	87.1	159.0
ERDA-76-1	Low	—		—	—	117.9
	High	42.6		26.0	67.0	135.6
EEI	Case C	25.6	24.5	20.3	39.1	109.5
Average		21.6	15.1	24.8	57.3	118.8
1975		15.8	9.3	18.6	27.4	71.1
1975–2000 growth rate		1.3%	2.0%	1.2%	3.0%	2.1%

3 • Exclusive Paths and Difficult Choices

sectors would, however, be substantially below historic levels; growth averaged 4.3% per year in the residential and commercial sectors from 1950 to 1973, and averaged 3.4% per year in the transportation sector.

Table IV presents a comparison of corresponding projections of U.S. primary energy supply in 2000 [30, 34]. Several points are noteworthy:

- Even with significant conservation, oil and natural gas demand are still dangerously close to today's levels (even in IEA's 101-quad scenario, lower than most current energy projections, oil and gas use is only 16% below 1976 levels). This spells trouble, particularly for oil, since imports currently provide in excess of 45% of demand. Faced with declining domestic production, imports would have to rise; by the year 2000 it is questionable whether such supply will be readily available, even from OPEC.
- Nuclear and coal production rise sharply, but are limited nevertheless by practical constraints to the rate of expansion, by limits to interfuel substitution (e.g., coal gasification), and by institutional barriers and environmental concerns.
- New energy sources (geothermal, wind, solar, and biomass) provide only about 5 to 6% of demand in the year 2000—constrained by limits to market development and by economics.

Domestic production of petroleum products currently account for only about 20 quads out of a 35-quad demand, and will likely be lower by the turn of the century. It is clear that oil represents one of the greatest of many difficult problems in future supply/demand matching. Certainly, it would be desirable to have 3 to 4 times as much energy from new sources (estimated at 6.1 quads in Table IV) that could be substituted for petroleum.

Resource availability and numerous constraints on the rate of expansion of new supply again argue in favor of moderated energy growth. Reduction of oil demand through conservation and substitution of coal, efficiently used electricity or renewable sources, is a central problem.

While these analyses vary widely in their projections of future demand requirements, they are not dramatically different in terms of the qualitative mix of energy supply (with the exception, perhaps, of the Energy Policy Project's zero energy growth scenario). That is, they are all predominantly based on existing and some new "hard" technology, with renewable energy sources entering quite rapidly around 1990 to 2000. It is primarily on the impact of efficiency improvements and conservation that they differ.

The "soft" energy path stands out in stark contrast to these analyses. Lovins's figure of 35 quads of oil and gas use in 2000 is substantially below the 42.5 quads of ERDA-48's imaginative "combination of all technologies" (COAT) scenario, the 46.3 quads of IEA's low projection, or the 53.8 quad average. His

TABLE IV
Comparison of Estimates of U.S. Energy Supply in the Year 2000 (Quads)

Estimate		Coal	Oil	Gas	Nuclear	Shale	Hydro	Other[a]	Total
IEA	Low	17.7	30.3	16.0	27.2	—	3.8	6.4	101.4
	High	31.2	37.5	16.0	31.0	—	3.8	6.4	125.9
Von Hippel and Williams	Low	22.7	42.5		12.4	—	3.7	7.9	89.2
EPP	ZEG	18[b]	39	29[b]	3[b]	—	4	7	100
ERDA-48	Imp. Effy.	22.9	40.3	22.8	20.4	—	3.7	12.4	122.5
	COAT (V)	39.1	19.7	22.8	24.3	8[b]	3.7	19.4[b]	137
EEI	Case C	16.4[b]	42.7	16.4	23.0	—	3.8	2.2	104.5
DRI	Med.	47.7	35.4	15.8	26.0	—	3.8	2.8	131.5
CAC, ETA	High	37.2	68.7		31.8	2	3.8	1.8	145.3
Oregon Conf. Est.		45	30	20	20	—	5	10	130
Average		32.4	35.3	18.5	22.7	0.4	3.8	6.1	119.2
1976		13.8	35.0	20.3	2.0	—	3.1	—	74.3
Lovins's soft path		25	35					35	95

[a] Geothermal, solar, biomass.
[b] Not included in column average.

3 • Exclusive Paths and Difficult Choices   **89**

figure of 35 quads from renewable, "soft" energy sources in 2000 is well beyond what other analyses project—even the 23 quads of ERDA-48's COAT scenario (not all of which is "soft").

3.4. Summary

From the foregoing discussion, it should be clear that our view of the future is neither the "hard" path nor the "soft" one. Rather, it is a moderate and viable path that utilizes feasible conservation to ease resource and supply constraints and costs; that can live within the constraints of available capital; and that comes closest to a reasonable match of supply and demand. By contrast, a view that sees the first and foremost purpose of each future quad of solar power to replace a quad of nuclear power fails to see the problem for what it really is—the awesome task of replacing the conventional fluid fuels on which our economy runs today. The moderate or appropriate path we advocate is a hard path only in the sense that it contains of necessity a predominance of what some call hard technologies. But it is not "business as usual"—rather, it recognizes that the problem at hand is so difficult that it requires the best effort of every possible approach.

4. ANALYSIS AND COMMENTS ON AMORY LOVINS'S SOFT ENERGY STRATEGY

4.1. Introduction

Soft energy path advocates, and in particular Amory Lovins [6, 7], have put forward the proposition that there are two energy paths that the U.S. can take:

- A "hard" path, resembling present federal policy, that relies on rapid expansion of centralized high technologies, especially electric generation
- A "soft" path that combines a prompt commitment to efficient energy use, rapid development of renewable energy sources matched in scale and quality to end-use needs, and special transitional fossil-fuel technologies

These two paths, he believes, are mutually exclusive. A summary of Lovins's projections for the hard and soft paths are contained in Table V.

A few of the points that soft energy path advocates raise are both valid and important. For example:

- That change and action are needed to solve the energy problems of the U.S. and the world at large

TABLE V
Hard and Soft Path Energy Projections
(in Quadrillion BTU per Year)[a]

	2000		2025	
Source	Hard	Soft	Hard	Soft
Coal	58	25	84	0
Oil and gas	68	35	16	0
Nuclear	35	0	130	0
Soft technology	0	35	0	63
Total	161	95	230	63

[a] Note: In 1975 about 75 quadrillion BTU were used, almost 80% supplied by oil and gas.

- That energy policy should fit the goals of society and not be developed by "incremental adhocracy"
- That growth does not necessarily produce societal equity
- That institutional problems present one of the greatest barriers to formulating a viable energy strategy
- That GNP is an imperfect measure of social well-being
- That environmental and public health risks should be incorporated in strategy formulation
- That efficiency improvements in energy production and use can provide large energy and cost savings
- That the development of renewable energy sources is desirable

These points are not new, of course. It is the melding into a cohesive formula for energy strategy and social change that is unique with Lovins—a strategy based on the implementation of a small-scale, decentralized, renewable energy supply. Arguments for a decentralized, no-growth society have been advanced previously, as for example in the *Ecologist's* "Blueprint for Survival" [35], and contrasts between the characteristics of hard and soft technology have been outlined by Robin Clarke, founder of the Soft Technology Research Community in Wales [36]. But none have approached the level of Lovins's analysis.

A central factor in his soft strategy is his belief in the ready availability and relatively low cost of the "soft" technologies, and concomitantly, in the riskiness and high costs of centralized "hard" technologies. His hard technology assessment tends to be pessimistic and his soft technology assessment tends to be optimistic. He ignores a number of critical factors, including: limits to the feasible rate of change in the supply system and in demand substitution; strong regional variations in supply and demand; historical increases in electricity and natural gas supply despite the continual availability of less expensive fuels. More detailed analyses of Lovins's propositions are set forth below. It is crucial, however, to keep in mind that discussion of specific statements or analysis in Lovins's work can easily distract one from concentrating on the value of the social issues that he clearly considers to be a most important aspect of his soft path.

Before proceeding, two comments are in order:

- Conservation and efficiency improvements should not be considered an exclusive feature of the soft path. Conservation is compatible with both hard and soft technology, and the necessity for its implementation should transcend the debate over hard versus soft.
- There is a distinction to be made between a commitment to the development and deployment of appropriate inexhaustible energy sources and an immediate commitment to a totally soft energy path. What we try to point out below are some of the problems of the soft *path,* including its uncertainties and its capital intensity. One can dispute the feasibility of Lovins's soft energy strategy and still believe strongly in the importance of solar and other inexhaustible energy sources [37].

4.2. What Is Hard? What Is Soft?

Lovins's claim of exclusivity for the "hard" and "soft" paths has puzzled commentators on his articles. As his hard path lies well above what we consider to be a reasonable projection of the future, we do not comment on it, other than to state that it is most likely unattainable. In more structured projections, there is ample evidence that "hard" and "soft" can and must coexist, as well as the clear demonstration of the compatibility of conservation and hard technology. There is no reason why a belief in the necessity of maintaining current hard technologies and the desirability of developing new ones should be incompatible with a belief in the necessity of conservation and the desirability of implementing appropriate soft technologies. Soft and hard have coexisted in the past, and a reasonable vision can look to peaceful coexistence in the future.

Besides, it is by no means clear that technologies can be separated into exclusively defined categories of "hard" and "soft." In fact, the boundary appears fuzzy. Alcohol refineries would appear to be as "hard" as conventional ones, and biomass on the scale envisioned by Lovins would require considerable collection and distribution infrastructure. Hydro plants presumably range from hard to soft according to scale. Nuclear power is definitely hard in current applications, but the joint Finnish–Swedish "SECURE" project for a 200-MW slightly pressurized underground pool reactor for district heating [38] appears little "harder" than fluidized-bed coal systems. It is well to note, too, that the definition of "hard" and "soft" is user-dependent; what is soft for industrial use may well be hard for the individual consumer.

In short, making "hard" and "soft" not only exclusive paths but exclusive in definition appears unnecessarily polarizing and inhibiting to innovative approaches to future energy supply and demand.

4.3. Comments on Lovins's Analysis

Our analysis of the economics of Lovins's "soft energy strategy" shows that:

- He overestimates the cost of nuclear power by about 78%, and the cost of coal-fired electric power by about 42%.
- He neglects the higher efficiency of most electricity end-use applications as compared to direct fuel use.
- He neglects the lifetime of energy resources and equipment, which is important in fossil fuel use and crucial in solar applications.
- He greatly underestimates the capital cost of "soft" technology, e.g., by a factor of ten times or more in the case of 100% solar heating.
- As indicated earlier, the "hard path" to which he compares his "soft path" is a straw man. It disregards the viability of efficiency improvements and the necessity of conservation in any mix of future energy supply and use.
- Lovins's "soft path" is more capital-intensive even than his "hard path." We estimate the 1975–2000 capital cost of the "soft path" to be around $3 trillion and that of the "hard path" to be around $2 trillion (in constant 1975 dollars). A "moderate path" employing significant conservation and primarily conventional technologies would have a 1975–2000 capital cost of 1 to $1\frac{1}{2}$ trillion dollars.

Lovins's estimates of capital costs for various energy technologies are given in a unit of his invention, the "dollars a barrel per day" ($/bbl per day). This is meant to represent the capital investment (in constant 1976 dollars) for the total energy system required to deliver the energy equivalent of a barrel of oil per day to the place of end-use. Some of Lovins's figures for "hard," "soft," and "transition" technologies are given in Table VI.

It is important to understand what this unit does and does not include. It is supposed to include all capital costs of extraction, processing, conversion, transportation, and delivery. It is a capital cost only; it does not include costs of fuel, operation, taxes, etc. Energy equivalents are calculated from the heat content of the energy as delivered to end-use. Thus,

TABLE VI
Lovins's Estimates of Approximate Capital Investment (1976 Dollars) to Deliver the Energy Equivalent of a Barrel of Oil per Day, Assuming Present Technologies[a]

Energy system	$/bbl/day
Traditional direct fuels, 1950s–60s	2–3,000
U.S. frontier oil and gas, 1980s	10–25,000
Synthetic fuels from coal or shale	20–70,000
Coal-electric with scrubbers	170,000
Nuclear electric, mid-1980s	235,000
Common industrial and architectural leak-plugging	0–5000
Most heat-recovery systems	5–15,000
Very thorough building retrofits	25,000
Industrial cogeneration, 1970s	60,000
Retrofitted 100% solar space heat with no backup, mid-1980s	50–70,000
Wind-electric Canadian design	200,000

[a]From Ref. 48.

1 barrel per day = 5.496 million British thermal units per day
   = 42 gallons of oil per day
   = 5383 cubic feet of natural gas per day
   = 1610.5 kilowatt-hours per day
      (heat value) or 67.1 kilowatts-average per day
   = about 10,000 square feet of solar collectors
      for water heating or about 20,000
      square feet for solar space heating

This careful choice of unit gives rise to several difficulties in understanding its significance:

- It neglects, in most instances, the capital cost of the end-use of energy.
- It neglects the considerable variations in the efficiency of energy end-use. Energy is never used at 100% efficiency.
- Electricity is converted to barrels per day as if it were used entirely for resistance heating; yet in the majority of applications, electricity is used at higher efficiency than direct fuel use (e.g., lighting and electric motors) and in some cases at efficiencies effectively higher than the resistance heating value (e.g., air conditioning, refrigerators, and heat pumps). Only about 20% of U.S. electricity is used in heat applications that could be

called inefficient (e.g., water heating, cooking, and drying), and even these can be improved with heat pumps, microwave ovens, etc.
- Most important, the $/bbl per day does not incorporate the lifetime of the equipment in the particular energy system. The useful life of a mine, power plant, or solar panel is an important factor in costs and in the rate at which capital investment is needed. For example, for a frontier oil or gas field with an equivalent full-capacity life of 12 years and a total capital cost of $17,500/bbl per day, the capital investment over a 36-year period (about the lifetime of an electric power plant) would be $52,500/bbl per day if costs remain constant. If costs were to increase at 5% per year in constant dollars, the 36-year cost would be $(1.0 + 1.8 + 3.2) \times 17,500 = \$105,000$/bbl per day.

Despite all these problems with the $/bbl per day, Lovins's energy costs even in this unit are substantially incorrect.

He overestimates the cost of nuclear and coal-fired power generation. Even adding in the cost of transmission and distribution equipment (as Lovins does), the total system capital cost associated with a nuclear power plant is about $1140 per kilowatt installed (constant 1976 dollars). Accounting for capacity factor and transmission losses, this translates to about $1966 per *average* kilowatt *delivered,* or about $132,000/bbl per day. This is considerably below Lovins's figure of $200,000 to $300,000/bbl per day. A similar estimate for a coal-fired power plant with scrubbers yields a figure of about $120,000/bbl per day, as compared to Lovins's $170,000/bbl per day.

Simultaneously, Lovins seriously underestimates the cost of solar technology. He estimates that a solar heating system supplying *all* heating needs with no backup could be retrofitted with *existing* buildings at a cost equivalent to $50,000 to $70,000/bbl per day (just $2700 to $3800 for a typical home!).

A *solar water heater* system would provide between 150,000 and 225,000 BTU per year per square foot of collector area, depending on location and efficiency. Daystar Corporation's complete water heating package includes 41.6 square feet of collector and costs $995; installation costs about $900. A similar Grumman system costs $900, f.o.b. factory. Thus, the *installed* cost of a *water heating* system should be about $1800 to $1900, or about $44 per square foot of collector area. This yields a capital cost, in Lovins's units, of $390,000 to $600,000/bbl per day.

Combination *solar space and water heating* systems deliver less energy per square foot of collector because of reduced heating needs in summer. A system designed to provide *about half* of home heating needs should deliver between 100,000 and 140,000 BTU per year per square foot of collector.

In the third cycle of grants in HUD's Solar Demonstration Program, the average installed cost is estimated at $31.74 per square foot of collector. The

3 • Exclusive Paths and Difficult Choices

Solar Utilities Company of California recently installed a 500-square-foot system at a cost of $12,000, or $24 per square foot. For a cost range of $24 to $32 per square foot, the capital cost of a conventional solar space and water heating system is $344,000 to $642,000/bbl per day.

These figures assume that existing problems of poor reliability, low efficiency, and short lifetime will be eliminated. Costs can be expected to decrease in the future; currently lower costs can possibly be found. However, projections of major cost reductions would be speculative. Life expectancy and maintenance costs are still largely conjectural.

Note that Lovins's figures are for a 100% solar system. In fact, a solar installation that required *no* backup would cost about *four* times as much as one supplying about 50% of heating requirements—or about *twice* as much per unit of heat delivered. Increased storage capacity could be substituted for some collector area but this would not significantly reduce costs, although we recognize an economically optimum trade-off between storage capacity and collector area exists.

Apart from optimistic cost figures, Lovins has apparently blundered also in the overall efficiency he uses for solar applications. Although collector efficiencies of around 60% can be achieved on clear days, the actual fraction of total annual insolation delivered to space and water heating by a 100% system is only about *15%*. First, 25 to 30% of insolation is too diffuse or intermittent to collect. Second, average collector efficiency will be significantly less than the maximum (going down as storage temperatures go up). Third, storage losses are about 15% (even with correction for loss of unwanted heat). Fourth, the duty factor for a $1(m^2 + m^3)$ system is only around 50%. Thus, the net fraction of total insolation that is used is approximately $0.72 \times 0.50 \times 0.85 \times 0.50 = 0.15$ not including pumping power. While the duty factor could be increased with higher storage temperature, this would result in a decrease in collector and storage efficiencies.

The current price of 100% solar heating in new homes appears to be about $600,000 to $900,000/bbl per day, professionally installed. For example, the Solar Heat Corp. house in Waltham, Massachusetts, designed for 100% heating though actually providing somewhat less, has been analyzed. This house is essentially what Lovins is describing—a well-insulated house (4.3 BTU/degree-day per foot^2 net space heat load) and moderately sized storage (0.55 m^3/m^2). Yet the capital cost, based on expected performance, is equivalent to *$615,000/ bbl per day*. Since this is a new house built by the president of Solar Heat Corp., its $20/foot^2 cost could be significantly less than for commercial retrofit.

As for the future, it is hard to accept that the capital cost of 100% systems can come much below $200,000/bbl per day. It is highly unlikely that a 1 m^3/m^2 system could achieve an overall annual efficiency greater than 20% ($0.75 \times 0.53 \times 0.85 \times 0.60$). At a cost of $100/m^2 in an average annual insolation of 180 W/m^2, this translates to

$$\frac{100 \times 5.825 \times 10^6}{180 \times 0.2 \times 3.412 \times 24} = \$197{,}600/\text{bbl per day}$$

At a more realistic 15% efficiency, the figure would be $263,500/bbl per day.

For *wind–electric* systems, Lovins gives a figure of $200,000/bbl per day, based on the 200-kW, vertical-axis wind turbine designed by the National Research Council of Canada and constructed by Dominion Aluminum Fabricating, Ltd. The first turbine was erected at Cap aux Meules, Magdalen Islands, in March 1977 and is being brought up to design power output. The contract price for this unit is $235,000, exclusive of foundation and grid connection [39]. Assuming a total installed cost of $280,000 ($1400/kW) a capacity factor of 30%, $150/kW for transmission and distribution, and 10% line losses, the capital cost is $5740 per average kilowatt delivered, or $383,000/bbl per day. Multiple-unit production cost could be as low as $260,000/bbl per day.

Costs for ERDA/NASA's horizontal-axis wind turbines are about two to three times as high.

Few U.S. locations could match the capacity factor of the Cap aux Meules site where the average wind speed is 19 mph. Because the output of a wind turbine varies as the cube of the wind speed, a site with 15-mph average annual wind speed would provide only half the power (15% capacity factor) of a site with 19-mph average wind speed—doubling the capital cost in $/bbl per day.

These systems do not include any form of energy storage for calm days. Thus, capital investment in a backup generator for part or all of the capacity of the wind turbine is required (at Cap aux Meules a diesel is used).

In short, Lovins overestimates some "hard" energy capital costs and underestimates "soft" energy capital costs (see Table VII).

TABLE VII
Total Energy System Capital Costs ($/bbl per day)

Source	Lovins	ERG
Nuclear-electric	200–300,000	132,000
Coal-fired electric	170,000	120,000
Solar		
Conventional	—	340–640,000
No backup	50–70,000	650–1,250,000
Wind-electric	200,000	260–750,000

4.4. On Electrical Generation

Comments on Lovins's analysis of electric generating costs notwithstanding, it is the case that electricity is generally more expensive than direct fuel in most intersubstitutable applications; early estimates [40] suggest this to be potentially true for synthetic gas. Inappropriate applications of electricity are costly and wasteful, and policy should attempt to avoid such through supply/demand matching and new end-use technology.

On the other hand, electricity since its introduction has been more expensive on the average than fluid fuels, just as natural gas has been more expensive than coal. Much of the dramatic growth in the use of these fuels must be attributed to their convenience—a factor that is hard to quantify. Future growth in electricity use can be expected for several reasons, including:

- Increased demand of electrical end-uses
- Convenience, particularly in interfuel switching
- Institutional problems for direct coal use
- Limits to nonelectrical energy source production

This raises important questions about efficiency. Certainly, the development of new fluid fuels, natural or synthetic, particularly for transportation, must be a central part of a cogent energy policy, as must be the early development of renewable sources for applications that would otherwise be supplied electrically (or the development of direct use of nuclear heat for process applications).

There are critical policy decisions required to avoid electric substitution in inappropriate applications. Industrial switching to electricity is continuing because of convenience in procurement and application. (Netschert [41] reports of factories using resistance heating for steam generation!) With natural gas in short supply and oil suspect since the embargo, many industries may prefer to switch to electricity and pay the price premium rather than become embroiled in fuel procurement problems or in federal, state, and local regulations over coal use—leaving utilities to act as their surrogate in procurement and regulatory matters.

Most current estimates project year 2000 electricity use to be 2.5 times or greater than today's. This represents a considerable onus to further improve the efficiency of electric generation, distribution, and end-use. Major efforts on the part of industry and government will be required to implement cogeneration and total energy systems, improve generating efficiency (topping and bottoming cycles, MHD, etc.), improve T & D efficiency (transmission, storage), and improve end-use efficiency (insulation, heat pumps, high-temperature heat pumps, solar-assisted heat pumps, storage, lighting, refrigeration, air conditioning, water heaters, electric motors, etc.).

To accomplish this will require innovative and flexible management—a sense of purpose tempered by a realistic view of what is achievable. As Pickering and Maxey have said: "practical conservatism goes hand in hand with abstract radicalism, because meaningful novelty is effectively postponed by both parties" [42]. The road we take can neither be impractical, nor the same stuff in larger doses.

4.5. Comments on Cogeneration

A major increase in use of cogeneration—whether electricity as a by-product of industrial steam generation or steam and heating as a by-product of utility electrical generation—is undeniably desirable for the increased efficiency and cost savings that result. Von Hippel and Williams [23] state that 60% of all industrial steam production (greater than 100,000 lb/hr) is suitable for cogeneration. Thus, if all of this 60% were produced in the year 2000 by fluidized-bed, gas-turbine (Stal-Laval type) cogeneration systems, they estimate an electric output of 3.6 quads/year (at 3413 BTU/kW-hr, or about 10^{12} kW-hr/year) could be achieved—about half the present U.S. electrical consumption. The potential for savings from utility industry waste heat are even greater.

The barriers impeding major deployment of cogeneration are (like much of the energy problem) institutional:

- Geographic mismatches of primary production and "by-product" demand (note in this regard that Lovins's suggestion to use only hydro and industrial cogeneration for electrical production involves serious geographical mismatches)
- The current policy of ex-urban siting of large power plants and resistance to urban siting of even relatively small plants (e.g., community resistance to Harvard Medical School's total energy facility)
- Regulatory problems (rates, environmental requirements, etc.)
- Inadequate financial incentives for cogeneration (beyond the producer's own needs) or for construction of district heating systems
- Motivating private industry to take on the increased burden of cogeneration production beyond its own needs

In the last regard it should be noted that industrial cogeneration, while more efficient, is hardly a "something-for-nothing" proposition. As Fig. 6 shows, the Stal-Laval cogeneration system requires 20% less fuel for steam and electricity production than the equivalent separate boiler and power plant (assuming that waste heat from the power plant is not utilized). However, in switching to this system, the industrial user is faced with an 80% increase in fuel procurement and a 130% increase in emissions as well as the increased capital cost. It is difficult to convince an industrial concern to accept this burden unless it plans to use a significant fraction of the electric power, or is given sufficient financial

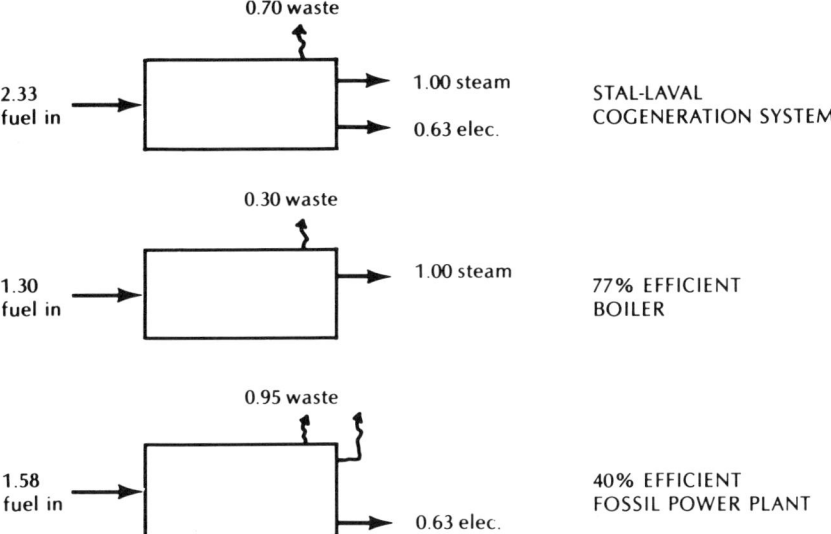

FIGURE 6. Comparison of cogeneration and separate generation.

inducement (such as a high tax credit or high rate of return), or has the utility cover part of the construction costs. In fact, given current institutional problems of coal use, it may be difficult enough to get the industrial producer to switch from fluids to coal, let alone to install cogeneration. Nevertheless, cogeneration is desirable and every effort needs to be made to remove the barriers to utility and industrial cogeneration.

As a final point, it should be noted that cogeneration and other total energy system concepts transcend the boundaries of "hard" and "soft," as does their desirability. A "hard" example is the proposed district heating system for Uppsala, Greater Stockholm, and Sodertalje using the Forsmark nuclear plant and heat pump feedback at the tail end [43].

4.6. Transportation

The transportation sector is given only light attention by Lovins. This is a crucial area because of the potential for efficiency, conservation (as Lovins point out), and increased mass transit, and because of the difficulties in meeting mid- to long-term demand.

Lovins's suggestion for a fuel alcohol industry 10 to 14 times the size of the present U.S. beer and wine industry does little to help. Szego and Kemp [44] estimate that land managed and harvested in a manner similar to a Southern tree farm would be sufficient to support 1 MW of electrical production per square mile. To produce one-third of current transportation fuel demand (19 quads) in this manner would require about 100,000 square miles—about 20% of current U.S. harvested farm acreage or about 35% of U.S. national forest acreage. Variations in crop, climate, insolation, water supply, and long-term productivity would likely increase the required area significantly. Biomass conversion (particularly of wastes) needs better attention, but the scale and manner of Lovins's approach seem inappropriate in an era when land use, especially for food problems, is critical.

Nevertheless, faced with declining oil supplies, a minor penetration of electric vehicles before the turn of the century, and uncertain prospects for synthetics, supply/demand matching in the transportation sector looms as a major difficulty.

4.7. On Nuclear Power

There are glimpses throughout Lovins's paper that his desire to halt nuclear power is not only intense but undoubtedly one of the seminal foci from which his statement grew. A strategy based on preoccupation with finding a rationale to prove nuclear power unnecessary, without thorough quantification of the criteria by which it is judged to be so obviously unacceptable, is likely to be flawed in its coherence and its adhesion to the practical. It would be fruitless to attempt detailed comment on such fundamental errors as "a nuclear temperature of millions," and diversionary to rerun the existing data on the safety and environmental impact of nuclear power relative to other energy sources [45]. The issue of future energy paths is more fundamental than the nuclear debate, although it has lain quietly behind that debate for many years.

But some points are in order. Lovins trots out a weary list of risks in operation, human fallibility and malice, risk through greed for profit, violence and coercion, abrogation of civil liberties, guarding of long-lived wastes, establishment of an elite priesthood, long-term effects of nuclear mishaps, and, in general and without evidence, a shopping list for a technology where everything is to be as bad as possible, in contrast to soft technology where everything is to be as good as possible.

More than 40 countries have now decided that nuclear power is an essential part of their future development. To suggest that "in almost all countries the domestic political base to support nuclear power is not solid but shaky" is contrary to the fact of a growing world commitment to nuclear power that is outpacing that of the U.S. It is ridiculous to quote Norway, which will not need nuclear power for some time, or Australia and New Zealand, which have not needed it until now, as examples of countries that have "rejected" nuclear power. International commitments to nuclear power will continue to grow of necessity.

3 • Exclusive Paths and Difficult Choices   101

International nuclear proliferation is a matter that requires very serious efforts for solution. But it is naive to suggest that a unilateral cessation of domestic *commercial* nuclear power activities (neglecting research and military programs) will bring a halt to proliferation. If the U.S. were to halt its nuclear energy programs, others might falter temporarily but it is doubtful that they would halt. The export of nuclear technology should not go uncontrolled, but it is well to note that no nation has yet produced a nuclear weapon through its commercial nuclear programs. As is often the case, the simple solutions sound attractive but they are unlikely to work. Control of international nuclear proliferation will mean hard work to develop comprehensive safeguards and international agreements, with the possibility of sanctions.

4.8. Environmental Effects

Embarking on a soft path with no immediate prospect of major input from clean, renewable sources means in essence the dispersed use of fossil fuels—oil in diesel generators, coal in basement furnaces, and a myriad of other polluting applications, including many present ones like the automobile. With dispersal, pollution control is more difficult, less amenable to regulation, and hardly more environmentally sound than the "hard" path. For example, I. Nisbet of the Massachusetts Audubon Society has stated [46] that:

> Wood fires are pleasant to sit by, but they are extremely inefficient sources of heat and they produce large quantities of particulates, carbon monoxide and other air pollutants. Stoves are more efficient and less polluting but nevertheless typically produce more particulates, carbon monoxide, and sulfur oxides per unit of heat than we permit from new fossil-fueled power plants. Wood fires may not do much harm if they are limited to dispersed suburbs and rural areas, but burning wood as primary fuel in urban areas conjures visions of a scrubber on every stove.

Nor would Lovins's path be totally "soft," since mining, refining, and delivery would still be essentially "hard." Without assurance of major solar energy input, the soft path would not necessarily be preferable in terms of our environment. Rather, it could be a return to negligent ways of the past in a much more heavily populated world.

On the other hand, the deployment of solar-based energy sources to the maximum extent feasible is obviously desirable in terms of their low environmental impact.

4.9. The Cost of Hard, Soft, and Moderate Energy Paths

Lovins has given no estimates of the total cost of his soft energy path. Since Lovins provides no detailed breakdown of his soft energy path (only a "shopping list" of possibilities), we have developed an illustrative one for the year 2000, matching the distribution of oil, gas, coal, and soft technology to meet the requirements of a 95-quad high-conservation energy demand and following Lovins's suggestions as closely as possible. It is clear that monumental changes would be required even by 2000—with just 37% "soft" energy as compared to 100% by 2025. For example:

- 60% of all industrial process steam boilers would be converted to coal-fired, fluidized-bed, gas-turbine cogeneration systems
- 12% of industrial steam boilers would be coal-fired and 27% would be solar-powered
- All industrial heating would be converted to gas
- All cooking and drying would be converted to gas
- Virtually all residential and commercial space would be equipped with high-efficiency solar heating (70% of space heat, 83% of water heat) with gas backup
- 55% of cooling would be solar with electric backup
- Electricity demand would be 23% higher than in 1975 (57% cogenerated, 15% wind, 15% hydro, 12% coal)
- 40% of transportation demand would be provided by biomass alcohol

Of course the matching of supply and demand and the mix of soft technology can be varied, but this does not substantially affect the pattern or scale of deployment. In fact, the changes required seem hardly less staggering than for Lovins's hard path scenario, particularly if the associated support facilities and infrastructure changes are included.

Table VIII gives our estimate of the total capital cost of the soft path to the year 2000—$2560 billion (1976 dollars) for the 30 to 35 quads of soft supply plus about $400 billion for conventional fossil supply and (very roughly) $140 billion for early retirement of central electric plant and transmission—or a total of *$3100 billion*. The uncertainty in this $3.1 trillion "soft path" cost to the year 2000 is probably on the order of plus or minus 1 trillion dollars.

This estimated capital cost is approximately $5\frac{1}{2}$% of cumulative GNP for the period 1975–2000 (at 3.1% real GNP growth)—a figure that undoubtedly precludes its feasibility in terms of capital formation. In fact, it is more expensive than Lovins's hard path, and much more costly than a reasonable moderate path.

Figure 7 repeats Fig. 5's presentation of estimates by the Institute for Energy Analysis [16], the Control Analysis Corporation [30], and the Ford Foundation's Energy Policy Project [24], of the cumulative capital investment in energy-related facilities required for a wide range of supply projections. Superimposed are our estimates of the costs of "hard," "soft," and "moderate" paths. Roughly speaking, very high year 2000 energy supplies of over 150 quads would require in excess of 3% of cumulative GNP, while only moderate to low energy requirements below 120 to 125 quads can stay within a level of about 2% of GNP. Even

TABLE VIII
Illustrative Estimate of Soft Path Cumulative Capital Cost to the Year 2000 (1976 Dollars)

Item	Unit cost ($/bbl/day)	Year 2000 supply (quads)	Delivery factor[a]	Cumulative cost to 2000 (billions)
Industrial cogeneration	50,000[b]	9.9 steam	0.75	298.7
Transportation biomass	30,000[c]	9.0	(1.0)	134.6
Solar space heating	400,000	8.0	0.7	1116.6
Solar water heating	200,000	2.5	0.7	174.5
Solar air conditioning	200,000[d]	2.5	0.7	174.5
Solar process heat	250,000	4.5	0.75	420.6
Wind-electric	380,000	1.3	0.9	221.6
Hydroelectric	130,000	1.3[e]	0.9	18.8
Conventional coal, oil, and gas supply				400.0
Retirement of central electric capacity				140.0
				3100

[a] Since unit costs are given in $/bbl/day delivered and supply is given in primary quads, this factor corrects for the efficiency of conventional supply. Cumulative cost is calculated as unit cost times supply times delivery factor times 498,500 bbl-yr/quad-day.
[b] Assuming a capital cost of $100 million for a 65-MWe, 105-MWt system, daily delivery of 1270 bbl-equivalent steam, 785 bbl-equivalent electricity (90% capacity factor, 5% loss factor).
[c] FEA high estimate.
[d] Assumed incremental cost for a building also equipped with solar heating.
[e] Only 0.3 quad of this is new capacity.

allowing for uncertainties and neglected cost items, it would appear that the cumulative 1975–2000 cost of a moderate (100 to 125 quad) path would be in the neighborhood of 1.0 to 1.25 trillion dollars, and certainly should not exceed 1.5 trillion.

This is not to suggest that solar energy is not a practical alternative for the future. Table IX compares the cost of several alternatives for home heating. Even with fairly optimistic assumptions about cost, performance, and lifetime, solar heating is the most expensive. However, with future improvements in solar applications and future increases in oil, gas, and electricity costs, solar energy has the potential to compete economically with conventional fuels. In some applications it is already competitive today. We believe that solar energy will be economically viable and will make an increasingly important contribution to U.S. energy supply. But our point is to emphasize that a "soft energy strategy" would be very capital-intensive, and that Lovins's soft economics are simply incorrect.

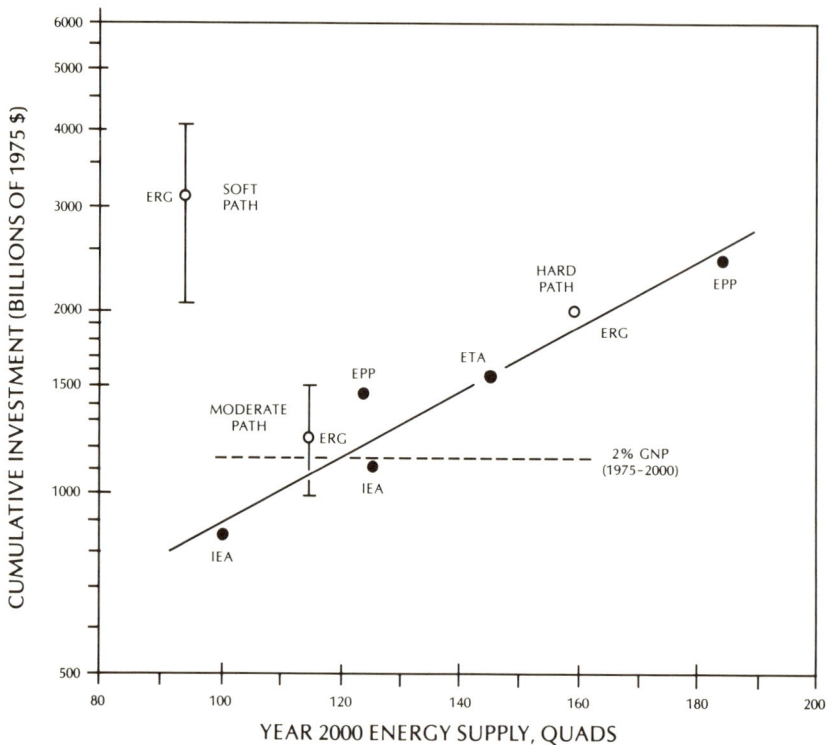

FIGURE 7. Estimates of cumulative 1975–2000 energy facilities investment, including estimates of hard, soft and moderate paths. (IEA) Institute for Energy Analysis, (ETA) Energy Technology Assessment Model, (EPP) Energy Policy Project of the Ford Foundation, (ERG) Energy Research Group.

4.10. Social Goals

Discussion of the technical fine points of Lovins's arguments probably has restricted value. Consider the following statements:

- "The distinction between hard and soft paths rests not on how much energy is used, but on the technical and sociopolitical *structure* of the energy system," or "Whether an energy system is high or low is derivative and less interesting."
- "A serious question can be raised whether economic calculations are particularly relevant today."

TABLE IX
Comparison of Home Heating Costs

System		Cost ($/MBTU)
Solar heating delivering 120,000 BTU/foot^2 per year and costing $25/foot^2		19.42[a]
Oil, at 50¢/gallon with 65% burner efficiency		5.55[b]
Electricity		
Resistance heat	3.5¢/kW-hr	10.25[b]
	5.0¢/kW-hr	14.65
Heat pump	3.5¢/kW-hr	5.13
(COP = 2)	5.0¢/kW-hr	7.32

[a] 9.32% annual fixed rate; based on 20% down payment, an 8½% 20-year mortgage on the balance, 1½% for annual maintenance, and full tax deduction of down payment and all interest charges.
[b] Cost of furnace, resistance heaters, or heat pump not included since one of these would be required for solar backup.

- "If nuclear power were clean, safe, economic, assured of ample fuel, and socially benign per se, it would still be unattractive because of the political implications of the kind of energy economy it would lock us into."

Thus, discussions of conservation, economics, rate of transition, nuclear safety, or the need for at least some hard technology must be seen as secondary to the social implications that Lovins considers "the guts of the issue" [47].

Social and institutional issues are discussed elsewhere in this report, and elegantly in a paper by Pickering and Maxey [42]. But some comments are appropriate here.

It is doubtful to us that energy can be used as the major instrument of social change to a decentralized society. Given that the "lead times" of the political and economic structures are shorter than for the energy system, it would seem that the changes would first have to be made in the structure of the economy and political thinking. Nor do we believe that energy strategy should be used as the means to manipulate a path for society; rather, energy policies must be developed in consonance with social objectives that have been openly agreed upon.

While one of the goals of soft energy path advocates, including Lovins, appears to be a decentralized society freed from the oppression of central bureaucracies, there is generally no real hint of the extent to which such an advocate would wish to see decentralization in government, nonenergy industries, or living patterns generally. Almost without exception, the advocate of an exclusively soft energy path does not consider the "hard" coercion that is likely to be required to implement the soft path, which is expensive and which limits individual choice in energy consumption patterns.

It is far from clear that the soft technologies avoid considerable infrastructure in manufacturing and distribution, particularly if the benefits of mass production are to be realized. To quote Pickering and Maxey again, "if the soft technologies are as commercially viable as Lovins maintains, they will attract investment and they will become absorbed in the web of commercial corporations competing in the manufacture, distribution, marketing and servicing of them. They will remain 'loyal' only to the degree that they are not generally marketable and, therefore, remain marginal to the supply of energy. What is more, Lovins does not seem to be aware that 'localism' is itself a cultural value whose appropriateness needs to be assessed."

It gives cause for concern when we hear that "soft technologies are ideally suited for rural villagers and urban poor alike" and that soft technologies "do not carry with them inappropriate cultural patterns or values" for poor countries. This seems hardly less than the "noble savage" concept resurrected in modern dress. We can point out to other peoples the mistakes we feel we have made, but we cannot dictate the path they should take, for the freedom of choice must be theirs.

5. OVERVIEW: MYTHS AND REQUIREMENTS

If one does not choose to consider explicitly the decision processes of those involved in the implementation of energy policy, there is always the danger that myths may form unstated assumptions in policy formulation. The following are examples of the kinds of myths that we have found to be alive and well and living in the middle of today's energy debate:

- The myth that conservation is "soft" rather than "hard." It is neither.
- The myth that electric utilities want rapid growth in electricity demand. In fact, profits can be better maintained in inflationary periods by avoiding any unnecessary construction programs and the frequent trips to the public utility commission that they require.
- The myth that technological fixes come quickly or easily. This is the "if we can put a man on the moon, then . . ." syndrome.
- The myth that scientists should agree. The future is uncertain; to the extent that scientists are futurists, one should expect them to disagree.
- The myth that there exist riskless technologies in energy supply.

In reviewing much of the current analysis on the formulation of energy policy, we have come to view the adoption of a certain type of strategy *process* as a prerequisite to responsible formulation. The basic criterion of this process is that it reflect the ethical responsibility of the strategist to create options for choice more than attempting to orchestrate the future. Rather than picking a future demand and advocating an exclusive set of energy supply options, energy strategy formation should focus on: (1) the strategic decisions necessary to create options that will enable future society to respond to a range of demand; (2) the kinds of options that ought to be created; (3) the deployment of options that have been created; and (4) the uncertainty that surrounds the entire process. In assessing viable strategy, the planner should therefore look to considerations of operational flexibility, institutional incentives (as a function of time), positive feedback to the political process, contingency analysis, and measures of successful implementation.

Our view of the advocacy of an exclusively soft energy future is that it flows from faulty strategic reasoning and is typically just plain wrong in assessments of cost, efficiency, and coercive social impact. Neither "hard" nor "soft" paths can accomplish such goals.

REFERENCES

1. M. W. Goldsmith, I. A. Forbes, J. C. Turnage, S. V. Weaver, and A. R. Forbes, *New Energy Sources: Dreams and Promises,* Energy Research Group, Framingham, Mass. (1976).
2. I. A. Forbes, *Energy Strategy: Not What But How,* Energy Research Group, Framingham, Mass. (1977).
3. *The Energy Daily 5,* No. 105 (1977).
4. *The Nation's Energy Future,* WASH-1281, U.S. Atomic Energy Commission, Washington, D.C. (1973).
5. C. Starr, The year 2000: Energy enough?, *EPRI J. 1,* No. 5 (1976).
6. A. B. Lovins, Scale, centralization and electrification in energy systems, *Symposium on Future Strategies for Energy Development,* Oak Ridge Associated Universities, Oak Ridge, Tenn. (1976).
7. A. B. Lovins, Energy strategy: The road not taken, *Foreign Affairs 55,* No. 1 (1976).
8. A. B. Lovins, *Soft Energy Paths: Toward A Durable Peace,* Ballinger Publishing, Cambridge, Mass. (1977).
9. C. L. Wilson, *Energy: Global Prospects 1985–2000,* Report of the Workshop on Alternative Energy Strategies, McGraw-Hill, New York (1977).
10. D. Bell, The year 2000: The trajectory of an idea, *Daedalus (Boston) 96* (1967).
11. V. C. Ferkiss, *Technological Man: The Myth and the Reality,* The New American Library, Inc., New York (1970).
12. T. Roszak, *The Making of a Counter Culture,* Doubleday, New York (1969).
13. P. C. Putnam, *Energy in the Future,* Van Nostrand, Princeton, N.J. (1953).
14. C. M. Summers, *Energy and Power,* Freeman, San Francisco (1971).
15. C. Marchetti, On strategies and fate, *Second Status Report of the IIASA Project on Energy Systems,* International Institute for Applied Systems Analysis, Laxenburg, Austria (1975).
16. *U.S. Energy and Economic Growth, 1975–2010,* ORAU/IEA-76-7, Institute for Energy Analysis, Oak Ridge, Tenn. (1976).
17. *Energy Balances of OECD Countries 1973/1975,* Organization for Economic Co-Operation and Development, Paris (1977).

18. J. Darmstadter, J. Dunkerly, and J. Alterman, How Industrial Societies Use Energy: A Comparative Analysis, preliminary report of a study prepared by Resources for the Future, Washington, D.C. (April 1977).
19. C. Marchetti, Primary energy substitution model on the interaction between energy and society, *Chem. Econ. Eng. Rev.* (Aug. 1975).
20. V. Smil, Renewable energies: How much and how renewable?, *Bull. At. Sci.* (Dec. 1979).
21. E. Hirst, Residential energy use alternatives: 1976 to 2000, *Science 194* (1976).
22. L. Schipper, in: *Annual Review of Energy*, Vol. 1, Annual Reviews, Inc., Palo Alto, Calif. (1976).
23. F. Von Hippel and R. H. Williams, Testimony on Behalf of the State of Wisconsin Before the GESMO Hearing Board, Docket No. RM-50-5, U.S. Nuclear Regulatory Commission, Washington, D.C. (March 1977).
24. *A Time to Choose—America's Energy Future*, Final Report by the Energy Policy Project of the Ford Foundation, Ballinger Publishing, Cambridge, Mass. (1974).
25. *Efficient Use of Energy*, AIP Conference Proceedings No. 25, American Institute of Physics, New York (1975).
26. E. Cook, *Study of Energy Futures*, Environmental Design Research Associates, Chapel Hill, N.C. (1975).
27. *An Input/Output Analysis of Energy Use Change from 1947 to 1967*, EPRI EA-281, prepared by Battelle Pacific Northwest Laboratories for Electric Power Research Institute, Palo Alto, Calif. (Nov. 1976).
28. R. Hall and D. Jorgenson, *Tax Incentives and Capital Spending*, The Brookings Institution, Washington, D.C. (1971).
29. *Economic and Environmental Implications of a U.S. Nuclear Moratorium, 1975-2010*, Vol. II, ORAU/IEA (M)-76-10, Institute for Energy Analysis, Oak Ridge, Tenn. (Sept. 1976).
30. *ETA Report and Analysis Service*, Control Analysis Corporation, Palo Alto, Calif. (1977).
31. *A National Plan for Energy Research, Development & Demonstration: Creating Energy Choices for the Future*, ERDA-48, U.S. Energy Research and Development Administration, Washington, D.C. (June 1975).
32. *A National Plan for Energy Research, Development & Demonstration: Creating Energy Choices for the Future*, ERDA-76-1, U.S. Energy Research and Development Administration, Washington, D.C. (April 1976).
33. *Economic Growth in the Future*, Edison Electric Institute, New York (1976).
34. *Proceedings of the Conference on Magnitude and Deployment Schedule of Energy Resources*, Oregon State University, Portland (July 1976).
35. E. Goldsmith (ed.), "Blueprint for Survival," *Ecologist 2*, No. 1 (1972).
36. J. P. Milton, Communities that seek peace with nature, *The Futurist* (Dec. 1974).
37. S. H. Butt, in: *Soft vs. Hard Energy Paths*, Charles Yulish Associates, Inc., New York (June 1977).
38. *Nucleonics Week 18*, No. 19, p. 14 (1977).
39. R. J. Templin and P. South, Canadian wind energy program, Vertical-Axis Wind Turbine Technology Workshop, Albuquerque, New Mexico (May 1976).
40. *The Energy Daily 5*, No. 87 (1977).
41. *Proceedings of the Workshop on Modelling the Interrelationships Between the Energy Sector and the General Economy*, EPRI-SR-45, Electric Power Research Institute, Palo Alto, Calif. (July 1976).
42. G. W. Pickering and M. N. Maxey, in: *Soft vs. Hard Energy Paths*, Charles Yulish Associates, Inc., New York (June 1977).
43. *Energy International 13*, No. 12, p. 19 (1976).
44. G. C. Szego and C. C. Kemp, Energy forests and fuel plantations, *Chemtech* (May 1973).

45. I. A. Forbes, M. W. Goldsmith, A. C. Kadak, J. B. Muckerheide, J. C. Turnage, and G. J. Brown, *The Nuclear Debate: A Call to Reason 2,* Energy Research Group, Framingham, Mass. (Sept. 1976).
46. I. C. T. Nisbet, Confessions of an environmental miser, *Technol. Rev. 79,* No. 5 (1977).
47. *The Energy Daily 5,* No. 72 (1977).
48. A. B. Lovins, "Invited Testimony for Hearings on the Costs of Nuclear Power," Environment, Energy and Natural Resources Subcommittee of the House Committee on Government Operations, Washington, D.C. (Sept. 1977).
49. *Mineral Resources and the Environment,* prepared by the Committee on Mineral Resources and the Environment (COMRATE) and the Committee on Natural Resources, National Research Council for the National Academy of Sciences, Washington, D.C. (1975).

4 An Energy-Deficient Society

Robert E. Uhrig

With the rapidly growing worldwide consumption of energy and the recognition of the finite supply of energy resources, we hear more and more about conservation, about a low-energy-use society, a reduced life-style, and a no-growth society. The U.S. is often cited as the worst offender of these concepts with its gas-guzzling automobiles, its poorly insulated buildings, its climatically controlled, windowless skyscrapers, and its alleged inclination for frivolous uses, such as electric toothbrushes, electric can openers, and air conditioning. Sweden is held as the model of a modern industrial society utilizing half the per capita consumption of energy.

We often hear the statistic quoted that the U.S. has some 6% of the world's population and yet uses 32% of the world's energy. There is an implication in this statistic that we, the people of the U.S., have been almost criminally wasteful, and that if we undertook the necessary conservation measures and changed our life-style sufficiently, we could reduce our consumption to our prorated share of 6% of the world's energy. Further, the advocate quoting this statistic implies that the elimination of wasteful practices, combined with the wide-scale use of mass transit systems, multifamily homes, and the general imposition of fuel-efficient practices, would allow us to solve our energy crisis without drastically affecting the quality of life. Those who defend the present U.S. situation, while conceding the desirability of sensible conservative measures, point out that the U.S. uses that 32% of the world's energy to produce some 31% of the world's GNP, maintaining that most of this energy is necessary to produce goods and services needed by people throughout the world. They further point out that this relationship between energy and GNP is true for other countries, even those that are reputed to be energy efficient. They cite the example of western Europe and

---

ROBERT E. UHRIG • Florida Power and Light Company, Juno Beach, Florida 33408.

Japan, which together produce a GNP equal to that of the U.S. and consume almost as much energy, even though together they have only 12% of the world's population.

And so the debate goes. It is carried on in the halls of Congress, in state legislatures, and at local government meetings; it is carried on the campuses of our colleges and universities, and it is carried on at cocktail parties, the local Rotary Club, and the local Chamber of Commerce. Predominantly, these discussions are carried on by well-educated, upper-class Americans, whose incomes fall within the upper 25% of our society, who live in comfortable surroundings, and who, in general, are beneficiaries of the present U.S. energy policy (or lack of such a policy). The debates are heavily dominated on one side by young idealists, often well-educated children of affluent parents, and on the other side by individuals that not only represent the "establishment," but, in fact, are part of it.

The energy crisis is also debated in the ghetto by those who the federal government euphemistically describe as "being below the poverty level." But the debate there takes a distinctly different character. Here the emphasis is on the high cost of gasoline and where the money is going to come from to pay the next utility bill, lest the electricity or gas be cut off. The debate is not concerned with esoteric issues like solar energy or environmental primacy; rather, conservation is viewed in the light of matching next week's paycheck against this month's utility bill, rather than such remote issues as balance of payments, or the politics of the Mideast. Here the gut issue is jobs, coupled with the rising aspirations of Americans in the lower economic strata, which can be fulfilled only with some reasonable approximation of full employment. To the laborer working on the Tennessee Dam being built by the TVA, it is inconceivable that some alleged "endangered species" such as the snail darter is of more importance than completing this energy-generating facility, let alone sufficiently important to cost him his job.

The NAACP's 1978 report at its National Energy Conference [1] addresses these issues head-on. A few quotations will suffice to illustrate this fact:

> We have examined the Administration's National Energy Policy in the light of the agenda for economic growth and development for America's black people . . . while we endorse the Plan's objectives of eliminating waste energy and improving utilization efficiency, we cannot accept the notion that our people are best served by a policy based upon the inevitability of energy shortage and the need for government to allocate an ever-diminishing supply among competing interests. . . .
>
> We are fearful that an energy policy with an overriding concern for the protection of the environment may cause government policy-makers in this area to lose sight of other more compelling economic and social objectives that are more important to Black Americans. The basic approach of the plan as it relates to energy supply reflects the absence of a Black perspective in this development.

4 • An Energy-Deficient Society

Underlying all of this debate is the issue of growth vs. no growth. Bayard Rustin, the black President of the A. Philips Randolph Institute, summarized a view often heard in the black community when he said [2]:

> [These days] one rarely hears of the "revolution of rising expectations." Instead we are confronted with a counter-revolution of lower expectations, a politics of limited horizons, and an advocacy of no-growth. This comes at a time when there is mass unemployment, a sizable and expanding black underclass of persons whose lives were barely touched by the civil rights revolution, and a declining standard of living for millions of working people. Yet, in this situation, many who profess concern about unemployment and poverty actively support the concept of limiting economic growth in order to protect the environment. The no-growth advocates fail to recognize that the adoption of their policy positions would significantly worsen the plight of the impoverished and disadvantaged.

Labor unions are also concerned about the effect of an energy shortage on jobs. Robert A. Georgine, speaking before the Edison Electric Institute's 45th Annual Convention, said [3]:

> The Congressional Budget Office told us what the embargo cost us in terms of jobs. When the embargo went into effect . . . unemployment rose by 1.1%. This was a loss of some one million jobs; this could be child's play compared to what could happen in the future.

It would appear that the black leadership and organized labor in the U.S. are reacting to the potential impact that current energy policies, particularly the conservation and "no-growth" syndrome, can have on the rising expectations of both groups. However, there has been no corresponding recognition on the part of the vast majority of American workers who are seemingly not aware that their livelihood is potentially threatened by the energy crisis. Recent polls have shown that Americans are beginning to recognize that the energy crisis is real, although a large number of them still harbor the suspicion that the energy crisis was contrived by the oil companies with a view to generating what many consider to be excess profits.

In 1979, the Federal Power Commission released a report indicating that certain regions of the nation "could experience significant shortages of electricity by 1985." The National Electrical Reliability Council was even more pessimistic and predicted "forced curtailments of electric power" in the near future. While these reports were widely disseminated in the newspaper, there is little understanding on the part of the public as to what this might really mean if it occurred. The most notable example of a serious electrical shortage in a modern industrial nation occurred in Great Britain immediately after World War II. For a period of several years, 10% voltage reductions, extended underfrequency operation, and 20% load-shedding were commonplace. There are those who attribute Britain's declining status among the industrial nations of the world to this important energy shortage at a very critical time in its history. More recently, the United Kingdom has suffered from energy shortfalls that have led to 4-hr rolling blackouts (where different groups of customers are systematically and sequentially disconnected for periods of up to 4 hr) and 3-day workweeks for certain industries. Needless to say, the economic and social impact upon the British people has been enormous.

Rather than attempt to unravel the complex relationship between energy and the economy, let us take a pragmatic look at the immediate results of an energy shortage, with special emphasis on a shortage of electrical energy. Perhaps the most comprehensive analysis of such a shortage is a 1975 study carried out for the California Energy Resources Conservation and Development Commission, entitled "Energy Shortage Contingency Plan" [4]. This 600-page report addressed the intermediate-term energy shortages (weeks, months, or a year or so) but did not address short-term (hours or days) or long-term (years) energy shortages. It concentrated on providing measures that might be initiated to bring the demand for energy in balance with the supply in a manner that would impose the least burden on society. It is useful to examine this report with a view to interpreting how an energy deficiency might be dealt with on a national basis and what its impact might be.

The California Plan for matching load with supply concentrated on reducing the energy used for those functions where it provided comfort and convenience before curtailing those functions directly associated with the production of goods and services. The plan was not a long-term energy resource allocation scheme, and it was not intended to substitute for or interfere with the capability of energy suppliers to deal with operational day-to-day emergency shortages. The plan consisted of many individual measures, in various degrees of severity. These measures were assembled into "packages" to meet a specific shortage level for specific energy sources. Individual contingency measures were classified by type of measures, affected use sector, and fuel type. The types of measures are fuel switching, efficiency improvement, use restrictions, space conditioning curtailment, and operations curtailment. The affected sectors are agriculture, mining, construction, manufacturing, commercial, residential, and utilities. The fuels considered are gasoline, kerosene, residual oil, LPG, coal, natural gas, and electricity.

The plan seeks to maintain economic (particularly production) activities, at as high a level as possible, because it recognizes that the basis of the State's overall welfare is in the fullest utilization of production and income-generating activities. Essential activities, such as public health, safety, and welfare in general, are exempt from the provisions of the use restrictions and operational curtailment measures.

In general, the plan relies heavily on public cooperation, voluntary compliance, and open decision-making, in the expectation that an informed public will more fully cooperate than an uninformed public. There are, however, provisions for mandatory compliance, and an appeals procedure is included to grant exemptions to those aggrieved by the mandatory provisions of the plan.

The plan provides a mechanism (a shortage index) for evaluating the extent of an energy shortage. While the shortage index itself is a numerical value based on expected energy use, incoming supplies, and inventory level, the actual evaluation of the shortage must be tempered with judgment. No separate "emergency energy date system" is provided, since the evaluation of the shortage index is an ongoing process. When a shortage situation is identified, a specific response to the shortage is then assembled, taking into account the relative shortages among the various energy forms and the time of the year in which the shortage occurs.

The California Public Resources Code requires that each electric utility, gas utility, and fuel wholesaler/manufacturer in the state prepare and submit to the Energy Commission an emergency load curtailment plan or an emergency supply distribution plan, setting forth proposals for identifying priority loads or users in the event of a sudden and serious shortage of fuels or interruption in the generation of electricity.

Each electrical utility was required to provide a load curtailment plan setting forth proposals for identifying priority loads or users in the event of a sudden and serious interruption in the generation of electricity. These load curtailment plans of the various utilities included the following components: (1) limiting lighting for outdoor signs, decoration, and billboards, reduction of indoor business lighting, and a virtual halt to space conditioning in homes; (2) voluntary load reduction plans implemented through a public campaign for energy conservation; (3) mandatory load reduction plans; and (4) sequential interruptions of service areas. Curtailment of some industrial operations would be expected but would stop short of a total plant shutdown. Commercial customers who utilize much of their energy for lighting and space conditioning would be expected to significantly reduce their consumption without ceasing operation.

Mandatory load reductions would be implemented only after voluntary steps prove to be inadequate. Most utilities feel that a state or federal declaration of emergency would be necessary for them to implement mandatory load reduction plans if they are to avoid lawsuits for consequential damages. Mandatory load curtailment would subject users to penalties or even loss of service if they fail to reduce their load by a specific fraction of their average usage over a base period. The utilities would also reduce voltage as a means of reducing resistive loads, primarily lighting and heating.

Rotating blackouts is the ultimate load curtailment measure that can be utilized by utilities. For most utilities, from 50 to 80% of the load can be subject to interruption on a prearranged sequential schedule. Unfortunately, many of the "critical loads" (such as hospitals, communications, police, water supply, etc.) cannot be isolated and inequities would undoubtedly occur.

Perhaps the ultimate question is whether the public would accept such measures on anything other than a short-term emergency basis. Clearly, the burden would not be shared equitably. If the emergency were due to causes beyond the control of the government, as was the case in the 1973 Arab oil embargo, there might be little choice. However, if the crisis were the result of a government-imposed reduction in generating capacity or due to construction roadblocks imposed for frivolous reasons (protection of some exotic endangered species) or some zealous moralistic imposition of a low-energy life-style (in the way prohibition was imposed), the American public can be expected to rise up in righteous indignation and take whatever steps they view as necessary to "correct" the situation.

In the past decade we heard much about a low-energy-use society, a reduced life-style, and a no-growth society. When neither industry, the U.S. government, nor its citizens responded to their pleadings, advocates for these concepts made concerted efforts to get them into both state and federal legislation, usually in the name of protecting the environment, and to use existing legislation for their own purposes. Among the requirements of NEPA is a requirement that "adequate" consideration be given to alternatives to the proposed action. Unfortunately, what constitutes "adequate" consideration can become a long legal battle. One nuclear plant was involved in environmental and safety hearings and subsequent litigation for over 6 years before receiving an unambiguous construction permit. The plant was 30% complete; the utility had over $300 million invested; and they were still in litigation (all the way to the U.S. Supreme Court), where the primary issue was whether, as required by the NEPA, the Nuclear Regulatory Commision (not the utility) gave "adequate consideration" to alternate sites, rather than putting the second nuclear plant on an existing site. Guidelines proposed by the Council on Environmental Quality in 1979 would have required that environmental considerations be the *only* criterion in choosing between the various alternatives. In effect, society is faced today with the existence of a legal framework, albeit in the name of the environment, which could virtually stop the construction of any future energy production facilities.

The situation with respect to siting coal-fired plants does not appear to be encouraging either. The Clean Air Act Amendments of 1977 put into law a plethora of new and complex concepts, such as Prevention of Significant Deterioration, Non-attainment Areas, Lowest Achievable Emission Rate, Best Available Control Technology, and others. In the implementation of the Act, additional new wrinkles were introduced such as the concept of "offset" whereby a utility might be required to clean up the emissions from an oil refinery before it was allowed to build a new plant, and requirements that a certain percentage of the sulfur be removed from coal regardless of how low the sulfur content was initially.

Each year Congress seems to impose more and more restrictions on what industry can do, seemingly without regard to the impact in terms of cost, time delay, resources expended, or benefits received. We have seen administrative and court actions implementing many conceptually good laws, such as the National Environmental Policy Act, the Clean Air Acts, the Clean Water Acts, the Toxic Substance Acts, the Resource Conservation and Recovery Acts, and others in ways that have virtually strangled the ability of any utility to build and operate new plants of any sort. There are many examples of where environmental concerns of a minor nature are holding up energy-producing facilities that have already cost hundreds of millions of dollars, and these costs must ultimately be borne by the public through higher prices for their energy.

It is clearly obvious that the utilities have stopped ordering new nuclear plants, at least for the time being. The reasons for this are complex, but the underlying fundamental reason is the financial risk associated with the uncertainties, the delays, and the escalating costs. As one utility's chief executive officer succinctly put it, "When you order a nuclear plant today, you play the game called, 'You Bet Your Company.' "

The Energy Plan introduced by President Carter in 1977, while espousing the current generation of light-water reactors as the "supply of last resort," demanded cancellation of the Clinch River Breeder Reactor Demonstration Plant, declared that it shall be U.S. policy not to reprocess spent fuel from commercial nuclear power plants, and unilaterally imposed (with congressional approval) new strict conditions on the export of all nuclear fuels and power plant equipment. Perhaps the most unfortunate aspect of this policy is that there is virtually no margin for error, and the rationale offered in support of the decision not to reprocess spent nuclear fuel has blatantly assumed that the actual uranium reserves in the U.S. are several times the known proven reserves reported by the Department of Energy and that they will be available at reasonable costs, like $40.00 per pound. The elimination of reprocessing ties up both remaining enriched uranium and plutonium in the spent fuel, thereby virtually guaranteeing high uranium prices and probably a uranium shortage before the turn of the century.

The Energy Policy seems to have assumed that all other nations having nuclear power will follow suit in abandoning both their nuclear fuel reprocessing and their breeder reactor development programs and would comply instantly with unilaterally imposed restrictions and access to nuclear fuel. Unfortunately, this is a delusion! Japan, which can supply only about 5% of its future energy requirements, has undertaken an extensive nuclear power program. Western Europe is well ahead of the U.S. in the development of nuclear breeder power plants and has operating prototype breeder plants in France and England today.

The Russians are expanding their nuclear power program, including the breeder, in spite of the seemingly large reserves of fossil fuels. We in the U.S. may be sufficiently well endowed with other natural resources that we can give up the option of nuclear energy for a few decades, but we do it at great peril to our economic and social well-being.

Adequate energy is essential to the well-being of the U.S. in both domestic and international affairs. The greatest economic hunger on the face of this earth is for energy, primarily in the form of electricity; energy that creates goods and services, energy that creates jobs, energy that raises the standard of living and improves the quality of life. In the final analysis, the economic strength of the U.S., the stability of its monetary system, and ultimately its security as a nation are at stake in the energy debate that is going on today.

An energy-deficient society is so inconsistent with the "American way of life" that it behooves us to take every possible step, including conservation and constructing new energy production facilities, including both nuclear and coal generating plants, to assure that our legitimate needs are met.

REFERENCES

1. Report of National Energy Conference, National Association for the Advancement of Colored People, Washington, D.C. (1978).
2. B. Rustin, Realization of Rising Expectations, presented at Edison Electric Institute 45th Annual Convention, June 13–15, 1977, Philadelphia, Pa.
3. R. A. Georgine, More Jobs More Energy, presented at Edison Electric Institute 45th Annual Convention, June 13–15, 1977, Philadelphia, Pa.
4. Energy Shortage Contingency Plan, Report to California Energy Resources Conservation and Development Commission, Document No. C78570 (1975).

5 Energy Shortages

The Downside Risks

A. David Rossin

The 1970s were a period of newfound concern for the environment and an awakening to the fact that natural resources are limited. Public concern was translated into laws. The laws require a vast array of regulations. Now, alternatives must be considered before a project can begin. The impacts and risks must be evaluated, reported, and, in some cases, debated.

The debate about energy—oil prices, synthetic fuels, solar energy, conservation—has become a battle when it touches nuclear power. Individual activists and interest groups raise questions and make charges about radiation, waste, safety, and weapons proliferation. Industry and government experts document their answers with thick reports. Whom can the people trust?

Nuclear risks are weighed against risks associated with coal (pollution, acid rain, the carbon dioxide layer); oil (spills, fires, tanker accidents, platform collapses, and dependence on the Middle East); natural gas (rising prices and limited resources); and new alternatives (unknown economics and unproven technologies). On the benefit side is the electric energy produced—the amount necessary to serve a nation that is learning to conserve.

The benefits are to be weighed against the risks. But rarely do people hear about the downside risks: the risks that come with not having enough electricity to go around. No one claims that nuclear power is the answer to all our energy problems. But nuclear power's contribution is vital. Without it the problem is tougher. It is time that Americans start to look at the downside risks of *not* having enough electric power plants.

A. DAVID ROSSIN • Electric Power Research Institute, Palo Alto, California 94303.

We have grown up with dependable and relatively economical electricity. Utilities, by law, have served all. But with too few power plants being built today (coal and nuclear) and the 8 to 10 years or more that it takes to build one, the odds of an electric generation shortage before the end of the 1980s are mounting. Even with *zero* growth in electric demand, by 1985 it would take 8000 MW each year (the equivalent of seven new large nuclear plants) just to replace the old plants that become obsolete. That is with *no* replacement of oil-burning plants and *no* growth in demand. Only 6000 MW was ordered in 1979. So the stage is being set today, not just for one of the three downside risks, but for any or all of them, perhaps even at the same time.

1. DOWNSIDE RISK ONE: THE SELF-FULFILLING PROPHECY

Utilities used to be accused (sometimes correctly) of building a new power plant, then advertising to promote demand for the power and justify its decision: the "self-fulfilling prophecy." Sometimes the advertising worked; sometimes it did not because national or worldwide events intervened. But that was back in the days of 3% interest rates when the new plant would produce power at less cost than what it replaced. Now there is no financial incentive for any utility to build anything. The large number of inflated dollars required will just raise total system generation cost.

But what is certain is that if a utility company cannot promise reliable electric service, the next factory or office complex will go somewhere else. With it go the jobs. Next the plans for expansion are canceled, followed by closing of companies that had been pillars of the community for years. That is the real self-fulfilling prophecy.

2. DOWNSIDE RISK TWO: GOVERNMENT TO THE RESCUE

If utilities cannot build, and citizens realize what is happening, they will not blame the citizen-activist groups that caused the delay, even if the leaders are still around. They will blame the utility company for failing to do its job. And some politicians will probably blast at the utility for not warning the public about what was in store.

If history repeats itself, the citizens will turn to government to take over the utility and build the plants its own regulatory process had stopped. Yet the risk remains that the same regulations and pressure groups may just stop the government too.

3. DOWNSIDE RISK THREE: PRIORITIES AND ALLOCATION

The theory of some who oppose nuclear power is that if electricity is restricted everyone will conserve, the right amount of energy waste will be skimmed away, and unnecessary growth, with all its environmental impacts, will be prevented. This may be the dream, but in the real world predictions rarely come out accurately, especially where energy is involved. Energy policy decisions sometimes produce results different from those promised.

If the power plant is ready but not needed, it does not run. Interest on its mortgage must be paid anyway, but at least the fuel is saved. If oil or gas can be saved and those plants are kept idle while coal or uranium fuel is used, the benefits are obvious.

However, on the downside, if there is not enough electricity to supply all users, priorities must be set. Over the short term, blackouts and brownouts due to storms or other emergencies will occur more often. But when there are only so many generating plants, and the various new demands turn out to be greater than the supply will be, priorities will have to be set.

What new use should have priority for a limited remaining amount of electric supply? A factory with its jobs? A new energy-efficient office building? A hospital? Apartments? One hundred four-bedroom houses? Should each residence have a limit? Should certain appliances be banned?

Just who should set these priorities? Not the utility company; that is not its role under the law, and a utility has no right to discriminate among users. Should the state order the utility to raise rates or should it tax energy use to depress demand, even though that would increase welfare payments? If not, that leaves allocation. The priorities would be set and enforced, not by the utility, but by government. But not one congressman, senator, governor, or other public official has called for public hearings on how to set priorities for electricity when there is not enough to go around!

A number of spokesmen are calling for *decentralized* energy sources; house by house or in each small community. Their stated objective is to free people from the big, centralized power companies. However, utilities do not decide what uses of electricity are valid and socially acceptable. Individuals and companies make those decisions themselves.

As long as there is enough utility electricity for reliable backup, anyone can build a windmill, a solar heater, or whatever he chooses. The centralized utility cannot stop anybody. The decisions are individual, local: *decentralized*.

Ironically, with shortages comes allocation of energy, and people may be left with no choice but to build and tend their own generators, even if they would rather go backpacking or read a book. Allocation is *centralized decision-making* by government: big, *centralized* government.

The disciples of decentralized energy sources offer no assurance that their alternatives can actually deliver. What they demand *now* is the commitment to stop nuclear power. Such a decision would make the likelihood of electric generating shortages before long very high indeed.

Nobody has a perfect crystal ball. No one can be sure just how much capacity would be right for 1990. But the risks of having too many power plants need to be compared with the downside risks of not enough.

A commercial airline considers downside risk. When a Chicago to San Francisco flight takes off, it is carrying a lot more fuel than it takes to make the flight. This means higher inventory costs and more weight on takeoff and landing because experience says there may be weather delays or other unexpected developments. Cutting fuel loads close to the line is tempting, but the downside risks of running short are well known, not only to pilots and executives, but to passengers and politicians.

If the debate about nuclear energy is to deal with risks, the downside risks of electric energy shortage had better become a feature of it. If a free society is to arrive at an energy policy, its people need to be well aware of the downside risks of all its options.

II Economics of Nuclear Power

A common theme in the energy debate is the allegation that nuclear power is uneconomic. Part II contains two papers that bear on this issue. In "Economics of Light-Water Reactors," A. D. Rossin addresses the issue of economics for light-water reactors, the type that represents the technology of today. He shows that nuclear power remains the cheapest baseload power among existing thermal stations and is, with only a few exceptions, the cheapest alternative for additions to baseload power. All of the so-called "add-on" costs for nuclear systems have not overturned this balance. What is euphemistically called "regulatory delay," i.e., the costs of fighting through the existing maze of hearings and interventions, is the principal reason for the absence of recent nuclear orders.

In "Fast Breeder Reactor Economics," C. P. Zaleski examines the economic prospects for breeders, which, in a nuclear energy future, must ultimately replace the reactors of today. What we can see in this paper is that, with breeders, nuclear power can deliver electricity almost indefinitely at prices that are only moderately higher than those of today's systems. Fast breeder reactor economics is not, of course, a hard fact today. What is presented is the way by which not merely breeders but any system with future prospects is planned for development through a process of incremental cost reduction through experience.

These two papers are not the only ones in this book that bear on nuclear economics. Chapters 9 and 10 by B. I. Spinrad consider cost-related aspects in the light of some nuclear weapon nonproliferation issues.

Our message on nuclear economics is a positive one. Yet, in this section introduction, we cannot help but comment on a negative aspect. We in the U.S. are not taking advantage of nuclear power for our maximum benefit. We are not building as many nuclear plants as we should, nor are we building them and developing advanced nuclear systems as expeditiously as we should. We are losing money now, and will lose more in the future, from these lacks. There is only one basic reason for this: a highly politicized opposition that has made building nuclear plants a political risk and thereby a financial risk. When one considers that a human being can be nurtured over a lifetime in dignity for less than 1 million dollars, how many lives have been sacrificed by the tactics of the nuclear opposition? The needless expense is now over 10 billion dollars; do your own arithmetic.

B.I.S.

6 Economics of Light-Water Reactors

A. David Rossin

1. ECONOMIC CONTEXT

In making an economic analysis, the performance of existing units is examined, detailed estimates are made of construction costs, and comparisons with alternatives are made. Allowances are added for contingencies, for the unexpected events that are bound to happen, without knowing just what and when those events might be. However, even if the economics favor nuclear power on paper, the decision might well be negative if there is doubt that the project can be carried through to completion. If that is the case (and it has been the case since 1980), the economic risks become prohibitive. There is huge risk in tying up a large amount of capital without obtaining any return on it. There is risk that state rate commissions might act in a punitive way, disallowing for rate-making purposes investments made in the project. But also there is recognition that delay or cancellation of a nuclear plant leaves the utility without that block of generating capacity; the years lost cannot be recovered, the funds cannot be used to build something else, and even the utility's financial structure may be hurt to the point where further financing becomes very difficult. The consequences may be expensive purchased power, deterioration of service, and public antagonism, the very things a well-managed utility strives to avoid.

On a nationwide basis, the risks include failure to reduce oil imports and the real possibility of electric energy shortages, just what a well-governed nation strives to avoid.

In 1980 three forces combined to make ordering of new LWRs unattractive, despite the favorable economics that would be expected if projects could be completed with confidence. The Carter Administration's "last resort" attitude on

---

A. DAVID ROSSIN • Electric Power Research Institute, Palo Alto, California 94303.

nuclear power and the policy against reprocessing and the breeder produced an atmosphere so negative as to discourage nuclear plant commitments in the U.S. and restricted U.S. vendors from competing worldwide. The regulatory situation, already negative and formidable, deteriorated to "impassable" since the TMI accident. Several major rule-makings were upcoming, any of which could result in serious problems for nuclear power. Meanwhile, the TMI-2 accident and failure of government to move effectively to dispose of nuclear waste increased public skepticism. Abetted by sensationalized media, public attitudes toward nuclear power deteriorated. While generally still favorable, the economic margins are narrow, and an unexpected event could turn the outlook negative. Thus, despite a very impressive economic history, the future for nuclear energy is clouded.

The record clearly shows that nuclear plants have generally been good investments. There are exceptions: TMI-2 is the obvious example. But it is important to recall that TMI-1 had produced hundreds of millions of dollars in savings for the residents of Pennsylvania before the accident and that the biggest cost is the more expensive replacement power that must be purchased until the units can be returned to service.

In contrast to charges by nuclear critics, the actual generation costs of nuclear power have been below those of the other available sources. The exceptions are large hydroelectric dams built before the inflation of the last decade and some mine-mouth coal-fired plants. This chapter presents cost data for the nation and illustrates the point with examples from several utilities. Next, factors that are changing nuclear power costs are discussed. In addition to sensitivity to price changes, discussions include the effect of delays, changes resulting from the lessons learned from TMI, costs of waste disposal and decommissioning, insurance, legal expenses, and research. Cost estimates are presented for future coal and nuclear plants, the only major viable options left at present. The conclusions examine the future prospects for reactor orders and the potential impacts.

2. PERFORMANCE NATIONWIDE

It would be attractive if a simple average of nuclear and coal-fired generation costs were published each year. However, this would not end the debate. Actual costs are averages of many plants built in different years, operated by different companies, and financed in different ways. Performance of both coal and nuclear plants varies widely from one year to another. Besides, plants already in operation will continue to operate in future years. Decision alternatives are confined to future plants, and new developments must be taken into account in estimating cost, as well as the effect of inflation and rising fuel prices over the life of the plant.

Data on operating performance of U.S. reactors are reported monthly by the U.S. NRC. In addition, *Nucleonics Week* reports performance of reactors throughout the world, except for the Soviet bloc nations. Thus, accurate, timely, and complete data on unit availability, capacity factor, and forced outage rates are available. Data for coal-fired plants are only reported annually to the Federal Energy Regulatory Commission and by the time they are collated and published, 2 years of delay or more is incurred. 400 MWe is used as the lower size limit for including coal plants. (Most nuclear plants exceed 600 MWe and plants under construction are in the 1100- to 1200-MWe range.)

The FERC data can be used to calculate fuel costs and operating and maintenance costs for all plants, but treatment of carrying charges on invested capital varies, and, therefore, only simple methods can be used to complete the economics calculation. Data are gathered by contacting individual utilities, and although some differences are substantive, the average is still a usable indicator and will be discussed below.

Year-to-year variations in costs between plants are expected. On the other hand, industrywide changes are important, and must be analyzed to see if they mark emerging trends with long-term implications.

1978 was an excellent year for U.S. nuclear power plants. But 1979 was marked by the shutdown of five units for seismic evaluations, three units began lengthy outages for steam generator replacements, and in the aftermath of TMI, extensive shutdowns for imspections and modifications were required. Table I shows the performance for 1977 through 1982. While 1978 shows what nuclear plants can do, 1979 shows the serious impact of these events. The penalty is 6.9 percentage points in availability and 6.3 points in capacity factor. The forced outage rate jumped 50%.

The exclusion of the two TMI units from the 1979 data would raise the average unit capacity factor for U.S. LWR units by 1.7 percentage points. Further exclusion of the five units that underwent reevaluations of some of their piping systems, during which the NRC required their shutdown, would raise the 1979 capacity factor average another 3.0 percentage points. Shutdowns for remedial

TABLE I
Performance of U.S. LWRs

	Performance index (%)					
	1977	1978	1979	1980	1981	1982
Average unit availability factor	73.4	74.8	67.9	65.3	67.8	65.7
Average unit capacity factor (DER)	62.4	65.2	58.9	56.0	58.8	56.7
Forced outage rate	7.6	9.5	14.3	11.5	12.7	12.9

action required by the NRC in the case of seven other operable nuclear units, which have Babcock & Wilcox nuclear steam supply systems, cost another 0.5 point. Short-term post-TMI safety measures adopted for other plants cost additional operating time. The first quarter of 1980 continued to show lengthy shutdowns, mostly for post-TMI-related and other modifications required by the NRC. The question is whether or not these units and the new ones that entered operation in 1981 and thereafter can demonstrate a high level of reliability in future years. One indication may be that even during 1979, 16 units achieved availability factors of over 85%: Browns Ferry-1 and 2, Cooper, Fort Calhoun, Haddam Neck, Hatch-2, Monticello, Oconee-2, Oyster Creek, Peach Bottom-2, Pilgrim-1, Point Beach-2, Prairie Island-2, Quad Cities-2, Rancho Seco, and San Onofre-1. The capacity factor of 12 units exceeded 80%. These high values for operating availability and capacity factor are indicative of the reliable service of which nuclear generating units are capable. Nationwide averages are shown in Table I. Despite longer outages for post-TMI backfits, capacity factors have begun to recover.

For large coal-fired plant performance nationwide, the average capacity factor over 20 years has been 59%.* Throughout the 1970s the annual national average was always between 56 and 62%. In 1979 it was 58.2%. Interestingly, the nuclear plant capacity factor from 1970 through 1976 was also 59%. In 1979 it was 58.2%. For the decade it was 59.8%. The real message is that for large steam-electric generating stations, capacity factors have averaged about the same (there are a lot of other parts of a power plant that must function besides the boiler); the average is just below 60%. Of course, there remains room for improvement.

3. ECONOMICS NATIONWIDE

Recognizing the difficulties in producing meaningful average generation costs, calculations have been made, but they should be taken *in context* and not as the deciding factor in energy policy decisions. The important thing is to be able to weigh economic factors carefully in choosing between coal and nuclear, if both can be kept viable alternatives.

To keep the analysis simple, power generation costs are separated into three segments: fuel, operation and maintenance, and carrying charges on capital. The cost of fuel to generate each kilowatt-hour is assumed to stay the same throughout a given year. On the other hand, the carrying charges are fixed, and the same amount must be paid, regardless of how many kilowatt-hours the plant produces

---

* Source: FERC.

during the year. The better the plant performs, the more kilowatt-hours there are over which to spread the fixed costs. Therefore, when capacity factors are high, energy generation costs are lowest. Operating and maintenance costs are part variable and part fixed. The sum of these three cost components is the net generation (bus-bar) cost per kilowatt-hour.

The AIF released a survey of 1978 economics[1] that quoted average nuclear generation cost at 15.4 mills/kW-hr for nuclear plants and 21.3 mills/kW-hr for coal plants. The survey was criticized by Komanoff[2]. Komanoff was correct in pointing out that only 39 of the nation's 70 eligible reactors were included, although it was made clear in AIF's publication that response by utilities was voluntary and noted the utilities that did not respond and showed the missing pieces of data. However, the quoted average is low, as Komanoff noted, because of the units that were omitted. (Charges of deliberate bias were made, but are obviously invalid.) Komanoff made assumptions for the missing units, raised the cost estimate to 18.9 mills/kW-hr, and then added 1.5 mills/kW-hr for waste disposal. Some utility figures already contain 0.5 to 1 mill/kW-hr for waste; some do not. The AIF solicited the missing data, and without an addition for waste disposal, recalculated the average to be 17.1 mills/kW-hr.

The AIF survey included only those coal plants operated by nuclear utilities, only 15% of total coal generation in the U.S. Komanoff lowered his coal cost estimates by including the American Electric Power Company and the Tennessee Valley Authority, but no others. These additions lower the coal cost to 21.1 mills/kW-hr. Then Komanoff reduced his figures by 2.1 mills more, savings only possible if the coal plants ran at 67% capacity factor. There are some large coal plants that are cut back in power during low system load periods, but Komanoff's adjustment is a gross exaggeration for plants with average availability over the years of only 74%. This effect is even greater in recession years when loads are lower than had been projected when the plants were built.

When a utility has an option of nuclear or coal for its next increment of generation, the economic choice could still be nuclear. The figures still give nuclear a several mill/kW-hr advantage because the fuel costs are lower. This can only apply if there is confidence that the plant can be licensed and built on schedule.

The DOE, confronting the same data problems, has attempted its own study[3]. The nuclear units in it generated 78.7% of all nuclear-powered electricity produced in the U.S. in 1979, and the coal-fired plants produced about 31% of all 1979 U.S. electric generation fueled by coal. DOE's summary of nuclear generation costs is shown in Table II.

The jump in generation costs in 1979 is dramatic. The reason is the drop in capacity factor because of TMI and the other problems discussed above.

The DOE estimated carrying charges on capital in a consistent way for nuclear and coal plants. Based on FERC data on capital investment, the DOE

TABLE II
National Averages of U.S. Nuclear Power Plant Generating Costs[a]

Cost category	Average costs (¢/kW-hr)					
	1977	1978	1979	1980	1981	1982
Fuel	0.3	0.3	0.4	0.5	0.5	0.8
O & M	0.3	0.3	0.4	0.5	0.6	0.9
Capital	0.9	1.1	1.4	1.4	1.4	1.8
Total generating	1.5	1.7	2.2	2.4	2.5	3.5
Capacity factor	66.0%	67.0%	60.0%	59.0%	61.0%	60.0%

[a] Table compiled using information from DOE, FERC, NRC (NUREG 0200 series), and Annual Economic Survey of Utilities, Atomic Industrial Forum, Inc. Costs are estimated to be accurate within ±0.1¢/kW-hr.

assigns a fixed charge rate (FCR) of 17% per year. In this way, the 1979 generation costs for nuclear and coal plants have been calculated (Table III). The difference (0.32 mill/kW-hr) is not large enough to be treated as statistically significant in light of the uncertainty in the various cost numbers. Although there are regional differences that could be analyzed in detail, the conclusion is that base-load coal and nuclear power plants produced electrical energy at about the same cost in 1979. As more new, high capital cost coal and nuclear plants come on line, the economics will depend more and more on how successful the utility is in keeping them in operation.

TABLE III
National Averages of U.S. Private Utility Nuclear and Coal-Fired Power Plant Generating Costs for 1979[a]

Cost category	Average unit costs (mills/kW-hr)	
	Nuclear	Coal
Fuel (actual)	3.92	12.52
O & M (actual)	4.25	2.19
Capital (estimated)	14.03	7.81
Generating (estimated)	22.20	22.52

[a] The average capital costs shown here are calculated using an assumed fixed charge rate of 17%. If an FCR of 18% were assumed, the estimated generating costs would be: nuclear, 23.02 mills/kW-hr; coal, 22.98 mills/kW-hr. If the FCR were 16% the results would be: nuclear, 21.37 mills/kW-hr; coal, 22.06 mills/kW-hr.

6 • Light-Water Reactor Economics

This will require improvement in the regulatory sphere. Regardless of whether the fault lies with the licensee (owner/operator) or the NRC itself, or even with clever adversaries who force the NRC to curtail a plant's operations, outages identified by the NRC due to regulatory restrictions were significant in 1979. Some were required inspections, others resulted from actual safety-related problems.

In any event, the energy not generated due to these restrictions was 30 billion kW-hr, or 12% of all the nuclear energy generated in 1979. Taken another way, this energy equaled 20% of all the outages, forced or planned. The oil equivalent is about four 1000-MWe oil-fired plants at 57% capacity factor. If all this generation were made up by importing oil, it would have averaged 141,000 bbl/day or well over a billion dollars in 1979.

3.1. Utility Examples

Nationwide statistics tell one part of the story, but individual experiences are very revealing. Most utilities have had good experience; some very bad. TMI, an economic miracle until the accident, could still bankrupt its utility if regulatory and political forces fail to take responsible, forward-looking views. Palisades and Davis-Besse have been plagued with problems. Trojan, a great performer when it has been running, has been saddled with more than its share of regulatory outages.

It would be prohibitive to analyze every utility's history. The examples that follow are informative, but no claim is made that they cover all scenarios: Commonwealth Edison is the largest nuclear power generator in the U.S. Its six large nuclear units were compared with its six large coal units through 1977[4]. This analysis has been updated. Virginia Electric and Power Co. and Florida Power and Light Co. have had extended shutdowns that affected their economics. Duquesne Power and Light had its Beaver Valley 1 unit shut down for a seismic analysis. New England, and particularly the Vermont Yankee unit, illustrate nuclear economics in a region heavily dependent on imported oil for its electricity.

3.1.1. Commonwealth Edison Company—A Case Study

Commonwealth Edison has six large nuclear and six large coal-fired units in operation, in addition to some 16 smaller coal- and oil-fired units and one smaller nuclear unit. This provides an opportunity to compare the economics and reliability of the different units under one corporate management and in a single geographic region, northern Illinois and the Chicago metropolitan area.

The system also has approximately 1200 MW of combustion turbine cause of the oil-fired peaking units, which run on No. 2 diesel oil. All scheduled gas use was eliminated in 1975 by order of the Illinois Commerce Commission.

The six large coal-burning units and the six large nuclear units provide the opportunity to compare performance of two sets of base-load power plants. Table IV provides the details on these plants. In brief, the nuclear units consist of four GE BWRs that went in commercial service between 1970 and 1972; two 1040-MW PWRs at Zion were placed in service in 1973 and 1974, respectively. The coal-fired units include four units of about 600-MW size commissioned between 1965 and 1968 and two large coal-fired units at Powerton Station of 850 MW each with service dates of September 1972 and December 1975, respectively.

Table V gives the actual bus-bar generation costs for these two groups of six base-load units. Note that Commonwealth Edison calculates its carrying charges at 20% of the gross plant investment per year compared with 17% in the DOE estimates above.

Figure 1 summarizes the generation cost of these two groups of units. Despite serious efforts to improve the capacity factors of these coal-fired units, their capacity factors have remained below 50% over a full 4 years. Four of the

TABLE IV
Commonwealth Edison's Large Generating Units

Unit	In-service date	Net capability[a] (MWe)	Type[b]	Construction cost ($/kWe)
Coal				
Joliet 7	9 April 1965	537	Western	113
Joliet 8	21 March 1966	537	Western	113
Kincaid 1	7 June 1967	606	Illinois	118
Kincaid 2	10 June 1968	606	Illinois	118
Powerton 5	30 September 1972	850	Illinois	231
Powerton 6	19 December 1975	850	Illinois	218
Nuclear				
Dresden 2	11 August 1970	794	BWR	147
Dresden 3	30 October 1971	794	BWR	147
Quad Cities 1[c]	16 August 1972	789	BWR	165
Quad Cities 2[c]	24 October 1972	789	BWR	165
Zion 1	2 October 1973	1040[d]	PWR	280
Zion 2	19 September 1974	1040[d]	PWR	280

[a] Net capability is the maximum dependable rating of the unit, i.e., what the utility expects it can get from the unit. This may be limited by design, license, or environmental conditions.
[b] For coal units, Illinois refers to high-sulfur Illinois coal and Western refers to low-sulfur Colorado, Montana, and Wyoming coal. For nuclear units, BWR refers to boiling water reactor and PWR to pressurized water reactor.
[c] The Iowa-Illinois Gas and Electric Co. owns 25% of these units.
[d] These units were limited by the NRC to approximately 850 MWe prior to 25 June 1976.

TABLE V
Generation Costs (¢/kW-hr): Commonwealth Edison Company[a]

	1977	1978	1979	1980	1981	1982
Six large nuclear						
Capacity factor	60.7%	70.3%	59.4%	61.0%	64.2%	59.1%
Fuel	0.35	0.37	0.42	0.36	0.46	0.56
Carrying charges on fuel	n/a	n/a	n/a	0.10	0.25	0.27
Operation and maintenance	0.21	0.22	0.38	0.40	0.42	0.51
Carrying charges	0.75	0.66	0.80	0.84	0.80	0.92
Total	1.31	1.25	1.60	1.70	1.93	2.26
Six large coal						
Capacity factor	46.2%	45.6%	43.2%	45.8%	45.1%	51.4%
Fuel	1.01	1.37	1.80	1.97	2.42	2.53
Carrying charges on fuel	n/a	n/a	n/a	0.15	0.32	0.26
Operation and maintenance	0.24	0.26	0.32	0.35	0.50	0.46
Carrying charges	0.84	0.86	0.90	1.14	1.42	1.30
Total	2.09	2.49	3.02	3.61	4.66	4.55
Oil—fuel cost only	n/a	n/a	n/a	6.63	8.13	8.86

[a] All data shown are taken directly from the company's books for 1977, except as follows: 1.3 mills is added to the nuclear fuel expense on the books to reflect the estimated cost of carrying charge on nuclear fuel in the reactor (0.8 mill), plus an additional allowance of 0.5 mill for the net cost of ultimate disposition of spent fuel. The per-books data already include about 0.5 mill for spent fuel disposition. The coal fuel expense figures per-books were increased by 0.6, 0.5, and 0.4 mills/kW-hr, respectively, to reflect the estimated carrying charges for maintaining a 90-day coal stockpile. Carrying charges were computed by applying a 20% annual fixed charge rate to the gross plant investment in the generating units in question and dividing by the number of kilowatt-hours generated (net) by such units.

six units (all of which were originally designed for Illinois coal) are now burning Western coal. This makes it difficult to reach design power levels. For example, Western coal contains more moisture and noncombustibles. The mills that pulverize the coal at the plant must grind 10–15% more coal to deliver the same heat content of coal to the boilers, and they generally cannot do it. The mine-mouth Kincaid units have suffered serious boiler tube problems and have required extensive repairs, maintenance, and deratings and a new coal washer has been built to meet environmental requirements. Meanwhile, coal prices have risen more than 25% in a single year. (Note the increase in coal cost from 10.1 mills/kW-hr in 1977 to 18 mills/kW-hr in 1979!)

Table VI lists the availability and capacity factors for these two groups of units. Piping repairs dominated the nuclear plants in 1974 and 1975, but then performance improved. In January 1977, Chicago experienced its coldest weather in 80 years. During this time, the large coal-fired units had severe problems with frozen waterways, and frozen coal would slide back down the conveyor belts.

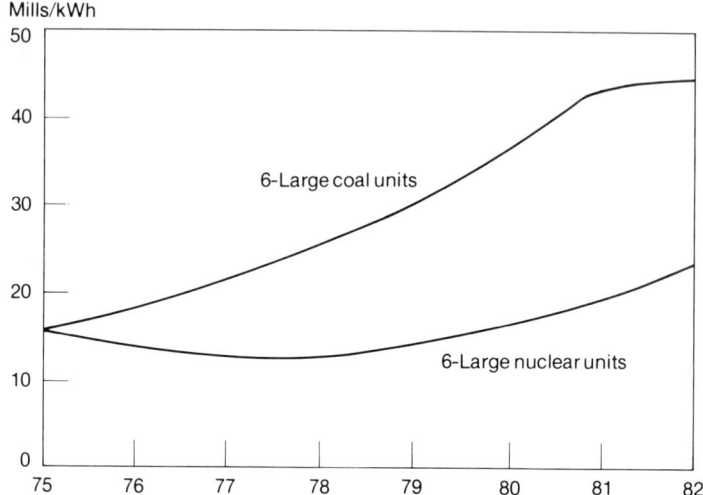

FIGURE 1. Actual bus-bar generation cost: Commonwealth Edison Co.

TABLE VI
Yearly Availability and Capacity Factors of Commonwealth Edison's Large Nuclear and Coal Plants[a]

	Large nuclear			Large coal		
Year	Number of units	Availability (%)	Capacity factor (%)	Number of units	Availability (%)	Capacity factor (%)
1970	1	60.6	31.5	4	77.0	58.1
1971	2	61.8	39.5	4	71.7	53.7
1972	4	68.6	61.1	5	71.0	55.5
1973	5	77.4	66.3	5	69.2	54.3
1974	6	59.5	51.1	5	72.6	58.0
1975	6	64.4	50.7	5	67.5	53.5
1976	6	71.4	57.3	6	60.0	44.7
1977	6	79.9	60.7	6	66.5	46.2
1978	6	83.6	70.0	6	72.0	45.6
1979	6	74.9	59.8	6	64.0	43.2
	Average:	72.6	58.8		68.5	50.3

[a]Availability and capacity factors were calculated beginning with the first full month after the in-service date weighted by the number of in-service months and net capabilities. The values for Zion 1 and 2 are based on net capabilities of 851 and 852 MWe, respectively, prior to June 25, 1976 and 1040 MWe thereafter.

The nuclear capacity, which at that time comprised approximately 30% of the system's capability, generated almost 50% of the total power.

The nuclear units also delivered during the heat of the summer of 1977. During August they were available 99.5% of the time.

January and February of 1978 brought a record snowfall of 83 inches, about twice the normal amount. On top of this, the 108-day coal miners' strike from December to March 1978 reduced coal stockpiles seriously, especially in the Midwest.

During the first 2 months of 1978, nuclear power accounted for more than 40% of Edison's output, despite the planned refueling of two of the company's seven reactors. Since about two-thirds of Edison's coal supply is from Western mines, it was not seriously affected by the coal strike. However, ice on the Illinois River, the main transportation route for coal, and a shortage of rotary-dump railroad cars caused fuel transportation problems. Only partial deliveries of coal were reaching Edison's generating stations.

With good nuclear performance, Edison was able to take care of its own energy needs as well as help its neighbors to the east.

Again during the summer of 1978, the nuclear plants provided 50% of the power. The nuclear unit showed availabilities of 92.1, 96.4, and 99.5%.

In 1978, the Quad-Cities Unit 1 broke the world record for BWRs in continuous operation (206 days), and posted the world's best BWR availability factor (94.9%). Its capacity factor was limited at times, however, by environmental regulations that govern its cooling water discharge. Dresden Unit 2 placed third in the worldwide availability ranking at 94.2%. The six nuclear units set a new company record by being ready for service more than 80% of the time during 1978.

In 1978 Commonwealth Edison's nuclear generating units produced a record 45.4% of its annual output for the first time, surpassing its production from coal-fired generation. Coal provided 43% of the company's 1978 production, while oil and other sources accounted for the remaining 11.6%. By contrast, just 15 year earlier this company produced almost all of its energy by burning Illinois coal. That fuel accounted for less than 7% of the total in 1979.

There are several ways to calculate cost savings. One method is to assume that instead of building the nuclear units, more coal units identical to the existing ones were built and operated in an identical pattern to produce the same energy as the six nuclear units. In 1977 the nuclear units generated for 7.6 mills/kW-hr less than the coal units. Their production was 26.6 billion kW-hr, so the calculated savings were $200 million, almost 10% of total revenues. This is more than the entire rate increase granted the company that year. The fact is that most of that saving goes to the consumer, and only a small fraction to the investors.

Comparable figures for 1978 show a 12.4 mill/kW-hr differential on a nuclear output of 30.6 billion kW-hr, a saving of $389 million. In 1979 the differential was 14.3 mills/kW-hr on 25.2 billion kW-hr for $360 million saving. As inflation increased in the years to follow, so did the savings.

Another approach is to look only at the fuel cost savings. Operating and capital costs are not included, since these are taken as sunk costs whether the plants operate or not. The figures can be calculated in a straightforward way. Table VII shows the calculated fuel saving for Commonwealth Edison's nuclear units if the nuclear fuel were replaced by low-sulfur Western coal or residual oil. Commonwealth Edison does not actually do this. In fact, power has been available for purchase from neighboring utilities on many occasions at prices below the fuel cost of residual oil.

In 1979 Commonwealth Edison operated two stations on crude and residual oil: Collins (five 500-MWe units) and Ridgeland (four units converted from coal in 1979 totaling 700 MWe). In general, these units were operated only on weekdays between the hours of 7:00 AM and 10:00 PM. Their use enabled the company to minimize generation from combustion turbines, which burn even more expensive No. 2 diesel oil. These units achieved an availability of 82.4% in 1979. Since they were only used as needed, their capacity factors were: Collins 21% and Ridgeland 45.4%.

Taken together, their average generation costs were:

Fuel	41.8 mills/kW-hr
O & M	3.0
Capital	21.9
	66.7

Recalculating the capital costs for a 60% capacity factor (unrealistic for oil on this system) reduces the cost to 51.6 mills/kW-hr. The high cost of oil fuel alone is almost three times the bus-bar nuclear cost and $1\frac{1}{2}$ times bus-bar coal generation cost.

The 2500-MWe Collins Station had not been part of Commonwealth Edison's system plan in 1971. But by that time, it became evident that interventions in the LaSalle County Nuclear Station licensing process would delay it at least from 1975 to 1978. With load actually growing at 7% per year, there was no time to build anything but oil-fired capacity. Construction delays and backfits to meet recent regulatory requirements continued to push LaSalle's service dates back. LaSalle 1 finally started up in 1982. Its total generation cost is over 40 mills/kW-hr. It is interesting to note that in mid-1980, Minnesota Power and Light Co. brought its new 500-MWe Boswell coal-fired plant into operation. The capital cost for this unit was $800/kW or 32 mills/kW-hr at 60% capacity factor.

TABLE VII
Fuel Cost Savings as a Result of Nuclear Generation

Established figures	kW-hr (millions)	%	Equivalent tons of low-sulfur coal (millions)	Equivalent barrels of residual oil (millions)	Savings in fuel expense ($ millions)	
					Low-sulfur coal	Residual oil
1973, 1974, 1975	Over 55,000		31	100	355	510
1976	24,525	40.0	13.6	43.8	234	520
1977	26,569	41.8	15	47	280	680
1978	30,628.2	45.4	17.1	54.7	365.9	797.6
1979	25,244.2	40.3	14.1	45.1	370.6	865.1

It is interesting to compare 1979 costs (Table VII) if the two groups of units both performed at identical capacity factors of 60%. The difference is still 11.44 mills/kW-hr in favor of the nuclear units.

Costs for 1979 reflect downtimes and delays from TMI-related and other regulatory actions. In 1979 Edison estimated that about 22 days of nuclear plant operations were curtailed by unanticipated regulatory demands. Replacement power cost $11 million and an estimated 450,000 barrels of oil were burned to make up these lost kilowatt-hours.

Cost estimates for TMI-related additions and backfits continue to change as the NRC issues new bulletins and orders. In early 1980, the industry estimated that changes already indicated by the NRC would cost between $1.5 and $8.2 billion and would require $40 to $280 million per year to implement. It was clear that experienced manpower necessary to engineer, install, and operate the new systems and equipment did not exist. The industry asked the NRC to set priorities among the various changes. The wide ranges result from plant-to-plant differences, depending on whether certain systems already exist or not.

Not knowing just what the investment will turn out to be, an arbitrary assumption of $50 million for a 1000-MWe unit is made for the purpose of analysis. Assuming a fixed charge rate of 20% per year and a 60% capacity factor, this investment would add 1.9 mills/kW-hr to the bus-bar cost.

Commonwealth Edison, with six 1100-MWe-class nuclear units under construction, believes firmly that its decisions have been the correct ones. Despite the rising costs of nuclear plant construction, the nuclear units still appear to be better investments than any alternatives available at the time or since. Faced with a serious cash problem in 1980, the company stopped all design and engineering work on its two Carroll County nuclear units scheduled for the early 1990s and on two coal-fired generating units that had been scheduled for 1989 and 1990.

3.1.2. Florida Power and Light Company

Nuclear generation in 1979 amounted to 26% of FPL's total. However, the remainder comes from burning oil, much of it imported. Nuclear operations saved 18.4 million barrels of oil. The differential fuel saving, which is passed directly to the customers because of the way in which FPL's fuel adjustment clause is designed, was $284.2 million.

FPL estimates that a "typical" 1000 kW-hr/month residential customer saved $86 in 1979. The total cost saving through 1979 of the two Turkey Point units and the one at St. Lucie has been $1.2 billion, $1\frac{1}{2}$ times the investment cost to build the three units.

3.1.3. Virginia Electric and Power Company

Virginia Electric and Power has operated Surry 1 and 2 since 1972 and 1973, respectively, and North Anna 1 since 1978. Both Surry units were shut down in February 1979. Since then, work has proceeded to replace the steam generator tubes and complete a seismic reanalysis and minor modifications to reestablish seismic capability. They returned to service during the summer of 1980.

The actual cost of steam generator tube replacement for both units is $133 million. Based on the simplified calculation in the previous section, this would add about 3 mills/kW-hr to Surry generating costs in future years.

North Anna 1 achieved a 62.5% capacity factor over its first 2 years of operation. Based on 1979 operation, its generation cost is estimated to be 31 mills/kW-hr. At the time of the TMI accident, North Anna 2 had already applied for its operating license. Permission to go to power was finally granted in August 1980, 17 months later.

Had North Anna 2 been available, its total generation cost would have been about 33–36 mills/kW-hr. This would have replaced coal generation at 39 mills/kW-hr and oil generation at from 52 to 70 mills/kW-hr. With both Surry and both North Anna plants available in 1981, Virginia Electric and Power can reduce overall generation costs significantly.

3.1.4. Duquesne Light Company

Beaver Valley Unit 1 is 47% owned by Duquesne. It was shut down March 9, 1979 due to a difference in calculated seismic loads. It should be pointed out that after completion of reanalysis of all five units that were shut down, only a few supports were modified. The design issue involved a design basis earthquake, so improbable that the units could have been permitted to operate while the reanalysis of piping and support design was carried out with no real increased risk to the public.

The outage lasted 162 days. Costs of replacement energy over that period averaged $123,000 per day. To the typical residential customer, the effect was approximately 5¢/day; to the company it amounted to $19,880,000. Once the unit was permitted to start up, it averaged 94.5% availability until its next refueling.

3.1.5. New England Electric System

For the 2-year period from May 1977 through May 1979, nuclear power supplied 14.3% of the system's generation. Oil-fired plants supplied about three-fourths as much as nuclear, the rest coming from hydroelectric dams. Practically no coal was burned. The oil-fired units achieved a capacity factor of 57.6% and generated at 26.1 mills/kW-hr. The four nuclear plants on the system generated at 14.4 mills/kW-hr, a saving of 11.7 mills/kW-hr. This equals $392 million saved and represents 40 million barrels of oil. (In 1981 the Vermont Yankee nuclear plant generated 78% of the amount of electricity used in the entire state of Vermont.)

The simple fact is that in New England all the hydroelectric power available is used. The rest, except for about 8% from coal, wood, and gas, is nuclear or imported oil. For every nuclear kilowatt-hour generated, the saving is translated directly into less oil that must be imported.

3.1.6. Maine Yankee

The Maine Yankee unit was shut down for 12 weeks in 1979 for seismic reanalysis. Calculated replacement power costs, shared among the plant's owners, totaled $41 million.

3.1.7. Ontario, Canada

Ontario Hydro set a record by exporting 11.7 billion kW-hr to the U.S. in 1979. Hydro's generation was 29% nuclear. This energy went to Vermont, New York, and Michigan, all states that could have more nuclear generation if proposed nuclear projects had been completed.

3.1.8. Electricité de France

The French national utility has embarked on a program of construction or standardized pressurized water reactors. By the end of 1982 a total of 23 new 900-MWe units were in service, in addition to several older units. Nuclear power supplied 40% of France's electric power and 12% of its total primary energy in 1982.

Electricité de France reported its average nuclear generation cost at 19.2 centimes/kW-hr, equivalent to about 2.8¢/kW-hr at the exchange rate current at the end of 1982. Its coal generation cost was 75% higher, while the cost of oil-fired generation was more than $3\frac{1}{2}$ times the cost of nuclear electricity generation.

3.2. The Fossil-Fuel Penalty of U.S. Nuclear Plant Delays and Cancellations

In the period January 1, 1977 to July 1, 1979, U.S. electric utilities slipped the schedules of their LWR nuclear power plants under construction and/or on order by a total of over 200 plant-years. During 1977, 1978, and 1979, 32 nuclear units were canceled by U.S. utilities, aggregating almost 35,000 MWe of generating capacity. More than 30,000 MWe have been cancelled since then, as shown in Table VIII.

One disadvantage inherent in the loss of nuclear generation can be expressed in terms of the fossil fuels needed for substitute generation. For instance, for the decade of the 1980s, the decline in scheduled capacity between utility reports as of January 1, 1977 and October 1, 1979 (see Fig. 2) means that the equivalent of about 3 billion barrels of residual oil will have to be burned to make up for the canceled or delayed nuclear capacity (see Table IX). Three billion barrels is about what the U.S. produces in crude petroleum in 1 year and represents almost 11% of proven U.S. crude reserves in 1978.

The oil equivalent of 226,000 barrels/day average for 1979 is almost one-fifth of the daily average of residual fuel oil imported into the U.S. during the first half of 1979. The lost nuclear generation of the 1980s represents a considerably higher percentage compared to expected imports. The delays and cancellations of nuclear units in the $2\frac{3}{4}$ years ending October 1, 1979 signify an average added use of fossil fuels during the 1980s equivalent to more than 800,000 barrels of residual oil daily. They represent a lost opportunity for drastically reducing residual oil imports.

The potential for replacing oil and gas has not disappeared. In 1982 more than one-fifth of U.S. electricity generation came from oil- and gas-fired power

TABLE VIII
U.S. Nuclear Power Capacity Ordered and Canceled

	Ordered		Canceled	
	No.	Capacity (MWe)	No.	Capacity (MWe)
Total through 1971	136	109,116	0	0
1972	38	41,509	6	5,738
1973	41	46,874	0	0
1974	26	30,961	8	8,290
1975	4	4,180	11	12,291
1976	3	3,790	2	2,328
1977	4	5,040	9	9,862
1978	2	2,300	13	13,333
1979	0	0	6	7,140
1980	0	0	16	18,085
1981	0	0	6	5,811
1982	0	0	7	8,558

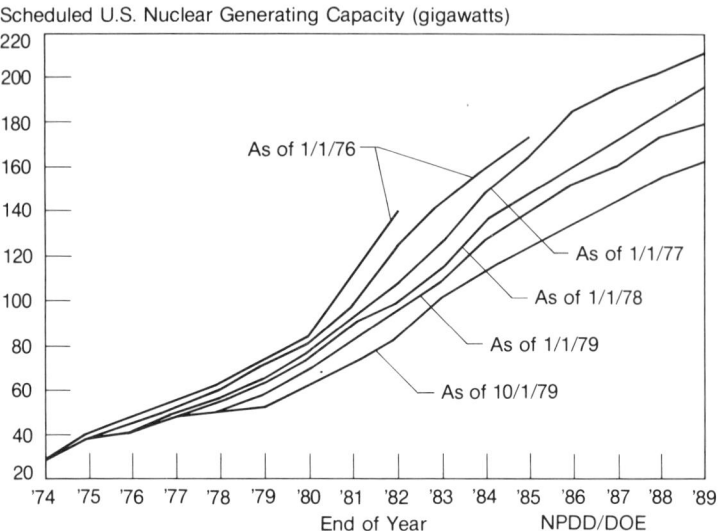

FIGURE 2. Nuclear electric generating capacity scheduled for commercial operation by U.S. electric utilities.

plants, and most of it was used to provide base-load generation. Five states with about 26% of the U.S. population accounted for two-thirds of the oil use in power plants: New York, Florida, Massachusetts, California, and Connecticut. Five other states, with about 24% of the population, accounted for more than 80% of the natural gas-fired generation: Texas, California, Louisiana, Oklahoma, and Florida[5].

3.3. Cost Factors

3.3.1. TMI-Related Costs

It has been estimated above that reasonable estimates may be $50 million investment per plant, which would add 1.9 mills/kW-hr in carrying charges, and $3 million in operating costs, which could add 20% or about 0.7 mill/kW-hr to O & M costs. These estimates may well turn out to be low for some nuclear plants. Large elements of cost include training simulators, classrooms, staff and trainee time, a remote emergency operations control center, and additional hardware for venting, sampling, and postaccident monitoring.

TABLE IX
Estimated Oil Equivalent Displacement Lost Due to U.S. Nuclear Power Plant Delays, Deferments, and Cancellations between January 1, 1977 and October 1, 1979[a]

During years	Barrels per year[b]	
	Millions	Thousands
1977	15.4	42
1978	37.6	103
1979	82.5	226
1980	127.6	350
1981	161.8	443
1982	199.5	547
1983	223.0	611
1984	271.1	743
1985	333.6	914
1986	404.5	1108
1987	455.3	1247
1988	441.8	1210
1989	434.1	1189
1977–1989	3187.8	672 daily av.
1980–1989	3052.3	836 daily av.

[a] The many delays and several cancellations from 1980 through 1984 make these figures much lower than actual.
[b] Based on displacement of 9 million barrels per year of residual fuel oil for electric energy generation by one 1000-MWe nuclear power plant.

3.3.2. Enrichment

The major area of government involvement is uranium enrichment. According to published financial statements, the U.S. government had a net income of $110 million from its enrichment operations in the fiscal year ending June 30, 1976. This was a rate of return of 13% on its invested capital. In addition, the government collected $50 million for depreciation of plants originally built for weapons material production, which it would not have recovered if commercial nuclear power had not been developed. Furthermore, although carrying charges on the enrichment plants, which were built years go, should have remained constant, the price of enrichment services has nearly tripled since 1976 and is predicted to go higher. To quote Gordon Corey, former Vice Chairman of Commonwealth Edison, "If this is a subsidy to industry, it is different from any other kind of subsidy I know."

3.3.3. Waste Disposal

There are uncertainties about waste disposal today, largely as a result of delays in government demonstration projects and the Administration's policy of deferring reprocessing. Utilities now charge an additional 1 mill/kW-hr on their books. This is required by the Nuclear Waste Disposal Act of 1982 and should be sufficient to provide for all disposal costs by the time final disposition is made of all nuclear waste. In any event, the waste management services the government provides will be billed to the utilities without subsidy, like the enrichment service. Even if costs were 100% higher, their impact on overall generation costs will not be enough to change the competitive position of nuclear power generation compared with coal.

Recent plans announced by the DOE have vastly stretched out the schedule for opening of the first repository. After the Carter Administration took office, the target date slipped from 1982 to 1997.

3.3.4. Decommissioning

It is estimated that the cost of decommissioning a nuclear plant will be about $40 or $50 million 40 years or more after start-up, depending on the criteria that are ultimately adopted. (Feasibility is not really in question; there are several options, with different price tags, from which to choose.) This translates into a cost of about 0.2 mill/kW-hr, and substantially less than this after present value discounting. In any case it will have a very small impact on utility rates or the viability of nuclear power.

3.3.5. Insurance

Even with TMI, there have been no claims on the government's part of the nuclear insurance structure. Utilities, like the owners of TMI, carry insurance on repairs to their own facilities. However, the TMI experience suggested that utilities should form an insurance pool to cover the huge costs of replacement power from a major extended outage. The method adopted involves contingent liability of $5 million per generating unit. It is conceivable that nothing will have to be paid out if operating experience is good. As for premiums to the government for Price-Anderson indemnity coverage, the premiums are too small to have an effect on nuclear power economics.

3.3.6. Legal

Although historically legal fees have been small compared to engineering and operating costs, they seem to be growing faster, and should be taken into account in calculating future plant economics.

3.3.7. Security

Guard forces at nuclear stations have jumped from about 25 to 100 persons in response to NRC security regulations. This adds $1 million a year to O & M costs at each nuclear station.

3.3.8. Industry Groups

Various problems have required utilities to form owners' groups or to subscribe to fund special programs. For example, the Nuclear Safety Analysis Center and the Institute of Nuclear Power Operations have annual budgets of over $30 million. A group of 20 utilities spent $30 million over 4 years on BWR stress corrosion and piping problems. Another group spent $50 million studying steam generator problems, and other groups are working on relief valves, containment and accident analyses. A number of groups (at much lower budgets) deal with regulatory and legal problems.

All of these costs will decrease the gap between nuclear and coal generation costs, but only as a small increment on large numbers.

3.4. Anticipated Costs of Plants under Construction

Few, if any, current nuclear plant projects can be expected to be completed on their original schedules or within budgets. Cost overruns have afflicted every project and the story of the problems, changes, delays, and cost increases for each nuclear project would make a book in itself. Articles, papers, and even television "documentaries" have been done on the subject. The causes are always complex, one delay building on another.

The industry has shown that regulatory delays have been the major factor. Regulators claim every delay has been necessary for the credibility of the process, that the law has required the actions that have been taken, and that construction problems and slow delivery of components have contributed even more delay. Nuclear critics claim the delays are inherent, that all they have done is remind the industry and the regulators of their responsibilities.

The reality is that every delay, no matter what the cause, opens the opportunity for more interventions, contentions, lawsuits, requests to reopen old issues, etc. The performance of nuclear critics, whether successful or not, has been obvious. They have taken advantage of delays to cause more delays. These delays, in an inflationary economy, force increased costs. Then they demand new Environmental Impact Statements because the cost–benefit balance may have changed.

All the added safety features and their costs notwithstanding, the largest factor in cost overruns has been delay. Prices of everything have risen with inflation. In addition, the cost of money has risen, sometimes precipitously, increasing the penalty caused by extended construction schedules. Interest on funds tied up during construction sometimes represents over half the total dollars by the time a long-delayed plant is completed.

As a result, plants now under construction, all of them delayed, some delayed more than a year when almost ready to operate, have had to increase their cost estimates. Figure 3 illustrates the trend of estimated capital costs for these nuclear units. Those expected to enter commercial service after 1980 reflect TMI-related delays and cost increases.

Cost estimates made when commitments are made are based on best judgment, and include some allowances for contingencies, but never assume that lengthy and unpredicted delays are bound to take place. This may turn out to be wishful thinking, but to assume such delays would be to arbitrarily inflate cost estimates.

One key to future commitments to nuclear power plants must be shortening of the licensing time and confidence that—barring emergence of new, real safety

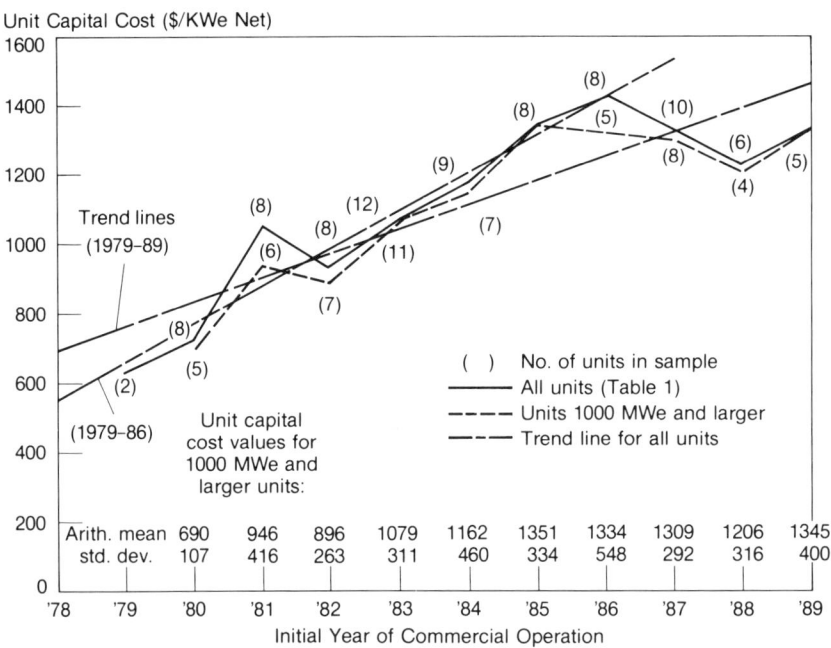

FIGURE 3. Average LWR power plant unit capital costs by year of entry into service (reported as of December 31, 1979).

issues—construction permits, once issued, will remain in force and schedules can be insulated from capricious interference by those who oppose nuclear power itself, whatever their reasons may be.

4. FUTURE ECONOMICS OF NUCLEAR AND FOSSIL GENERATION

Changing regulations and other factors require continued reevaluation of available alternatives. The results of such a recent reevaluation for new nuclear and coal-fired base-load units scheduled for service in the late 1980s, based on estimates made in 1979, are summarized in the following pages. The figures are no longer realistic, but they illustrate the cost of components and the methods for estimating costs. History demonstrates how unpredictable costs are, particularly when lead times are long.

In order to make valid comparisons of nuclear and coal-fired plant construction costs, one must either use comparable time periods or else express all costs in constant dollars. The constant dollar approach (e.g., all costs in 1980 dollars) is acceptable in making rough cost comparisons, because there is little evidence that future costs of one type of plant will escalate more or less than the other. However, the use of constant dollars for fuel cost estimates tends to penalize the nuclear alternative. Therefore, cost comparisons are usually developed both with and without inflation, so that judgment can be exercised in making plant investment decisions.

Nuclear plant construction costs at a single utility (Commonwealth Edison) have increased from $149/kW for Dresden units 2 and 3 (which were initially contracted for on a fixed-price turnkey basis in 1965 and placed in service in 1970 and 1971) to estimates of over $700/kW for the Byron and Braidwood units, which will not go in service until the mid-1980s, at perhaps twice the original estimates. At the same time, coal-fired plant costs have increased from $113/kW for Joliet units 7 and 8, which were placed in service in 1965 and 1966, to $225/kW for Powerton units 5 and 6, which were placed in service in 1972 and 1975, to which another $150/kW or more (in 1977 dollars) must be added if stack gas scrubbers are required, and several times those amounts for plants committed in 1980.

These upward cost trends, approaching 15% a year in either case, have resulted from general construction cost increases combined with increasingly stringent environmental and safety requirements. These are nationwide phenomena affecting both nuclear and coal-fired plants.

The fact is that there are uncertainties about the cost of all new generating stations. It isn't a matter of whether or not conservation investments are cheaper.

If, in spite of conservation, a new plant is needed, hard decisions have to be made about what to build, when, and where.

The conclusions of cost estimators are consistent:

- Nearly all find generation from nuclear will be less costly than from coal-fired plants over the life of the plant.
- Both nuclear and coal energy generation costs will rise.
- The extent of the rise depends largely on inflation.
- Both coal and nuclear power are viable options on an economic basis alone.
- Detailed cost estimates are essential for any specific project decisions.

Despite the fact that a nuclear plant, if it could be built on schedule, would be a better investment for both customers and stockholders, the regulatory and political atmospheres are so negative that the necessary confidence does not exist in the U.S. to make new nuclear commitments in 1983. Even if it did, there are few utilities that can handle the huge capital commitment based on their current capital structure. The investment community views the numbers of dollars that ultimately have to go into the project, the number of years until returns on the investment begin to be realized, and the uncertain prospects for fully adequate rates in future years, and weighs these concerns against the size of the company and its own growth prospects.

Even consortia of utilities will find it difficult to finance new nuclear units in this atmosphere. The irony is that it is almost as difficult to finance a coal-fired plant, and prohibitive to commit to a full commercial-scale solar, wind, hydro, geothermal, or even conservation project *unless* adequate revenues from rates or government subsidies are *assured*. This is the reason, despite lowered projections of electricity demand, that so few new plants of any kind are being announced.

However, with regard to nuclear plants, the economic uncertainty stemming from uncertain, unstable, and politicized regulation, and the lack of public understanding of energy realities, relative risks, and waste disposal technology have created an economic atmosphere in which investments cannot be made. Economic costs cannot even be estimated because the uncertainties are so great. Investment risks that are open-ended cannot attract the type and volume of investment necessary to construct nuclear power plants. Thus, the economics are not simply bad, they do not even exist. Nuclear power plants have been removed from the marketplace as far as new energy capacity decisions are concerned.

In time, the slowdown in construction of new generating capacity will mark a limit on the rate of economic growth that the nation can sustain. A continued hiatus on new orders makes this future inevitable, its time-frame being governed by the extent of conservation, population growth, and the expectations of people. It means that the risk of prolonged energy shortages has to be faced.

If the average age at which a power plant is retired were 30 years, the number of megawatts of electric generating capacity to be lost in 1985 matches the new capacity scheduled for operation that year. (It is unrealistic to expect any new plant committed in 1985 to be on line before 1995.) Thus, the nation's capacity picture could soon become static, with no new capacity for growth or for replacement of oil- or gas-fired plants. The problems of the future may not only be inevitable, but their onset may well be predictable.

4.1. Conversion of Nuclear Plants to Coal

In response to political pressures, a number of utilities have been required to make calculations of the cost of converting partially completed nuclear stations (even TMI) to coal. The results have been cost figures that are extremely expensive, generally billions of dollars more than the cost of completing the nuclear plant.

For example, the conversion cost for the Shoreham plant of the Long Island Lighting Company is estimated at $9.9 billion, considering construction, operation, and fuel costs over 30 years and the costs of delay in terms of replacement power. For Diablo Canyon, the estimate is over $6 billion. GPU's estimate for converting TMI-2 is $1.7 billion, not including replacement power, since the delay in restarting TMI-2 is unknown.

The results of all of these studies indicate a huge financial penalty involved in conversion. It is clear that if there is an advantage to nuclear in the first place, the cost of converting a plant designed for nuclear to coal would be substantial. The addition of all kinds of coal-handling facilities, sludge-handling facilities, ash-handling facilities, and pollution control equipment means that the only thing that could really be saved in the conversion would be the electric switchyard, a generator, and a turbine optimized for a steam supply with very different characteristics than those obtainable from an efficient coal-burning plant. Sites suitable for nuclear plants might even be unacceptable environmentally for coal, and the logistics of handling unit trains at those sites might be prohibitive.

4.2. Energy Alternatives

Without question, effective conservation of energy will increase, and the nation's success in achieving energy conservation will affect future electric loads. This will impact the scheduling for the construction of additional coal and nuclear plant capacity. However, it will not affect the economics of the choice between the two.

The possibilities for solar power remain unclear with regard to the extent that solar energy can reduce the need for new electric generating capacity.

However, either direct conversion of solar energy to electricity, or the direct use of solar energy to reduce electricity demand, requires that solar energy be produced and used during daylight hours. This is the period when most utilities experience peak demand, and many utility systems have to call upon their peaking capacity (some of it oil-fired) in order to meet those demands. As a result, solar energy really competes with oil burning (and perhaps some coal burning) rather than with nuclear power.

In fact, nuclear power and the future use of solar energy are not only compatible, but nuclear power is a necessary condition for achieving significant solar input on a national basis. Only if utility systems can operate reliably and economically is there any chance that sufficient capital will be available to invest in even more highly capital-intensive solar energy systems. However, solar would integrate well with utility systems and make low fuel cost energy available at peak load times. If energy storage has to be added to permit the use of solar energy at times other than peak daylight hours, the likelihood of its being economically competitive is really very small.

Other energy alternatives are all of interest. No nuclear proponent claims that nuclear is the only answer. They explain that nuclear energy is an essential part of the mix of viable energy sources and that as base-load capacity, it plays its proper role.

5. CONCLUSION

This analysis shows that if utility decisions could be made on a clear economic basis, nuclear plants would be attractive to utility decision-makers today. The lack of orders can be attributed largely to uncertainties in the regulatory process and the way in which political pressures influence it. Utilities can cope with bad regulatory decisions and stringent regulations. They are not well equipped, however, to cope with indecision, interminable delay, and indefinite deferral. The lack of objectivity and stability on the part of regulatory bodies is most frustrating to a rational utility decision-making process.

Perhaps recurring crises in the Middle East will reawaken the American public to the vulnerability of this nation to imported energy supply. This would suggest that it is imprudent to delay operating licenses for nuclear plant, let alone construction permits and commitments to new ones. It is ironic that while the nation is obtaining economical and reliable power from some 80 licensed nuclear power plants, the accident to one of them has created an atmosphere in which small groups can use the regulatory process to delay the operation of the newest and safest units. At the same time, political leaders who profess their lack of enthusiasm for nuclear power are neither eager nor willing to call for shutdown of operating plants. These positions are inconsistent.

The economic impact of nuclear decisions (or lack of decisions) is not quickly reflected in the political process. However, the nation is paying a serious long-term price of huge magnitude that it will not be able to write off easily. The real danger to the nation comes in the fact that by the time energy capacity shortages become clearly evident to the average citizen, there will be a lead time of 10 years or so before new nuclear generating capacity can be brought on line. Experience shows that the cost of building long-term projects in a crisis mode is high indeed, and in a technology where safety is the prime importance, it should also be recognized to be in the nation's interest to plan and build nuclear plants with utmost care, inspection, and quality control.

The construction of more nuclear power plants is not inevitable. However, the nation should be fully aware of the consequences if it chooses not to build them. If society recognizes those consequences, it is likely that a very strong political consensus would emerge that would call for the building of these plants. If they are to be built, they should be built on prudent schedules.

The National Environmental Policy Act almost forces decisions to be made only when there is no other alternative. This is a far cry from the kind of decision-making that was possible in years past, in which those with responsibility could look ahead and try to protect the nation against possible contingencies by timely decisions and prudent scheduling.

Even if nuclear power were to have a slight cost disadvantage, nuclear plants would be essential for diversity of supply to guard against the depletion of domestic oil and gas and the debilitating effect of huge oil imports, and to compete with coal in order to keep all fuel prices from skyrocketing further. In fact, for diversity of supply or for geographic and system planning reasons, a utility might choose an option that appears to have the higher bus-bar cost if the differential is small, should it appear possible that the choice might result in lower long-term costs to its customers.

If projected bus-bar costs differ by as little as 20%, there is an essential role for both technologies, coal and nuclear. There is simply too much uncertainty to claim that any projections will be wholly accurate over the next 40 years. Furthermore, uranium and plutonium have virtually no other use than for energy. This is not true for oil, coal, and gas, which have important nonenergy uses for plastics, chemicals, and other purposes.

Central-station generating unit commitments for the foreseeable future will be nuclear and coal. Any policy that precludes or restricts either technology would be unwise for the U.S. as a whole.

REFERENCES

1. AIF INFO (May 14, 1979).
2. C. Komanoff, *Power Propaganda,* Environmental Action Foundation (March 1980).
3. DOE "Update," pp. 54–67 (May–June 1980). Data include nuclear units larger than 400 MWe and coal units larger than 500 MWe with in-service dates from 1965 on.
4. A. D. Rossin and T. A. Rieck, *Economics of Nuclear Power* Vol. 201 (Aug. 1978).
5. D. Bodansky, A strategy for saving oil, *Journal of Contemporary Studies* (Spring 1983).

7
Fast Breeder Reactor Economics

C. Pierre Zaleski

1. INTRODUCTION

The long-term incentive for the development of breeders is well known: better use of natural uranium in order to extend the potential of nuclear energy by a factor of 50 with given uranium resources, and in fact by a much larger factor, since also uranium resources from low-grade ores can be used economically in breeder reactors. Nuclear energy without breeders represents a relatively modest contribution to the world's total energy supply; with breeders, it allows thousands of years of high energy consumption.

The question one may ask is: taking into account the relative merits of different sources of energy, will breeder reactors be deployed, and if so, when? The relative merits of importance are: environmental impact, safety, public acceptance, relation to nuclear weapons proliferation, effect on balance of payments, energy independence, and economics.

Assuming that for all the aspects except economics, breeders could achieve at least a neutral score against their competitors, the development of this type of reactor will depend on their economic competitiveness. Economic competitiveness may be different for different countries. However, if it is achieved in one country, one can reasonably assume that, with some delay, it will also be achieved gradually in more and more countries. We may, therefore, concentrate our analysis on one country, France, as it has one of the most advanced programs of breeder reactor development in the world.

Reactor economics has to be measured with a yardstick. One can talk about absolute values of energy cost, but these absolute values, during years of high inflation, do not have a very lasting meaning.

C. PIERRE ZALESKI • Center for Geopolitics of Energy and Raw Materials, University of Paris–Dauphine, 75775 Paris Cedex 16, France, and University of California, Los Angeles, California 90024.

Therefore, we prefer the use of relative values in measuring costs; we will try to relate the projected cost of electricity produced by breeders to that of electricity produced by their competitors.

In fact, the closest competitors of breeders in France are presently light-water reactors (LWRs).

Therefore, if competitiveness with LWRs is achieved, one then can expect that breeders will be built. Their operation in symbiosis with LWRs will improve the overall performance of nuclear energy through a stabilizing effect on natural uranium prices.

The overall aspects of breeders, other than economics, seem at least in France to compare favorably with those of LWRs. We can, therefore, try to approach the question of breeder deployment by comparing the costs of electricity produced by breeders and by LWRs, as a function of time or rather as a function of uranium prices. Indeed, the cost of electricity produced by breeders is extremely insensitive to the price of uranium. Breeder reactors consume negligible quantities of uranium as compared to their competitors, LWRs. In addition, for a very long period of time they can use as feed material, instead of natural uranium, uranium depleted in the isotope 235, which is produced in large quantities as a useless by-product of thermal reactor operation and weapons material production. On the other hand, the cost of electricity produced by LWRs is strongly dependent on natural uranium prices. For example, if the price of natural uranium increases from \$60 per lb. U_3O_8 to \$240 per lb. U_3O_8, the cost of a kilowatt-hour produced by an LWR will increase by more than 50% (all other conditions unchanged and based on the current French cost context). We can therefore see that starting with a noncompetitive liquid-metal fast breeder reactor (LMFBR), one can reach competitivity just by a sufficient increase in natural uranium prices.

We will now discuss the question of the relative economics of light-water and breeder reactors in four sections as follows: a brief description of methodology; discussion of construction cost; discussion of operating cost; kilowatt-hour cost calculation and conclusion.

2. METHODOLOGY

Our objective is to compare the cost of producing electricity by two different nuclear power stations: the LWR and the LMFBR. In considering this cost, one should take into account three components:

- *Construction* cost incurred before the plant is put into operation; interest expenses prior to operation
- *Operating* cost, including maintenance and fuel cycle costs that are incurred during the useful life of the plant

7 • Fast Breeder Reactor Economics

- *Taxes* other than those included in construction and operating costs, and *other costs* resulting from engaging in a commercial enterprise.

In fact, the first two components are the major ones; in addition, one can assume that the third is similar for electricity produced by each of the two types of power plant. Therefore, when examining the competitivity of the two nuclear power plant types, we can limit our discussion to the first two components.

2.1. Present Value

To make our analysis, we will use the concept of "present value." This concept is derived from the fundamental notion of a return on investment or an interest one pays when one borrows money for investment. If we assume that the interest rate is constant in time and equal to i, then we can write the following equation:

$$E_{pr} = E_{pa}(1 + i)^n$$

where E_{pr} is the present value of a past expense E_{pa} that has been incurred n years ago. Conversely:

$$E_{pr} = \frac{E_{fu}}{(1 + i)^n}$$

where E_{pr} is the present value of future expenses E_{fu} that will be incurred in n years.

The present value concept may also be applied to revenues:

$$R_{pr} = \frac{R_{fu}}{(1 + i)^n}$$

where R_{pr} is the present value of future income R_{fu} that will occur in n years.*

2.2. Comparison of Cost of Production

We want in fact to compare the cost of production of a kilowatt-hour by two systems, LWRs and LMFBRs. We may assume that during the lifetime of

---

* We have used as the unit of time scale 1 year, and to be consistent we must express the interest rate per year. However, if for some calculations, for example fuel cycle calculations, we need a finer description as a function of time, we can very well use another unit for the time scale, for example 1 month or 1 week, and then of course we must express the interest rate per month or per week.

the plant the deflated cost of a kilowatt-hour would be constant and, as we have mentioned before, the deflated interest rate would also be constant.

We use the notion of deflated interest rate and deflated cost so as to isolate our calculation from the problem of erosion of the value of money. We will, therefore, calculate the present value of all expenses incurred (construction cost, operating cost, fuel cycle cost, etc.) at a given point in time, and then equal their value to the present value at the same point in time of the revenues generated by sales of electricity during the lifetime of the plant. (In this equation we will not take into account either profits or taxes on the revenues.) From this equation, we can easily extract the cost of the kilowatt-hour produced by the plant. For convenience, we may decide to choose as the given point at which we are performing the present value calculation the moment at which the power plant begins operation. In fact, however, one can use any point in time, provided one treats consistently all past and future expenses and revenues.

A detailed description of the economic calculations and methods that we are using, including the effect of taxes, may be found in *Fast Breeder Reactors* [1]. Although the methodology of calculation is quite straightforward and simple, the difficulties of an economic appraisal of LMFBRs are great and lie in the calculation and selection of correct values for the parameters to be used in the formulas. Often, these values must be found by extrapolating from present, insufficient experience.

3. CONSTRUCTION COST

Although the technical feasibility of the LMFBR concept has been well demonstrated by the successful operation of EBR-2, Phenix, and a number of other French, British, Russian, and U.S. breeder reactors, its economic competitiveness has yet to be demonstrated. Breeder reactor fuel cost is expected to be low because of its almost negligible demand on uranium resources, but there is considerable concern that LMFBR capital cost could be so high that electricity production by breeder reactors might not be economic in this century.

An estimate of the capital cost of a fully developed commercial "series" of LMFBRs is still difficult and includes some uncertainties. The main uncertainties could result from safety-related regulatory requirements, which in many countries are not yet clearly defined for LMFBRs.

One can note that very stringent safety requirements were applied to the near-commercial breeder reactor plant, Super-Phenix, under construction in France. The safety requirements for LMFBRs appear to be more stringent than for LWRs, highlighting an increased concern about reactor safety and reflecting the technology evolution of recent years. In fact, the operators and designers of Phenix and Super-Phenix (Electricité de France, Novatome) believe that recent and more accurate safety studies demonstrate that some regulations applied to Super-Phenix were too conserative, and could be relaxed for subsequent breeder reactor models without compromising the overall level of safety. Indeed, French safety authorities recently accepted this point of view and gave preliminary approval to a less conservative design for Super-Phenix II. Essentially, their present position is to consider that whole core meltdown should no longer be taken as a design basis accident.

All this indicates that the capital cost of large breeders, or at least of follow-on models of the Super-Phenix system, should not increase due to changing regulations, and may even decrease. This should hold even more for the relative capital cost of breeders versus LWRs.

This being said, we feel that the most objective and sound approach is to compare the capital cost per kilowatt installed of LMFBRs already built or under construction to the same cost for comparable LWRs, which is now relatively well established, and then to try to extrapolate this result to fully developed commercial breeders and LWRs.

We have followed this approach in a study published in *Nuclear News* [2]. This study was based on comparison of early U.S. and French LWRs and LMFBRs. The results indicate that the capital cost per kilowatt installed for LMFBRs appears to be in range of 115 to 140% that for LWRs.

An independent study was made about 8 years ago using French conditions for both reactor types. Its rather optimistic assumptions are listed in the Appendix. Consequently, the results are lower than the ones obtained below. The authors distinguish a first and a second generation of breeders for which they obtain kilowatt-hour cost ratios of 1.34 and 1.16 (LMFBR/LWR), respectively.

After 1976, the important new step in the world development of LMFBRs has been the construction of the Super-Phenix power plant.

We will try to analyze below the additional experience brought by this new power plant.

3.1. The Near-Commercial Plant Super-Phenix Mark I

The third stage (after Rapsodie and Phenix) in the French breeder program is Super-Phenix Mark I. The experience with this reactor, for the time being, is restricted to design, testing of components, and construction; first power generation being expected in 1985.

The Super-Phenix plant at Creys-Malville is the culmination of the breeder R & D phase. With its 1200 MWe, it will represent industrial confirmation of the technology [3, 4].

Super-Phenix project engineering and construction work have shown clearly that a plant of such a design can be engineered, manufactured, and built without major difficulties. This is all the more remarkable as Super-Phenix has been preceded only by a first-of-a-kind project, Phenix.

Between Phenix and Super-Phenix, size increases and cost considerations led to certain modifications. The fuel elements are slightly larger. They are designed for higher burnup (the design target is 100,000MWd/t). Refueling and maintenance is scheduled only once a year. The main difference, however, concerns steam generator design: the Phenix modular type was replaced by the more economic helical-tube steam generators (Figs. 1 and 2).

The contract for Super-Phenix was awarded by NERSA, the owner of the Creys-Malville plant, to NOVATOME-NIRA in April 1977; the start of construction was authorized in May 1977, and no work was done before this date except some preliminary site preparation. Power operation is scheduled for 1985, 8 years after the contract award.

At the time of this writing, practically all the Super-Phenix components have been ordered, involving some 70 main supply contracts awarded to 35 different European firms. These contracts, in many cases, imply a combined effort on the part of manufacturers from several different European countries—backed, however, by assistance from members of the Rapsodie or Phenix engineering teams. Due to their very large size, certain components could not be

FIGURE 1. Super-Phenix: Helical winding of steam tubes on the central body of a steam generator. (Courtesy Novatome.)

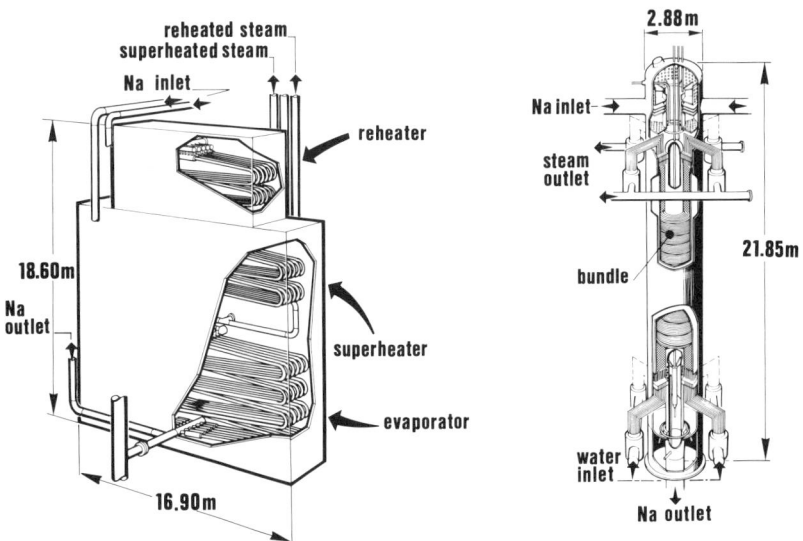

FIGURE 2. Comparison of Phenix (A)(186.3 milliwatts) and Super-Phenix (B)(750 milliwatts) steam generators. Same scale. (Courtesy Novatome.)

FIGURE 3. Super-Phenix: Work in progress in the site shop, January 1980. (Courtesy Novatome.)

delivered in one piece, and it was necessary to provide special assembly facilities on-site (Fig. 3).

Construction is now practically completed and testing of components has started; filling of secondary sodium loops should be finished by the end of 1983 (Figs. 4 and 5).

The status of the Super-Phenix project has been described above to indicate that the capital-cost figures for this reactor, which are subsequently used to estimate capital costs of future FBRs, are based on actual experience.

One may question whether the breeder technology used in Super-Phenix and to be used in future breeders is sufficiently demonstrated. We can make the following comments on this subject.

The successful operation of many experimental reactors and three demonstration plants gives a first tentative answer. More specifically, the excellent performance of Phenix has given confidence to French engineers. For example, in 1979 Phenix had an availability of 93.5% and a capacity factor of almost 80.5%. On the average, Phenix's load factor for its first 10 years of operation

FIGURE 4. Super-Phenix: Creys-Malville site, reactor building, August 1982. (Courtesy Novatome.)

FIGURE 5. Super-Phenix: General views of the Creys-Malville site, April 1982. (Courtesy Novatome.)

exceeded 50%, which compares favorably with that of comparable water reactors. One may also ask whether the Phenix technology is representative of commercial breeder technology.

As it is reasonable to assume that the French commercial breeder Super-Phenix Mark II will have a design very similar to that of Super-Phenix Mark I, the previous question can be answered by comparing the Phenix and Super-Phenix Mark I design and technology.

Without going into details, we can say that fuel, core, primary loops and components, fuel handling, reactor vessels, etc. all present a moderate size extrapolation when going from Phenix to Super-Phenix Mark I (much smaller than extrapolation from Rapsodie to Phenix) (Figs. 6 and 7). Essentially the same materials and technology are used.

The only large extrapolation, actually a change in technology, exists for the steam generators.

However, mock-ups of the new steam generators were subjected to a very extensive testing program for prolonged periods under nominal as well as abnormal conditions. All tests were successful (Fig. 8).

Thus, the technology of Super-Phenix Mark I is rather well demonstrated, and we can expect it to be representative of later reactors.

The above description of Super-Phenix status and discussion of the technology used is intended to show the limits, but also the soundness, of a capital-cost evaluation of future commercial breeders based on this project.

3.2. Construction Cost Comparison

First, what is important to define is the term of reference—in other words, which LWR costs will we select to compare with Super-Phenix costs? To obtain the most consistent and logical comparison, we will try to select LWR power plants of similar nominal output and, as far as possible, built during the same period of time as Super-Phenix. These criteria point to the selection of the series of eight 1300-MW reactors called P4, which includes four reactors at the Paluel site, two at Saint-Alban, and two at Flamanville. The most contemporary to Super-Phenix are the four units at Paluel, which should start in operation in 1984 and 1985. However, the site work at Paluel (cooling by seawater) is particularly expensive. On the other hand, the fact that this site hosts four units, while on the other sites there are only two, at least partially compensates for the site expense at Paluel.

The cheapest site is Saint-Alban, with once-through river water cooling. However, the variation of the cost of a kilowatt installed between the most expensive site, Paluel, and the least expensive, Saint-Alban, is still not very great—around 15%. Therefore, it seems to us that the best term of comparison should be the average value of a kilowatt installed for the eight P4 units. This

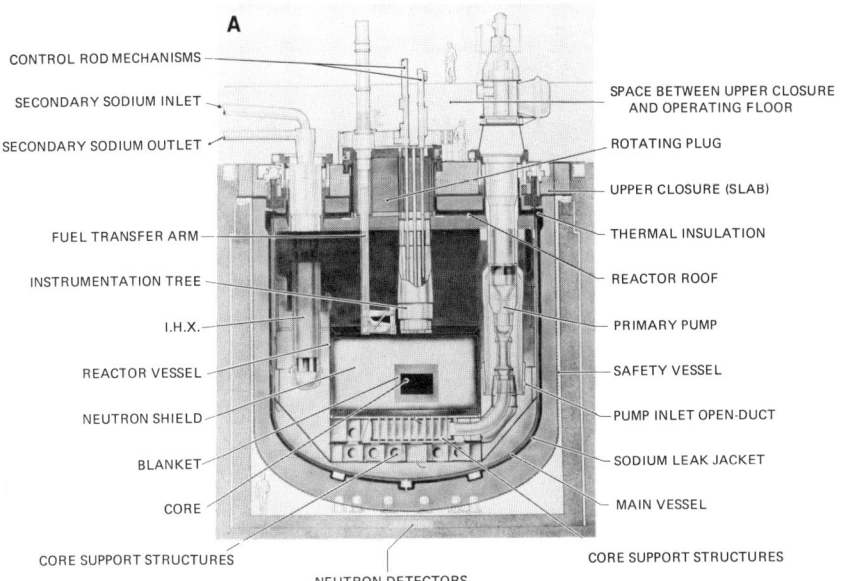
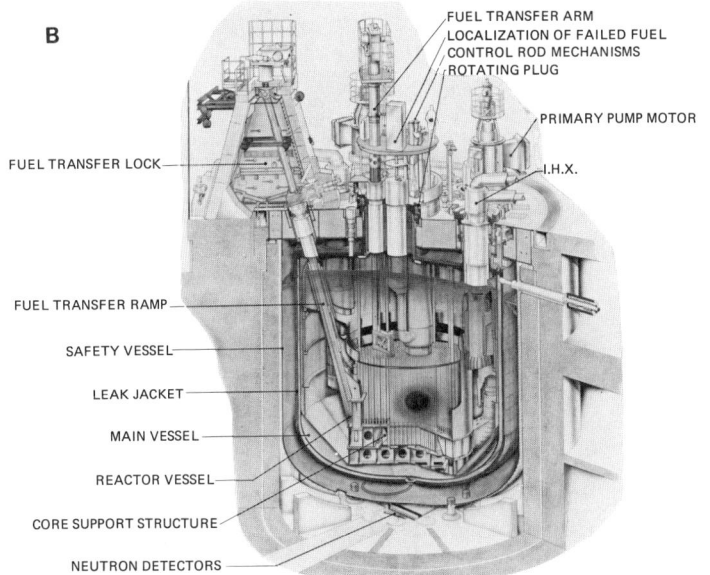

FIGURE 6. Phenix: (A) Reactor and primary circuits; (B) section through pump and IHX.

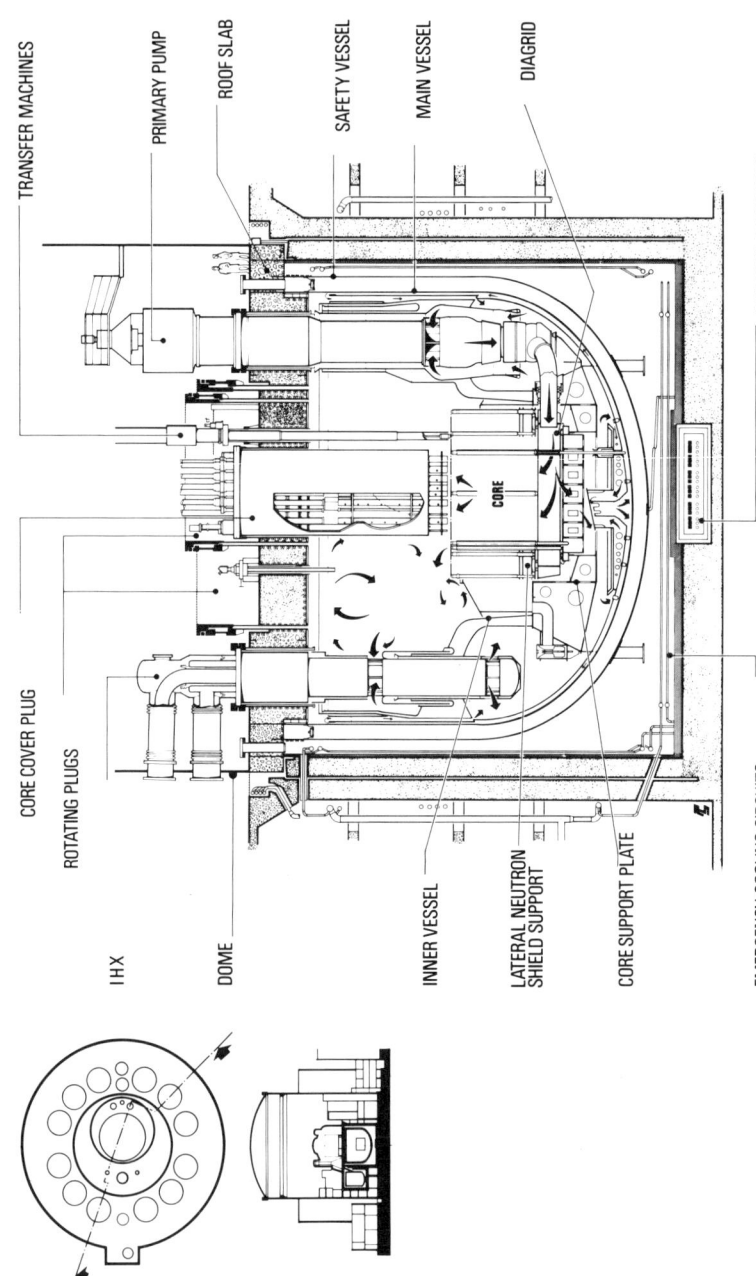

FIGURE 7. Super-Phenix: Arrangement of main primary circuit components. (Courtesy Novatome.)

FIGURE 8. EDF steam generator test facility at Les Renardieres.

value for total construction cost is equal to 4626 francs/kWe in constant francs of January 1982.

Here, we do not use the formulation of "present value" introduced in the previous section and which we will use for the general calculation, but rather real expenditures in constant money (deflated to January 1982 French francs). Therefore, the total construction cost includes all expenses except interest during construction and fuel (the first core or even the first two cores which sometimes are included in construction costs).

As regarding Super-Phenix, the most recent cost estimate of October 1983, which should be very close to the final one, gives a value of 10,750 francs/kW installed in January 1982 francs. This sum, again, is the total construction cost not including interest during construction, fuel, or a special fuel storage facility (a sort of "AFR") that was decided recently because of delays in establishing the off-site fuel reprocessing facility. This last item clearly has no equivalent in the P4 series and should not be taken into account in the cost comparison from our perspective.

The simple division of these two values for Super-Phenix and P4 gives a value of 2.32:

$$\frac{\text{SPx I}}{\text{LWR-P4}} = \frac{10,750}{4,626} = 2.32$$

This is the value of the ratio between the construction cost per kilowatt electric of Super-Phenix I and the average construction cost per kilowatt electric of the P4 series of 1300-MWe LWR power plants.

It is now necessary to apply some corrections to this number if we wish to compare the potential cost of an industrial series of LMFBRs to the cost of the already industrial developed series of LWRs. Let us first consider the study now being done by Novatome and Electricité de France of a possible next French breeder, Super-Phenix II. This study of a breeder reactor of 1450 MWe, versus 1200 MWe for Super-Phenix I, indicates that practically the same size and weight of primary components will be sufficient for the new breeder, which has 22% higher net electricity output. Therefore, one can expect a saving of some 22% on the cost of the nuclear island between Super-Phenix II and Super-Phenix I when expressed as cost per kilowatt electric installed. (We must remember that the nuclear island cost for Super-Phenix represents more than 50% of the total plant cost.) This saving is due mainly to improved knowledge of the technology, allowing a decrease in the design margin of most components and permitting a much more compact overall design. In addition to this saving, one can expect some additional saving due to a more rational approach taken in the safety area, as we have mentioned previously. Altogether, one could expect that Super-Phenix II built under the same conditions as Super-Phenix I—that is, a unique design

7 • Fast Breeder Reactor Economics

(prototypical), one unit on the site, in a complex international management context—should cost 20% less per kilowatt electric installed than Super-Phenix I. This would bring the capital cost per kilowatt electric ratio of Super-Phenix I to P4—2.32—down to a ratio of Super-Phenix II to P4 of 1.86.

Now, we must consider the possibility of cost reduction for LMFBRs when and if they are developed in series like LWRs. In fact, we are comparing the cost of Super-Phenix I and II to the P4 series of eight reactors built by pairs (at least two on the same site) and within an optimal time frame to allow the best use of personnel and site equipment. In addition, the P4 series was preceded by and gained experience from three series of 900-MW LWRs built in France: a preliminary series of 6 units; the CP1 series of 18 units; and the CP2 series of 10 units, for a total of 34 units.

We have therefore to consider three corrections:

- *The series effect*—for a series of reactors, one will have lower expenditures per kilowatt electric installed for studies, tests, spare parts, and some manufacturing equipment, due to the effect of sharing.
- *Two-unit site*—in the P4 series we have considered at least two units per site built in an optimal time frame, instead of one reactor on a site for Super-Phenix.
- *International context*—an international organization, NERSA, which subcontracts most of the engineering responsibility to EDF, certainly costs more than the normal EDF organization; the desire to distribute the contracts among different nations in proportion to national shares in the project may also add some costs.

Altogether, these three corrections may reduce the cost of construction per kilowatt electric installed by some 22%, or an average of 5 to 9% for each correction. This would bring the ratio between the capital cost per kilowatt electric of an industrial series of LMFBRs and that of an LWR series to 1.45:

$$\frac{\text{SPx II}}{\text{LWR-P4}} = 1.86 \xrightarrow{22\%} \frac{\text{industrial LMFBRs}}{\text{LWR-P4}} = 1.45$$

We will use this value as a basis for our general calculation, at least for the *realistic* version of the calculation. We will, however, take into account some more optimistic evaluations of LMFBRs to calculate a lower—in our opinion, very *optimistic*—value for the above ratio by applying an additional 10% reduction, which yields a ratio of 1.3:

$$\frac{\text{industrial LMFBRs}}{\text{LWR-P4}} = 1.45 \xrightarrow{10\%} \frac{\text{optimistic LMFBRs}}{\text{LWR-P4}} = 1.3$$

Finally, we will use these two ratios—1.45 "realistic" and 1.30 "optimistic"—for our calculation of the relative cost of kilowatt-hours produced by LMFBRs and LWRs.

4. OPERATING EXPENSES

4.1. Cost of Fuel

Some input data of importance for calculating LWR and FBR fuel cycle costs have changed drastically during the past decade. For example, the price of natural uranium over the last decade increased by a factor of 8 and then decreased by 30% (spot market prices at least). Another example of a steep increase is the estimated cost of LWR spent-fuel reprocessing. Nine years ago, most of the estimates for LWR fuel reprocessing and transportation were in the range of $150/kg. Presently, the only commercial operation of LWR fuel reprocessing is at La Hague, France; the most recent French estimate is $750/kg. This includes transportation of fuel, reprocessing, waste solidification, and retrievable storage of solidified waste for 100 years.

The increase in natural uranium cost can be rationalized by remembering that we are dealing with an essential raw material whose resources and reserves are not inexhaustible; as for other raw materials, the general trend is one of cost increase (in constant money). Superimposed on this general trend there are fluctuations due to changes in the demand/supply situation, which are often influenced by political actions (U.S. embargo on natural uranium imports, position of the Australian Government on uranium mining, actions of OPEC countries) or by independent economic and political considerations (e.g., sharp slowdown in world nuclear plant ordering, especially in the U.S.).

It is, therefore, reasonable in forecasting long-term trends in fuel cycle costs to assume a further increase in the uranium price. The question is, how much of an increase should be expected, and when? A few percent average annual increase? Or, like oil (and a few years ago uranium), an increase that could reach or even exceed 100% in some years? We will not try to answer this difficult question; rather, we will present LMFBR/LWR economic performances as a function of uranium price.

The variation of LWR spent-fuel reprocessing cost is of a different nature. We are dealing with a technology that is just starting first commercial operations. Early cost estimates were extremely uncertain, usually underestimating technical difficulties and the trend toward ever stricter regulations (safety, safeguards, radiation exposure limits).

One can note that the prices of services for which an industrial commercial

7 • Fast Breeder Reactor Economics

operation is well established (UF_6 production, LWR fuel fabrication, uranium isotope enrichment) are stable or even slightly decreasing in constant money.

On the other hand, fuel fabrication and reprocessing for LMFBRs are, at this stage, not industrial or commercial operations. The technology is known and demonstrated, but commercial-size plants are not even under construction. Therefore, bearing in mind the experience with evaluating the cost of LWR fuel reprocessing, one must be cautious when evaluating LMFBR fuel reprocessing and fabrication cost.

Selection of Costs for Different Fuel Cycle Components

The selection of correct values is both essential and difficult. Commercial breeder reactors (competitive with LWRs) are not expected in operation by even their most optimistic supporters before the year 2000. Their useful life may be 30 years. Therefore, the time interval for which we would need to evaluate the cost of fuel cycle components is 2000–2030.

We will express all values in constant 1983 U.S. dollars, and we will use an exchange rate of $1 = 8 French francs.

1. *Cost of Natural Uranium.* 1983 spot prices of U_3O_8 (yellowcake) are about $26/lb. Short-term trends (within 2–4 years) do not show a rapidly increasing price, but it is likely that longer-term trends (within 10–20 years) will show a steady price increase. To evaluate the value of the price of uranium during the first 30 years of the next century is a difficult task. Without going into detailed discussions, we will select two values that may be taken as limits of a possible range of "weighted average" values of the price of uranium during the lifetime of the first commercial breeders. These limits are $60/lb. U_3O_8 and $240/lb. U_3O_8.

2. *Cost of Conversion of Yellowcake to UF_6.* Conversion cost does not have a very great impact on the final results. In addition, there are no major reasons for this cost to change in the future. We will therefore adopt the current value of $5/kg.

3. *Cost of Enrichment of Natural Uranium.* We will select the value of $120/SWU (separative work unit), which is close to the present Eurodif price, expecting that in the longer term an increase in energy cost above the inflation rate will be compensated by improvement of the technology.

4. *Cost of Fuel Fabrication for LWRs.* The cost of fuel fabrication for LWRs is rather well known and in the last few years has tended to decrease in constant money.

There are some differences between manufacturers (mainly in the U.S. and Europe) but a value of $130/kg seems reasonable.

5. *Cost of LWR Spent-Fuel Reprocessing.* The cost of LWR spent-fuel reprocessing, including transportation, waste vitrification, and retrievable storage

of solidified wastes, was estimated recently at $750/kg. Some experts feel that this cost could still increase, even in constant money, mainly due to stricter regulations. On the other hand, one can expect that future reprocessing plants will take advantage of operating experience with the first commercial La Hague plant, and therefore will allow lower prices.

These two trends may compensate each other in the long term. We will therefore adopt $750/kg as a reasonable value.

6. *Fabrication Cost of Fast Breeder Fuel.* There are large uncertainties concerning this cost. Three factors contribute to a much higher cost for FBR fuel fabrication than for LWR fuel fabrication. They are:

- FBR fuel contains a higher proportion of fissile material.
- FBR fuel is designed for much higher specific power than LWR fuel, and therefore FBR fuel pins have much smaller diameter than LWR fuel pins.
- The throughput of the fabrication plant will, at least initially, be much smaller for FBR fuel than for LWR fuel.

Preliminary studies based on experience with fabrication of fuel for Phenix and Super-Phenix indicate that the fabrication cost of FBR fuel may be eight times the fabrication cost of LWR fuel (realistic value).

In the longer term, one can expect a relative decrease in FBR fuel fabrication cost (experience gained, higher plant capacity), allowing us to take as a possible lower value a factor of six between FBR and LWR fuel fabrication costs (optimistic value).

This gives us the following two values (six or eight times $13/kg):

$780/kg (U + Pu), $1040/kg (U + Pu)

7. *FBR Fuel Reprocessing Cost.* The indication for the first FBR fuel reprocessing plant is that the cost of this service may be as high as six times that for LWR fuels. Subsequent plants, by taking advantage of improved technology and larger size, may allow a reduction of this factor to four.

We will consider two values: $3000/kg and $4500/kg.

8. *FBR Blanket Fabrication Cost.* This cost should be similar to the fabrication cost of LWR fuels. We will therefore take the value of $130/kg.

9. *FBR Blanket Reprocessing Cost.* Here again, the cost should be similar to the reprocessing cost of LWR fuel. We will therefore take the value of $750/kg.

10. *Price of Plutonium.* This price may vary during the deployment of breeder reactors. It is clear for many reasons that the use of plutonium in breeders, if they are available, is much more efficient (cost-efficient) than use of that plutonium for recycling in thermal reactors [6]. However, large inventories of civilian plutonium, mainly coming from the La Hague reprocessing plant, are

beginning to accumulate, and there will not be enough breeder capacity to use this plutonium for some time yet.

Therefore, the economic advantage of recycling plutonium in thermal reactors is a valid question today. We do not intend here to give a detailed discussion of the equivalent value of fissile plutonium burned in thermal reactors—i.e., giving the same cost of a kilowatt-hour produced as enriched uranium fuel. Nevertheless, we may take as a reasonable value under present economic conditions that of $20/g fissile plutonium.

This value does not mean that reprocessing of LWR spent fuel exclusively for reuse of recovered uranium and plutonium in LWRs is economic under present conditions. The economic calculation may depend, among other things, on what cost assumptions one makes for permanent storage of unreprocessed fuel, reprocessing, and so forth.

In any case, we can assume that in the early stages of LMFBR introduction, there will be a large stockpile of civilian plutonium available from existing reprocessing plants (La Hague) that will be operated—if not for other reasons—for optimal waste management.

This plutonium stockpile could be used in thermal reactors (recycle) to replace enriched uranium even if the value of plutonium burned in thermal reactors is much lower than that of ^{235}U.

But what is more important, the recycling of plutonium in thermal reactors is limited to a few cycles by the buildup of the isotopes 240 and 242.

Therefore, even if one decides to recycle plutonium in thermal reactors, there will be at the beginning of the next century a stockpile of plutonium inappropriate for further thermal recycling, and therefore useless and even posing a problem of the disposal of highly radioactive, very-long-lived material.

This plutonium could be burned in breeder reactors; thus, it seems reasonable to assign it a zero value at least at the beginning of introduction of LMFBRs. One can note that later on, when the competitiveness between LWRs and FBRs is fully achieved, the price of plutonium can be calculated so that the price of a kilowatt-hour produced by LMFBRs and by LWRs remains identical.

We therefore will use for our basic calculations a plutonium price of zero, but also make some calculations with a plutonium price of $20/g to see the influence of an increasing plutonium price on the competitiveness of LMFBRs versus LWRs.

4.2. Balance of Operating Costs

Operating costs other than fuel costs are relatively well known for LWRs in France. They include two components. One is independent of the number of kilowatt-hours produced and includes salary of operating crew, part of maintenance costs, etc.; the other is proportional to the number of kilowatt-hours produced and includes such things as certain taxes, consumption of certain products, and the rest of maintenance costs.

The constant part may be expressed in dollars per kilowatt electric installed per year and is equal to $33/kWe/year.

The variable part is equal to 0.16 mill/kW-hr. For the "realistic LMFBR case" calculation we will take for the operating cost the LWR operating cost multiplied by 1.25. This yields: $41/kWe/year and 0.20 mill/kW-hr.

For the "optimistic" case we will take the same operating costs as for LWRs.

5. KILOWATT-HOUR COST CALCULATION AND CONCLUSION

5.1. General Parameters

There are still three general parameters that we must consider:

- Interest rate
- Total time and payment schedule for construction (or the present value of construction cost)
- Useful life of the plant and load factors achieved (or the present value of all full-power operating hours)

1. *Interest Rate.* In France, the government authorities define the interest rate to be used in calculating the economic performances of different investments (this official interest rate is in fact used to eliminate investments one should not make).

The rate presently is set at 9% per year in constant (deflated) money, which of course is a rather high interest rate. In the U.S., a typical utility may finance a power plant with 55% capital raised by debts or bonds and 45% by equities or stocks. A deflated bond interest rate of 2.5% and a deflated equity interest rate of 7.0% are often used in economic calculations. In this typical case, the effective deflated interest rate would be 4.5%:

$$i = (0.55 \times 0.025) + (0.45 \times 0.07) = 0.045 \text{ or } 4.5\%$$

We will, therefore, perform our calculation with the French deflated interest rate of 9%, but we will also study the influence of a more realistic (for U.S. conditions) deflated interest rate, which may be as low as 3%.

2. *Construction Time and Payment Schedule.* Super-Phenix construction time will be some 7.5 years, while most of the 1300-MWe LWR plants in France are scheduled for a construction time of roughly 6 years.

However, construction of the first 1300-MWe plant, Paluel-1, took over 7

7 • Fast Breeder Reactor Economics

years. Considering the prototypical character of Super-Phenix and the international management of this enterprise, we can say that for the industrial series of LMFBRs one should be able to attain at least the same length of construction time as for LWRs, if not a shorter time.

As regarding the schedule of payment within this construction time, we do not see any reason for a meaningful difference between LMFBRs and LWRs.

Therefore, we will select the same construction time—6 years—and same schedule of payment for LMFBRs and LWRs.

3. *Present Value of Full-Power Operating Hours*. The current assumptions for French PWRs are a useful lifetime of 25.11 years and a present value of full-power operation of:

$$61{,}788 \text{ hr at an interest rate of } 9\%$$

or

$$107{,}072 \text{ hr at an interest rate of } 3\%$$

We will select for LMFBRs practically the same values (the small difference is due to different techniques of fuel management).

The useful lifetime will be 25.51 years and the present value of full-power operating hours will be:

$$60{,}718 \text{ hr at an interest rate of } 9\%$$

or

$$107{,}600 \text{ hr at an interest rate of } 3\%$$

This assumption seems realistic based on the operating experience of Phenix, the first French demonstration plant, which during its first 10 years of operation achieved a plant capacity factor of over 50% and in some years (notably 1979) exceeded 80% capacity factor.

This result for a demonstration plant of a new technology compares favorably with similar results for LWR plants.

5.2. Kilowatt-Hour Calculation

We will perform two series of calculations for LMFBRs, one *"realistic"* that tries to use our best estimates for all values to be taken into account, the other *"optimistic"* and giving in our opinion the possible lower limit for a favorable evolution of LMFBR economics.

We will also perform these calculations for two prices of natural uranium, $60 and $240/lb. U_3O_8, to show the variation of the competitivity of LMFBRs versus LWRs for different uranium prices. Finally we will study two other effects:

- Value of interest rate: 9 or 3%
- Value of plutonium: $0–$20/g Pu

The key hypotheses of our calculations are summarized and listed below.

1. **General Hypotheses**

Deflated interest rate (two values)	3% and 9%
Plutonium (two values)	$0 and $20/g fissile
Natural uranium prices (two values)	$60 and $240/lb. U_3O_8
Conversion $U_3O_8 \rightarrow UF_6$ (one value)	$5/kg
Enrichment (one value)	$120/SWU
Power plant construction time (one value)	6 years
Useful lifetime of power plant (one value)	25 years
Present worth of operating hours (two values)	61,000 hr—9%
	107,000 hr—3%

2. **LWR Key Parameters**

PWR (P4 industrial series) unit size	1300 MWe
Construction cost	$692/kWe
Fuel fabrication cost	$130/kWe
Fuel reprocessing cost	$750/kWe

3. **LMFBR Key Parameters**

Super-Phenix-type industrial series unit size	1450 MWe
Design and fuel cycle optimized for low plutonium prices	$0–$20/g fissile
Date of introduction	After 2000

	Realistic	Optimistic
Construction cost	$1004/kWe	$900/kWe
Fuel fabrication cost	$1040/kg	$780/kg
Fuel reprocessing cost	$4500/kg	$3000/kg

The results are summarized in Tables I and II.

For "realistic" LMFBRs:

$122/lb. U_3O_8—9% and Pu = $0

7 • Fast Breeder Reactor Economics

TABLE I
LWR Kilowatt-Hour Costs

Key parameters			Results (mills/kW-hr)					
							Total	
Interest rate (%)	U cost ($/lb. U_3O_8)	Pu cost ($/g fiss.)	Invest-ment	Fuel cycle	Other operating	Below 25	25 to 35	Above 35
9	60	0	11.2	9.55	5.75		26.5	
9	120	0	11.2	14	5.75		30.95	
9	240	0	11.2	22.9	5.75			39.95
9	60	20	11.2	9.2	5.75		26.15	
9	120	20	11.2	13.65	5.75		30.6	
9	240	20	11.2	22.55	5.75			39.50
3	60	0	6.4	8.7	5.6	20.7		
3	240	0	6.4	19.0	5.6		31.0	

$157/lb. U_3O_8—9% and Pu = $20/g fissile
$142/lb. U_3O_8—3% and Pu = $0
$159/lb. U_3O_8—3% and Pu = $20/g fissile

For "optimistic" LMFBRs:

TABLE II
LMFBR Kilowatt-Hour Costs

Key parameters			Results (mills/kW-hr)					
							Total	
Interest rate (%)	Realistic/ optimistic	Pu cost ($/g fiss.)	Invest-ment	Fuel cycle	Other operating	Below 25	25 to 35	Above 35
9	Realistic	0	16.6	7.2	7.1		30.9	
9	Realistic	20	16.6	9.6	7.1		33.9	
9	Optimistic	0	14.9	5.2	5.75		25.85	
9	Optimistic	20	14.9	7.6	5.75		28.25	
3	Realistic	0	9.5	8.6	7.0		25.1	
3	Realistic	20	9.5	9.5	7.0		26.0	

$57/lb. U$_3O_8$—9% and Pu = $0
$88/lb. U$_3O_8$—9% and Pu = $20/g fissile

From these results we can calculate six values for natural uranium cost, which give the equivalence of kilowatt-hour cost between LWRs and LMFBRs depending on the hypothesis used.

We may make some observations based on these calculations. First, increasing the plutonium price has a modest effect on LWR kilowatt-hour cost, but a substantial effect on LMFBR kilowatt-hour cost—an effect that is also enhanced by a high interest rate.

Second, decreasing the interest rate from 9% to 3% for a zero plutonium value has—surprisingly enough—a more favorable effect on LWR kilowatt-hour cost than on LMFBR kilowatt-hour cost.

This could be explained by the fact that a large fraction of expenses for the LWR fuel cycle are incurred in advance of power production, e.g., purchase of natural uranium and enrichment of fuel. Therefore, a decrease in interest rates has a very favorable impact on LWR fuel cycle cost, more than compensating for the greater impact of a decreasing interest rate on LMFBR investment as compared to LWR investment.

As soon as the plutonium price increases, the interest rate on advance purchase of fuel (Pu) for LMFBRs moderates this trend, and the decrease in interest rate becomes less unfavorable to LMFBRs.

Third, for the realistic cases, the value of natural uranium that assures the competitivity of breeders with LWRs lies between $119 and $157/lb. U$_3O_8$, which are certainly very high values.

For the optimistic cases the values of $53 to $88/lb. U$_3O_8$ seem quite reasonable for the time period considered (2000–2030).

In any case, selecting a proper value for the price of natural uranium during the years 2000–2030 and beyond is a difficult matter, and the decision-maker considering the development of breeders must make some guesses. He must, however, also keep in mind the risks of not having this source of energy if it is needed versus those of subsidizing its development and then not needing it, or needing it only later.

There would seem to be a clear economic advantage for all countries interested in LMFBR development to join their efforts in developing the technology. In this way, they minimize the risks by sharing expenses and then sharing the benefits. In fact, this approach has already been taken in Europe, where the Super-Phenix power station is being built by France, Italy, Germany, Belgium, the Netherlands, and Great Britain. This kind of cooperation should be extended to other countries and to a broader scope of activities.

6. SUMMARY

A comparative analysis of the power generation cost of a fast breeder reactor (LMFBR) and a light-water reactor (LWR), both built in France during the past seven years, is presented. The analysis is based on the separate evaluation of capital, operating, and fuel cycle costs.

The reference LWR, a 1300-MWe PWR, is one of a series of eight built around this time that did benefit from the experience of about 34 PWRs previously built in France. The LMFBR, the 1200-MWe Super-Phenix I, is the first of a kind, with only a much smaller prototype (Phenix) as precursor, except for the even smaller experimental fast reactor Rapsodie.

The basis of the analysis is a comparison of the actual capital costs. Taking a modest or a more optimistic allowance for cost reductions that should be expected if the LMFBR is built in a series as the PWRs shows that the eventual cost could be 45 or 30% higher for the LMFBR than for the PWR. However, there is a compensating effect resulting from the uranium costs: Electricity from the PWR becomes considerably more expensive with increasing uranium prices, as compared to a negligible effect for the LMFBR. The capital cost disadvantage may then be overcompensated, even taking into account the higher reprocessing and fuel fabrication costs.

Tables I and II indicate that the eventual electricity costs in mills per kilowatt-hour are in the same ballpark for both reactor types. Depending on the future cost of uranium, the LMFBR power costs could be somewhat higher or somewhat lower than the PWR power costs.

APPENDIX

An independent study was made 6 years ago for French conditions [5], based on the following assumptions (all costs are expressed in constant 1979 French francs):

- The nuclear part of a 1300-MWe PWR built presently in France represents 38 to 39% of the total direct construction cost; it corresponds to 900F/kWe.
- The nuclear part of a unit in the first series of the Super-Phenix Mark II breeders (four to six units) will cost 1.9 times that of the nuclear part of a PWR, or 1712F/kWe.
- The nuclear part of a completely developed breeder (after four or five Super-Phenix Mark II plants are built) will cost 1.45 times that of a PWR, or 1304F/kWe.

- The balance of the plant cost will be the same for breeders and for PWRs; the other expenses, including interest during construction, will be proportional to the direct cost in both cases (same duration of construction and same spending schedules).

These hypotheses yield a total construction cost for first-generation LMFBRs of 4700F/kWe, and for second-generation breeders 4100F/kWe, compared to 3460F/kWe for PWRs.

These rather optimistic assumptions led Baumier *et al.* [5] to a capital cost ratio of 1.34 for the first generation of breeders versus PWRs and 1.16 for the second generation of breeders. These numbers are to be compared with the values of 1.45 and 1.3 given above.

ACKNOWLEDGMENT. I wish to thank Mr. Jacques Charles for performing some of the fuel cycle calculations.

REFERENCES

1. P. S. Owen and R. P. Omberg, in: *Fast Breeder Reactors* (A. B. Walter and A. B. Reynolds, eds.), pp. 53–100, Pergamon Press, Elmsford, N.Y. (1981).
2. M. Levenson, M. Murphy, and C. P. Zaleski, Economic perspective of the LMFBR, *Nucl. News* (April 1976).
3. G. Vendryes, Super Phenix: Full scale breeder reactor, *Sci. Am.* p. 26 (March 1977).
4. Creys Malville Nuclear Power Station, *Nucl. Eng. Int.* (June 1978).
5. J. Baumier, J. Charles, and M. Labrousse, Les tendances économiques á long forme des surgénérateurs et leur place dans les contextes énergétiques mondiaux et francais, *Revue Générale Nucléaire* No. 6 (1979).
6. J. Bussac, A. Cania, P. Clauzon, A. Meyer-Heine, G. Vendryes, and C. P. Zaleski, Utilization of plutonium in fast and thermal reactors, Seminar of the European Society for Atomic Energy, Baden-Baden (Sept. 1973).

III Recycling and Proliferation

Part III represents a collection of papers on the issue of nuclear fuel recycling and its relation to the nuclear weapons proliferation problem. Few will dispute that nuclear weapons proliferation is a serious problem; but we point out here that it is not a problem that is significantly aggravated by the existence of a nuclear power industry, nor even by the use of plutonium as a fuel for that industry.

In the first paper, "International Cooperation in the Nuclear Field: Past, Present, and Prospects"* by B. Goldschmidt, we review the history of attempts to create international institutions so that the benefits of nuclear power can be shared by the world while avoiding the risks of nuclear war. Then, B. I. Sprinrad offers two papers: "Nuclear Recycling: Costs, Savings, and Safeguards" and "Alternative Fuels, Fuel Cycles, and Reactors: Are They Useful? Are They Necessary?" These are didactic papers; their main purpose is to explain the interplay between the continuing development of new reactors and fuel cycles and the real issues of nuclear safeguards.

Following this, B. I. Spinrad and E. L. Zebroski discuss the weapons proliferation issue in depth. They point out that the only true defense against widespread proliferation is a national and international consensus that this is in no one's interest. Such a consensus is demonstrable in the roster of countries who could have weapons any time they want to, but do not have them. To the extent that nuclear power, and particularly the breeder, alleviates world tensions arising from controlled oil supplies, it makes a positive contribution to maintaining this consensus. These same points are reinforced by Zebroski's "Paths to a World with More Reliable Nuclear Safeguards," which follows.

---

* Reprinted with editorial modification from *IAEA Bulletin* Vol. 20, No. 2 (1978), with the permission of the International Atomic Energy Agency, Vienna, Austria.

The last paper is a very brief note on "The Homemade Bomb Issue" by B. I. Spinrad and E. L. Zebroski. In it, the obstacles to making a clandestine bomb are explained. The conclusion is that, given these obstacles, the homemade atomic bomb should be ranked very far down on anybody's list of things to worry about.

The last point having been detailed, we confirm as our message that nations can best be persuaded to abjure nuclear weaponry by giving them, in exchange, free access to the benefits of nuclear power (which is exactly what the original Non-Proliferation Treaty offered).

We would be less than candid if we did not emphasize that large-scale nuclear war is *very much* to be feared. The big threat, against which all other threats pale, is the "balance of terror" between Russia and America. We will never be free of fear so long as the two superpowers keep themselves super-armed with nuclear weapons to the ridiculous degree that both have world-overkill capability.

B.I.S.

8 International Cooperation in the Nuclear Field

Past, Present, and Prospects

Bertrand Goldschmidt

1. THE FIRST EUROPEAN COOPERATION

Early in 1950, the Norwegian physicist Gunnar Randers found himself in a predicament: he was in charge of the construction, already at an advanced stage, of one of the first nuclear reactors to be made outside the Anglo-Saxon world, but he had not managed to find the requisite uranium. Indeed, it had been on his instigation that the Norwegian government, encouraged by its national production of heavy water (the first such industrial production in the world), had granted just after the war the appropriations required for the construction of a heavy-water research reactor of low power, quite reasonably assuming that the uranium needed for its operation would be found on Norwegian territory in the course of the few years required to build the reactor.

This was still at the height of the period of secrecy, for the Anglo-Saxon wartime Allies—the United States, the United Kingdom, and Canada—had decided in November 1945 to abide by the agreement reached at the Quebec Summit Conference in August 1943, which was, first, to keep secret the technology developed jointly by them during the conflict and, second, to buy up between them all the uranium available in the Western world. This policy, which blocked access to the two ingredients essential for any nuclear development, i.e., uranium and technical know-how, was designed to mitigate the consequences of the overlap between the military and scientific aspects of atomic energy.

In principle, it was a temporary measure pending, as they said at the time, implementation by the United Nations of international control of the new source

---

BERTRAND GOLDSCHMIDT • French Atomic Energy Commission, Paris, France.

of energy, or as we would say today, pending establishment of an effective world policy of nonproliferation.

France was at the time the principal Western power not bound by the policy of secrecy. A number of French scientists had played a prominent part during the war, first in the British team and later, after 1943, in the Anglo-Canadian team, in particular by contributing 180 kg of heavy water purchased from Norway at the beginning of 1940, which represented at that time the whole of the world's stock. It was, moreover, precisely this transfer of heavy water from France to England in June 1940, and then to Canada, that led the Canadians to specialize in that particular branch of reactor technology and to champion today the cause of heavy-water power plants.

It was further purchases of heavy water from Norway after the war, once again, that made possible the construction of the first two French research reactors. Because of this close relationship, the French did not withhold from Randers the main results obtained with their first research reactor, completed at the end of 1948. By 1950 the construction of the Norwegian reactor was fairly far advanced, but uranium exploration had proved most disappointing. Randers went to Paris to ask Frédéric Joliot, the leading French nuclear scientist, to give him the necessary uranium, since France had done better than Norway in her first attempts at prospecting.

Convinced that he was the only one who could help his Norwegian colleague out of the difficulty, as the rules of the Anglo-Saxon "condominium" were opposed to virtually all exports of uranium, Joliot decided to drive a hard bargain: he agreed to provide the uranium, but without the technical information relating to the essential processes of purification and transformation into the metallic form, insisting, furthermore, that the reactor should be considered a Franco-Norwegian project. Randers rejected these terms, which he considered too harsh. Although he had warned Joliot that he could overcome his difficulty in another way, Randers was thought to be bluffing. Such was not the case, however, for shortly afterwards an agreement was concluded between Norway and the Netherlands, which had in its possession 10 or so tonnes of uranium bought in 1939 on the advice of a university professor and hidden away during the war; the existence of this material had been kept secret up until then.

Realizing their mistake, the French approached the newly created Netherlands-Norwegian group with an offer to come in as a third partner, this time on rather generous terms as far as the exchange of information was concerned, and with an offer to purify and transform the uranium required for the reactor. But it was too late; the Anglo-American allies were now alerted to the proposed arrangement and were opposed to Franco-Dutch-Norwegian collaboration in which France would inevitably play the dominant role; Washington advised against it, and the United Kingdom contributed part of what Randers had asked for from Joliot by taking over the task of purifying the impure Dutch oxide and converting it into metal. This is how the first international cooperation in the civil nuclear field came about. The Dutch-Norwegian reactor was completed in 1951 at the Kjeller Centre, the first nuclear establishment to open its doors to technicians from other countries.

2. THE POLICY OF ISOLATIONISM

A much more important agreement might have seen the light of day, however, at an earlier stage, following an undertaking given to Churchill by Roosevelt at the end of 1944 in a memorandum stating that close atomic cooperation between the American and British governments would continue after the war in the military and commercial fields unless the two parties put an end to it by mutual agreement. But having accepted, at the end of 1945, the principle of effective and complete collaboration in the civil field, and then having evaded the issue for a period of several months, Truman finally told Attlee that the U.S. could not possibly conclude an agreement with the United Kingdom enabling it to start up a complex of industries that might be exploitable for military purposes just at a time when the three Anglo-Saxon allies were embarking, at the United Nations, upon the search for world agreement on control. Nonproliferation was thus used for the first time as an argument for reappraising a previous commitment.

Indeed, over the period between mid-1946 and mid-1948 almost 200 meetings at the United Nations were devoted to what might be called today the first "international evaluation" of the technical stages of the industrial development of atomic energy from the standpoint of controlling it. The point was to study what was intended to be the practical application of the first nonproliferation plan, namely, the American Lilienthal–Baruch Plan. This plan was based on safeguards devolving from multinational administration; a supranational body was to be the proprietary possessor of the world nuclear industry and to exploit and develop it in the name, and in the interest, of all nations. In short, as an example of world government in a matter of worldwide concern, it was undoubtedly the last chance for mankind to live in a world free of nuclear weapons.

The political constellation at the height of the cold war was hardly compatible with a project as revolutionary as this, and in the spring of 1948 the United Nations Atomic Energy Commission announced it had reached a deadlock and broken off its deliberations. The first attempt at nonproliferation had failed, and the policy of secrecy and isolationism could only be intensified as a result, lasting in fact until the first break: the joint Dutch-Norwegian undertaking described above, which was launched in 1950 and continued until 1960, bringing together the nuclear research teams and activities of the two countries. But the irony of fate is that on account of the Norwegian hydroelectric resources and the discoveries of natural gas in the Netherlands, and more recently of the North Sea oil, these two countries have in the end found themselves under less pressure to generate nuclear power, and the long-term importance of their early collaboration has thereby been reduced.

If I have dwelled at some length on these now only dimly remembered episodes in the prehistory of the Nuclear Energy Agency of the OECD, which celebrates its 20th anniversary this year, it is because we can already see in them the seeds of the problems that face us today; we can see the advantages of the concept of international administration of nuclear facilities, and a demonstration at once of the benefits a country can reap from the possession of basic materials and advanced technologies and of the limitations of a policy of withholding such materials or technologies.

3. THE POLICY OF OPENNESS

The first phase of the history of nuclear international relations, i.e., the period of isolationism, came to a close in the mid-1950s following the famous "Atoms for Peace" speech by Eisenhower at the end of 1953, the relaxation of Anglo-Saxon policy and the lifting of secrecy during the United Nations Conference on the Peaceful Uses of Atomic Energy, held at Geneva, in 1955.

The second phase lasted until the beginning of the 1970s and was marked by the disappearance of the uranium monopoly, the availability of enriched uranium from America on the world market, and growing international trade first in research reactors and then in nuclear power plants, with scientific and technological secrecy vanishing from the scene (except as regards enrichment) and giving way to conventional industrial secrecy. This phase was also marked by the institutional and broad-scale acceptance for the first time in history of genuine international control, namely control of the peaceful uses of atomic energy, the safeguarding of which had now become one of the essential features of any nuclear collaboration between nations.

It was a period of openness marked by a more liberal attitude that was not, of course, entirely free from the political constraints inevitable in connection with the applications of uranium fission, but one characterized by a certain stability in those political constraints and by complete freedom with regard to technological and industrial options; in short, a relatively happy period favoring the growth of the nuclear industry. The resulting climate, at the end of the 1960s, enabled a large number of countries to accept the renunciation and discrimination inherent in the Non-Proliferation Treaty, made palatable as it was by a guarantee of free development for all techniques, and to place their entire nuclear activity under international control.

This period began with the golden years 1955–58, followed by years of readjustment, which lasted until the middle of the 1960s (industrial development having got off to a false start, or at least having moved forward too rapidly), and finally, from the mid-1960s onwards, was witness to the development of nuclear power on a grand scale in more and more countries. Since the beginning of the 1970s, more specifically since 1974, we have entered a third phase in world nuclear development—a time of reappraisal. It is a period of instability marked by constant shifting of the administrative and political constraints, even during the implementation of contracts, and for some countries fraught with uncertainty as to the future of a nuclear industry caught by such changes in the midst of expansion.

From the political standpoint this change can be explained both by the emphasis placed on things nuclear by those opposed to our technological way of life and by the importance given to the problem of nonproliferation at a time when there is need for added recourse to nuclear power in order to satisfy the energy requirements of a world shaken by the oil crisis. Within the industrial context, the change stems from reactions to the emergence of competition in the area of enrichment and power plant construction, and more recently from the expression of doubts as to the political and economic wisdom of reprocessing and breeder reactor operations.

This period has seen the return, as in the years of isolationism, of restrictions on the transfer of materials, equipment, and technology as the result of a lack of confidence in international commitments and in the degree of protection afforded by international control. It is too early for us to predict when and how we shall recover the much-needed stability in national and international regulations and restrictions pertaining to the new technology, or for us to know whether, having suddenly gone from the "Dark Ages" of secrecy and isolationism to a period of "Renaissance," with all its openness and stability, we are witness today to a lasting change, or whether, on the contrary, we are living in a cloud destined soon to pass and enable us to develop international nuclear collaboration once again in a climate as favorable as the one we knew in the past, by which I mean the climate essential for an effective policy of nonproliferation so adversely influenced by the present mood of increasing distrust created, paradoxically enough, by the very measures adopted for the sake of nonproliferation.

4. SCIENTIFIC COOPERATION

Since 1955 a veritable plexus of international relations, both bilateral (with the most advanced countries, especially the U.S.) and multilateral, has grown up in Europe and the world as a whole through the Nuclear Energy Agency (NEA), with 23, originally 17, Member States; EURATOM, with 9, originally 6, Member States; and the International Atomic Energy Agency, today with more than 100 Member States.

Within this framework of collaboration we can clearly single out two categories (which we shall discuss in turn) according to whether the collaboration is or is not predominantly of the industrial and commercial type. Within the vast field of international relations free from direct commercial implications, mention should first be made of theoretical research, which, moreover, often lies on the fringes of nuclear energy properly speaking. We could refer here to high-energy physics and the success of CERN at Geneva; the joint Franco-Soviet project for the bubble chamber in the Serpukhov accelerator; German-Anglo-French work on the high-flux reactor at Grenoble; joint European activities in the field of controlled fusion, sponsored by EURATOM; and the recent decision to build the JET device in Great Britain.

The training of engineers specializing in the nuclear field has been a broad-scale international activity; the Argonne and Harwell schools, like those at Karlsruhe and Saclay, have played a major part in the specialized training of engineers and technicians from, first, industrially advanced countries and, more recently, less developed countries. Technical assistance, especially in the application of artificial radioisotopes, radiological protection, and efficient utilization of research reactors, has assumed growing international proportions, first with the support of the advanced countries and now mainly under the aegis of the IAEA. These efforts have made it possible to instill in many countries not yet ready to embark upon nuclear power the confidence that they have not been completely left out of the nuclear venture. An example we might mention here is the international food irradiation project under joint IAEA/NEA sponsorship.

To establish connections between experts in all countries is also essential. The four United Nations Conferences on the Peaceful Uses of Atomic Energy held at Geneva, the recent IAEA Conference on Nuclear Power and Its Fuel Cycle, held at Salzburg, the meetings and symposia organized by the European Atomic Energy Society (a "club" made up of the national commissions of 14 European countries and set up in 1954), the American and European nuclear societies, FORATOM and the various national industrial forums—all have helped in their respective spheres to popularize nuclear energy by removing the final traces of the era of isolationism.

The part played by the NEA in connection with basic constants and nuclear data, like that of EURATOM and the IAEA in the realm of safety standards, deserves mention here, as does the establishment of safeguards systems, for which we are indebted to the IAEA, and also the EURATOM, though on a geographically smaller scale.

5. MULTINATIONAL UNDERTAKINGS

Finally, four European multinational projects going back 20 years, one set up by EURATOM and the other three by NEA, have contributed to the acquisition and sharing of information, each in its own distinctive way. The EURATOM Joint Research Centre, composed of four establishments—Ispra, the largest,

Geel, Karlsruhe, and Petten—suffered initially from the fact that under the Treaty it was assigned too broad an area of competence; although the Member States with modest national nuclear programs favored this multidisciplinary approach, countries like France wanted it to be geared to activities with a technical risk or cost justifying joint action.

These difficulties were smoothed out when the Member States reached an agreement, at the beginning of the 1970s, on a certain number of activities, mainly of public interest (reactor safety, waste management, etc.), requiring a staff of 2500 and an annual budget of 100,000,000 European Monetary Agreement units of account. NEA was spared difficulties of this kind by virtue of the optional participation of Member States in its three joint projects. Two of these were concerned with reactors: (1) the boiling water reactor plant at Halden Island, Norway, first built to provide steam for a pulp factory; only the current running costs of this reactor are charged to participating countries interested in fuel element behavior and power plant safety studies; (2) the Dragon high-temperature reactor, built and operated jointly at Winfrith in England; this plant was shut down in 1976, having been for a long time a highly advanced reactor of its kind and having furnished valuable information on a reactor type that, in spite of the difficulties encountered, even today still has numerous supporters and has not lost its chance of competing on the world market, especially where possible use in the chemical or metallurgical industry is concerned.

The third joint NEA undertaking—Eurochemic—was intended for the acquisition of industrial know-how and experience in irradiated fuel reprocessing, France being the only one of the 13 participating countries to have had such experience at the time. The pooling of plutonium extraction technology, considered today one of the most sensitive operations from the standpoint of proliferation, seemed at the time to be highly desirable within the context of European cooperation and did not give rise to any political difficulties. Built at Mol in Belgium, thanks to a good understanding between the principal European chemical industries concerned, the plant continued operating until it closed down in 1974, i.e., for nearly 10 years, under satisfactory conditions. Now it is soon to be bought up by the Belgian government, an action that runs counter, paradoxically, to the present political trend in favor of internationalizing the administration of such plants.

Though a technical success, this undertaking was a serious failure economically, for three main reasons:

- The inadequate capacity of the plant (100 tons/year), which represented a poor compromise between a pilot facility and an industrial-scale plant
- A depression on the world reprocessing market during a large part of the life of the undertaking
- The company's articles, under which it was not obligatory for shareholders to contribute to the increased capital outlays necessitated by a too low

initial estimate of the cost of the plant, or to the repayment of growing annual deficits, but which allowed defaulting shareholders to reap the benefits alone, namely the acquisition of technical information

The fact that there was a surplus of reprocessing capacity at the end of the 1960s which was responsible for Eurochemic's difficulties even led the United Kingdom and France, already in possession of large plants, and Germany, which was about to build one, to join forces in 1971 under an agreement (at first commercial and later on technological) giving rise to United Reprocessors (UNIREP) with a view to avoiding the risks entailed in an uncontrolled development of reprocessing capacity prior to saturation of existing plants. A sudden reversal of the situation, so typical of the development of nuclear energy, which has now shown up in an alarming shortage of available capacity, has led the three UNIREP partners to reconsider the role of this industrial undertaking.

And so, passing from Eurochemic to UNIREP, we have arrived at international relations that are predominantly industrial and commercial but that are still subject, to a lesser or greater degree, to political constraints.

6. COMMERCIAL RELATIONS

The activities we have just reviewed fully deserve the description of international collaboration, involving as they do in each case a joining of the forces of several countries in pursuit of a common aim. But the terms "collaboration" or "cooperation" have also been used often enough, no doubt in contrast to the paralysis in relations during the period of isolationism, to describe bilateral commercial arrangements in which the forces are not truly pooled, but in which, on the contrary, the position of strength of the country selling, i.e., the possessor of the nuclear materials or technology or both, is used to impose political conditions on the purchasing country.

This has often been the case with the supply first of research reactors and then nuclear power plants, which has become a conventional transaction in international relationships in the field; here the more advanced supplier decides on the political constraints and provides at least the most important components, fuel elements and training facilities for the teams of operators, while the purchaser, on whose territory the facility is to be set up, contributes as much as he can afford to the project. Between 1956 and 1958 it was the British, Canadian, and French advocates of the natural-uranium-fueled and graphite- or heavy-water-moderated reactor type who captured the first markets. But American industry, financially and politically backed by the U.S. government, reacted promptly, profiting from its work with research reactors and submarine engines based on enriched uranium; it managed to gain the upper hand in the market, having first tried out in Europe—through the USA-EURATOM Agreement of 1959, which included joint financing of industrial research—what were called proven power plants although they were not yet proven in actual fact.

The joint construction and administration of nuclear power plants are obviously an ideal area for collaboration; through such collaboration it has become possible, in one case after another, to bring together the engineers and technicians, nuclear industries and electricity companies of a number of countries. This is true of the Franco-Belgian power plants at Chooz and Tihange, the Hispano-French power plant at Vandellos, completed in the 1960s, the German-Belgian-Dutch SNR breeder reactor under construction, and the Franco-Italian-German prototype plant Super-Phenix—evidence of the broad collaboration existing between European countries in this technology so important for the future of nuclear energy.

7. THE URANIUM MARKET

The provision of natural uranium and enrichment services is likewise an important aspect of international nuclear relations. Their availability under stable as well as economically and politically acceptable conditions is essential for the regular implementation of nuclear power programs. Where natural uranium is concerned, the relations that involve international cooperation more than any other are those binding one country with another on whose territory the former carries out, or takes part in, exploration, prospecting, and possibly mining operations, usually in exhange for the right to take away some part of the uranium produced. If the mining zone is an important one, we may find a number of mixed and multinational groups, each responsible for a deposit. This is the case, for example, with the Niger deposits discovered by France in the middle of the 1960s, in the mining of which, besides the organizations from those two countries, there are now participating German, American, British, Spanish, Iranian, Italian, and Japanese companies.

Nevertheless, in the Western world the uranium market has been subject, on many occasions in the past, to sharp fluctuations reflecting a complete lack of any spirit of international collaboration in this field. The de facto monopoly of the Anglo-Saxons on uranium purchases meant a dearth of uranium for the other Western countries during the 1950s. Then, starting from the early 1960s, a slowdown in American purchases and the discovery of large deposits in the U.S. resulted not only in nonrenewal by the U.S. of some important contracts still running in Canada and in Africa, but also in protectionist action involving an embargo on American uranium imports.

The outcome of this was swamping of the market and a slump in prices at a time when an increased prospecting effort, such as France alone of the European countries was then making in Africa, would have been necessary to cope with a demand in keeping with the veritable blossoming of ambitious power plant programs, i.e., at the end of the 1970s and the beginning of the 1980s. Furthermore, the chief manufacturer and exporter of power plants helped, at the beginning of the 1970s, to perpetuate the stagnation by capturing and neutralizing part of the domestic and foreign market in the U.S. Indeed, the manufacturer offered along with the sale of power plants all the uranium needed to fuel them for their 30 years of operation, without covering itself by buying the requisite stocks, and thereby artificially reduced demand. The concerted efforts of producer countries other than the U.S. aimed at remedying the situation were slowly beginning to produce results when suddenly, in 1974, the situation was reversed and the shortage emerged again accompanied by its inevitable counterpart—a considerable increase in the price of uranium.

A more or less simultaneous occurrence of several factors was responsible for this reversal: the oil crisis; the decision taken by some electricity companies to cover themselves on a long-term basis and even to stock up with uranium—which they had never done before; the persistent failure of the Australian producers to come on the market, owing to the influence of trade unions hostile to nuclear energy, at a time when large deposits had just been discovered there; the slowdown, and later discontinuation, of Canadian exports as a result of constant doubt as to the political terms of sale; the emergence of the U.S. as a buyer on the foreign market as a consequence of the import embargo gradually being lifted; and finally the revelation that many sales thought to be firm were without security.

It is to be hoped that Canada* and Australia, two of the biggest Western producers, will soon get their exports going again and that the doubts regarding constraints associated with nonproliferation will be satisfactorily dispelled at the same time. Although it is normal and desirable that natural uranium or slightly enriched uranium sales for civil projects should be subject to conditions associated with peaceful utilization and to the international safeguards required for that purpose, it is in an equal degree unacceptable, for the vast majority of importing countries, that such conditions should also be accompanied by demands constituting direct interference in their national fuel cycle strategy.

8. URANIUM ENRICHMENT

Similarly, the fact that the U.S. government had until recently a virtual monopoly on the supply of enriched uranium required for fueling light-water reactors and power plants—by far the most common type—gave it an invaluable

*Editor's note: the matter has already been favorably decided.

political and commercial advantage. It was then inevitable that the more advanced countries should aspire first to building the power plants themselves—as was done by Germany, Sweden, and France—and then to acquiring a certain degree of independence as regards enrichment.

A brief account of international relations in connection with the isotopic separation of uranium, an amazing example of collaboration and competition, will be a good way to finish this review. In 1941 the British leaders were convinced of the importance, and the difficulty, of constructing a national plant for the production of uranium-235, for which the gaseous diffusion technique had already been selected. Potential siting of the plant soon became a bone of contention, first between the supporters and opponents of independence vis-à-vis the Americans, and later between Churchill and Roosevelt. The war finally came to an end without the British plant seeing the light of day—it was not built until 10 years later at Capenhurst—and without the British being associated with the technology used for the American gaseous diffusion plant at Oak Ridge.

In 1954, negotiations got under way for the construction in France, with the assistance of British industry, of a plant similar to the one at Capenhurst, which had just started operating. The French proposal was at first well received by the British officials concerned, who were in favor of exporting their most advanced techniques, but the affair was dropped in the face of official opposition by the U.S., consulted under the Anglo-American Agreements of 1943 on secrecy in atomic matters. At the end of 1955, during the initial negotiations for EURATOM, the French delegation described the isotope separation plant as a priority project to be carried out jointly, without waiting for the Treaty to come into force; and this viewpoint was adopted by the Committee of the six heads of delegation.

Since the Organization for European Economic Co-operation (OEEC) had already included a plant of this kind in its list of possible joint undertakings, a study group was set up with the participation, besides the six, of Denmark, Sweden, and Switzerland. The French members were advocates of gaseous diffusion, a process for which the research had already reached an advanced stage, while the Germans, at first advocating a jet process that they had discovered, then joined forces with the Dutch and supported the ultracentrifuge technique, whose economic advantages and imminent development they vouched for.

In 1957, following the Suez crisis, the study by Armand, Etzel, and Giordani, the three "Wise Men" of Europe, on Europe's nuclear energy requirements helped to bring about the downfall of the initial project for a European enrichment plant. Their report, prematurely prophetic as it envisaged immediate and broad-scale recourse to nuclear energy to combat an imminent shortage of energy and oil, and the concomitant drain on foreign exchange, as a necessity, was based on an idea intended to unite the six and gain more from the American achievements, especially from the highly favorable price of enrichment in the U.S. Thus, at the beginning of 1958 the study group for the European isotope separation plant, of which I was chairman, ceased to exist, France not having been able to convince a single one of her eight potential partners of the urgency of the project.

It was not in fact until 10 years later, after the start-up of the French plant at Pierrelatte, that the European experts first from FORATOM and then from EURATOM proclaimed the need in 1967 to free Europe from its complete dependence on the U.S. in the matter of fueling power plants with enriched uranium. They were even willing to accept, if necessary, a higher price for a European product.

It was against this background that there came, in 1968, the sudden and dramatic announcement by the German, British, and Dutch governments that they would be cooperating on the ultracentrifuge technique in view of the progress the three countries had made independently in that process, which consumed far less energy than gaseous diffusion. Their aim was to build pilot facilities, at Almelo in the Netherlands and at Capenhurst in England, and construction was to start by the beginning of the 1970s.

The negotiations ended on 5 March 1970, the very day of the entry into force of the Non-Proliferation Treaty, with the signing by the three governments of the Almelo Treaty, stipulating the political conditions for industrial cooperation between them aimed at creating an improved model of the centrifuge. It was not until last year that the two prototype facilities were officially started up at Capenhurst and Almelo, each one operating with different centrifuge models. Through extension of these plants it is expected to attain 2 million separative work units (SWU) by 1982, as opposed to the 100 thousand SWU at present in operation.

One might have thought that the American reaction to this European initiative would have been to ease up the commercial conditions governing enrichment services. On the contrary, they were made at that time much stricter: the terms now included payment in advance for part of each order, several years' notice of consignments, and a penalty in the event of cancellation of a consignment, the price for which in any event was to be fixed unilaterally by the U.S. authorities, and only at the time of delivery.

At the beginning of 1971 the Soviet Union made its appearance on the market by concluding an enrichment contract with France. The reaction of the U.S. to this breach of their monopoly was an offer, made at the end of 1971, to make their gas diffusion technology available to multinational undertakings safeguarded by the IAEA and open to participation by the U.S., but not in a position to compete commercially with American production. This offer, the terms of which were dictated not just by a concern for nonproliferation, was not taken up by the advanced Western countries to whom it was made.

8 • International Cooperation

Just the reverse, it helped to consolidate the European projects, more especially the interest in a French proposal for a joint venture based on gas diffusion, which led to the creation in 1972 of a study group composed of industrial organizations from Germany, Belgium, the United Kingdom, Spain, France, Italy, the Netherlands, and Sweden. The industries of URENCO Member States and Sweden withdrew from the association in 1973 and 1974 when it was being converted into an industrial company and was about to move on to the construction stage; Iran later joined. In this way, at the end of 1974, just at the moment when the Americans made it known that they could not accept any more enrichment orders, work was begun on the Tricastin plant, which is to have an output of approximately 11 million SWU and is to start operation next year. The possibility of extending this work by the construction of a second plant, CO-REDIF, with the same partners, is also being considered.

EURODIF and URENCO should constitute by 1985 one-third of the Western enrichment capability. Unlike URENCO, the industrial operation of which was preceded by an intergovernmental treaty, EURODIF is an industrial company based on French law, with a French majority interest, but with multinational participation. It was, so to speak, in at the start of this dual nuclear operation—doubly nuclear in as much as the plant is to be supplied with electricity from four pressurized water reactors each with an output of 900 MWe.

9. INTERNATIONAL NUCLEAR FUEL CYCLE EVALUATION

Finally, to end this survey of multinational activity in the area of fuel enrichment, mention should be made, within the context of nonproliferation policy, of a program that received its impetus from President Carter—the expert studies to be conducted in connection with the International Nuclear Fuel Cycle Evaluation (INFCE), designed to find ways of providing supply guarantees to importing countries (reinsurance, banking, etc.) and thereby encouraging them to abandon any plans they might have had to develop their own enrichment capability. INFCE will also be tackling the important problem of supplying the numerous research reactors in the world fueled with highly enriched uranium, the one and only supplier of which so far—the U.S.—is becoming more and more reluctant on grounds of nonproliferation. One solution at present being considered would be to use uranium with less than 20% enrichment in a fuel specially designed for the purpose.

The proposal for multinational development of a "nonproliferation" enrichment process announced by France in the spring of 1977 could also be a means of solving this problem in a politically satisfactory manner.

INFCE is in fact like a very broad, continuing international symposium on all aspects of the fuel cycle, seen mainly from the standpoint of nonproliferation.

As a most unusual exercise in the friendly exchange of information, it has 2 years in which to study, as its main function, the best ways of promoting the development of nuclear energy while curbing proliferation as much as possible. It will be called upon, among other things, to reply to the objections leveled against fuel reprocessing and breeder reactors, considered by some to be both too hazardous from the standpoint of proliferation and not really justified in the economic sense. It will act as an honorary board of judges that could eventually, or so many of us think, rehabilitate these parts of the cycle by demonstrating that they are capable of being protected against the risks of proliferation and are also indispensable for the full-scale development of nuclear energy.

Although now, in 1978, 20 years has passed since the establishment of the international organizations, NEA, EURATOM, and IAEA, it is only 40 years since the discovery in Europe—in laboratories at Rome, Berlin, Paris, and Copenhagen—of the fission of the uranium nucleus.

The distance covered since then is staggering: the dream of the alchemist realized, with the production of tons of a new element; radioisotopes in almost unlimited species and quantities now at the disposal of research, medicine, agriculture, and industry; the weapon that enabled us to end the Second World War and the arms on which depends the balance of peace that has followed it; the modest but remarkable success in propulsion of ships, including the nuclear submarine; and finally, at a time when our civilization is confronted with the oil crisis, a new form of energy that already accounts for almost 10% of the electricity generated in the industrialized countries (even close to 20% in some of them) and which by the turn of the century will be absolutely essential in overcoming underdevelopment.

After meeting and surmounting the many difficulties built into this technology, in most cases through cooperation between countries, let us hope that man, who so often destroys what he creates, will not set up any more formidable obstacles in this, his triumphant path.

9 Nuclear Recycling

Costs, Savings, and Safeguards

Bernard I. Spinrad

1. INTRODUCTION

Everybody is in favor of recycling. Steel, paper, aluminum, and countless other materials take less energy to recycle than to produce anew from virgin resources; also, this saves resources. With the energy and resource saving there is almost always a saving in money, too. Not only that, but there is also a saving in garbage disposal if garbage can be turned to use.

Enthusiasts of recycling are quick to point out economic advantages, and indeed to argue for them even when the figures indicate the contrary. Nevertheless, environmentalists who loudly endorse recycling reverse their position when it comes to nuclear fuel. Even more curiously, although the recycling of nuclear fuel will lead to reduced demands for freshly mined natural uranium and reduce land requirements and long-term risks of nuclear waste disposal, environmentalists remain opposed to nuclear recycling. Could it be that the label "environmentalist" is a misleading label? Preempting a good word is an old propaganda trick—witness the "democrats" and "republicans" of the German Democratic Republic, otherwise known as East Germany.

In the face of charges that nuclear fuel recycling would not be economic—charges that we will refute by careful explanation—we have to contend with the fact that this recycling has not been widely practiced, although it is technically proven to be feasible. History records that nuclear industry plans for widespread recycling were frustrated by antagonistic government actions, stimulated by

---

BERNARD I. SPINRAD • Department of Nuclear Engineering, Iowa State University, Ames, Iowa 50011.

pressure from environmentalist organizations, however; so that this is by no means an argument against the economics of nuclear recycling.

The economics of recycling nuclear fuel involves two separate matters: first, the actual costs and savings of the recycling operation in terms of money spent, made, and saved; and second, the impact of the recycling on the future cost of uranium. It is hard to quantify precisely these latter savings, which arise from reducing the demands that will be made on our finite resource of natural uranium. Qualitatively, however, they will always add to the economic desirability of recycling.

To make this discussion of nuclear economic matters more intelligible to nonprofessionals, we shall begin with a review of the relevant physical and chemical processes.

Today's power reactors are of the light-water reactor (LWR) type. They are fueled with uranium that contains 2–4% by weight of the fissile isotope ^{235}U. The rest of the uranium is ^{238}U, an isotope that is not fissile, but is the source material for plutonium that is made in the reactor. Plutonium is also fissile, i.e., easily fissionable in any reactor. The concentration of ^{235}U in an LWR is higher than it is in natural uranium. The increased concentration is achieved by enriching the uranium. This is done in enrichment plants at Oak Ridge, Tennessee, Paducah, Kentucky, and Portsmouth, Ohio, by a process known as gaseous diffusion. Natural uranium, which contains only 0.71% of ^{235}U, is fed in and the enriched product, along with depleted "tails," are withdrawn. It takes about five units (pounds or kilograms) of natural uranium to make one unit of 3% product.

This enriched uranium is chemically converted and fabricated into nuclear fuel assemblies. These are then loaded into a reactor, where energy is produced by the fission (splitting of the atoms, converting part of their mass into energy) of ^{235}U. The process continues until about 3% of *all* the uranium originally loaded has fissioned. This means that one fission has taken place for each atom of ^{235}U that was loaded. But this does not mean that all the ^{235}U has fissioned. The reason is plutonium.

Plutonium is produced in the reactor as a by-product. It is formed by capture of neutrons in ^{238}U. It is more easily fissioned than ^{235}U, so a great deal of the plutonium formed is actually fissioned in place. In a typical LWR, about half the plutonium formed is fissioned, and by the time the fuel is ready for removal, about two-thirds of the fissions take place in plutonium. The formation of plutonium in the reactor also contributes to our ability to control the chain reaction, as it makes it possible to keep the reactor running with much less motion of the control rods; technically, the reactivity change is smaller than it would be without the plutonium.

The final result of the conversion of part of the ^{238}U to plutonium is that the spent fuel still has valuable materials in it. In fact, spent fuel from LWRs

9 • Nuclear Recycling

has almost the same amount of ^{235}U in it as natural uranium does. Let us recall that it took about 5 lb. of natural uranium to get 1 lb. of enriched fuel. With recycle of this uranium, we could save 1 of those 5 lb., save the mining that was needed to get it, and leave more of the uranium resource available for the future.

But uranium is not the only useful nuclear fuel material in the spent fuel. There is also plutonium—a little more of it, in fact, than there is of ^{235}U. It, too, can be used in LWR fuels; recycling of the plutonium would save still another pound of the original 5 lb. we have used for illustration, and in addition less enrichment would be needed. However, there is a competitive use in prospect: the fueling of fast breeder reactors. It turns out that while plutonium is a useful fuel in LWRs, it is the premier fuel for fast breeders, for reasons that we shall discuss subsequently. Thus, with plutonium, the technical question is not whether to use it, but how.

To recover uranium and plutonium, the spent fuel must be chemically reprocessed, and this is the step that has now become controversial. It is alleged that the plutonium, when it is recovered, might be "diverted"—stolen to make private atomic bombs. To try to show how this might be done leads the inventor of schemes into James Bond scenarios of the most outlandish sort, for plutonium is intensely radioactive, even when it is purified, and as a result its theft is more difficult than the theft of gold from Fort Knox. But that is another chapter of the story. For now, we just want to describe the uranium and plutonium recovery process.

In reprocessing, the once-used uranium fuel is chopped up into small pieces and exposed to concentrated nitric acid. This dissolves the uranium, the plutonium, and fission products, but leaves intact the pieces of metal cladding that originally contained the uranium. The uranium and plutonium are then separated from the fission products by a process called solvent extraction. An organic solvent (a light oil, such as kerosene) that doesn't mix with water or nitric acid is contacted with the acid solution. The solvent contains a compound which combines with uranium and plutonium. The reaction "complex" containing the uranium and plutonium is much more soluble in the oil than in the original solution. The latter now contains only fission products and a trace of the uranium and plutonium originally present, and is ready for disposal as high-level waste. The organic solution can be "stripped" of the heavy metals in a number of ways. Some permit separate stripping of the uranium and plutonium in a step known as partitioning; others strip them together. In any case, the stripped solution (or solutions) is (are) purified to get rid of the very radioactive fission products, and the final product can be prepared by routine chemistry. If the stripping of the uranium and plutonium are done more or less together, and if the final removal of fission products is omitted, one has the proposed CIVEX process whose product is a usable fuel material, but one that is so radioactive that it can only be handled in specially built industrial facilities.

The radioactivity from uranium is so feeble that it can be safely handled in the open. The only precaution that is necessary is to keep radioactive dusts from forming. Consequently, the fabrication of uranium fuel assemblies is an established, not very expensive process. Plutonium is much more radioactive, even when purified—particularly the plutonium recovered from power reactor fuel. Even the less radioactive plutonium used by the military for weapons must be handled only in sealed "glove boxes" that maintain a plastic barrier between the material and the workers. The more radioactive reactor-grade plutonium must be handled in facilities that shield against penetrating radiation as well. For this reason, the fabrication of fuel assemblies containing plutonium is much more expensive than is uranium fabrication. We expect that ultimately the cost differential between the two processes will be small—perhaps as little as 25 or 50%. To achieve this, however, a green light for the necessary research and development is needed. Economies will come from achieving a high degree of process automation and process reliability—specifications that do not necessarily mean high cost but do require testing and experience.

2. INDIFFERENCE COSTS

The strictly economic analysis of whether or not to reprocess spent fuel for recovery of its values obviously hinges on whether one makes money or loses money on the transaction. If there is no reprocessing, there is a cost for spent fuel management and disposal, and there are also no rebates. If there is reprocessing, there is a charge for reprocessing and for high-level waste disposal, but there are rebates from sale of the uranium product and of the plutonium. We have at least some data on which to base all the items of expense and income—except for the value of plutonium. Thus, we must find a way to estimate the latter, since there is no market in it at present. What we attempt to evaluate is its "indifference price."

An indifference price is obtained by considering two alternatives. The first is to buy natural uranium, enrich it, and fabricate the enriched product into reactor fuel. The second is to mix plutonium with depleted uranium (obtained virtually free of charge from the rejected "tails" of the enrichment plants) and fabricate the mixture into an equal amount of equivalent fuel. The indifference price is the price we would pay for plutonium that would make the costs of these alternatives a toss-up.

We use as a standard the money that has been spent to have 1 kg of 3%

enriched uranium in fabricated fuel assemblies. In 1978 dollars, this cost is $1100/kg.*

Now we must work out the costs of equivalent fuel using plutonium. Physics calculations indicate that a fuel with 33 g of plutonium in 1 kg of depleted uranium will perform in an LWR about as well as the reference 3% uranium fuel. The cost of this equivalent fuel is then, in 1978 dollars per kilogram,

$$33V + F$$

where F is the cost of fabricating plutonium-containing fuel and V is the value of the plutonium in dollars per gram. We can take values of F varying between $150 and $450 /kg, i.e., we can vary the fabrication cost from 50% more than that of uranium fuel to three times as much. The indifference price is that obtained by solving the equation

$$V = (1100 - F)/33 \qquad \$(1978)/g$$

The plutonium value, V, then turns out to vary between $20 and $29/g. A nominal value of $24/g is often used to represent this number, in the realization that uncertainties about fabrication costs create about a 20% uncertainty in plutonium value.

3. RECYCLING IN LWRs

What, then, is the balance sheet for fuel recycling in LWRs? Table I tells the story. Where do the numbers come from?

The charge for fuel assembly disposal is one suggested in 1978 by the DOE and never changed. It includes interim storage, fuel encapsulation, and costs of permanent disposal in a geological repository. This charge is almost certainly more than the charge would be for disposing just the fission products from this same fuel, as the unreprocessed spent fuel is bulkier, produces more radioactive decay heat, and has much more plutonium in it, requiring more stringent geological containment. Of these disadvantages of spent fuel, the most severe is probably the extra heat generation, because this parameter determines how much land area must be committed to disposal of high-level waste.

The charges for disposal of high-level waste are set slightly below those for spent fuel disposal, at worst. Storage and recanning costs are eliminated and

---

* The cost of the reference nuclear fuel cycle has tracked the Consumer Price Index quite closely over the past 6 years, making its "constant dollar" cost a very stable number. We use figures from 1978 because they are readily available.

TABLE I
Costs of Reprocessing and Not Reprocessing

No reprocessing	
Waste disposal charge on fuel	$250/kg
Net cost	$250/kg
Reprocessing	
Reprocessing charge	$150–300/kg
High-level waste disposal	$150–200/kg
Total charges	$300–500/kg
Uranium credit	$100/kg
Plutonium credit[a]	$160–225/kg
Total credits	$260–325/kg
Net (credit) or cost	($25)–$240/kg

[a] Assuming 8 g of fissile plutonium per kilogram of spent fuel.

ultimate disposal has fewer restrictions. The lower number is actually an upper limit to what developers of waste-disposal methods have suggested.

Reprocessing costs cover a range, arising from the great number of estimators. The higher number is about what is being charged by the French reprocessing plant in La Hague, and is what the owners of the Barnwell reprocessing plant in South Carolina (which has never been operated) have suggested they would charge. The lower number is what Oak Ridge estimators have calculated as the cost of a utility-financed plant. In between is the original cost suggested by Barnwell management for 1975 operation; part of the factor-of-two difference is inflation, part is the desire to recover interest accrued while the plant has been idle, part is undoubtedly the result of a supply (no plant available)–demand (many utilities want reprocessing) imbalance.

The basis for the uranium and plutonium credits has already been given.

What Table I clearly shows is that, at very worst, reprocessing costs slightly less than throwing the spent fuel away, and at best might return a small profit to the fuel owner. The condition is that *the plutonium must be reused*.

4. PLUTONIUM IN FAST REACTORS

We have mentioned that plutonium in fast reactors is a more valuable fuel than in thermal reactors of the LWR type. Let us explain and quantify this point. Plutonium, when it fissions, produces on the average many more neutrons than does ^{235}U. The ratio is about 2.9 : 2.5. As a result, many more neutrons become available per fission to maintain the neutron chain reaction and to enter into nuclear reaction with ^{238}U so as to form more plutonium. But in a thermal reactor, plutonium also enters into a side-reaction with neutrons. The process is known as radiative capture, and results in the loss of an atom of the fissile isotope of plutonium, ^{239}Pu, by formation of the nonfissile isotope ^{240}Pu. Per ^{239}Pu atom destroyed, we only get 1.9 neutrons in the thermal reactor because of this process. (The ^{240}Pu that is formed makes it far more difficult to use the resulting "reactor-grade" plutonium for weapons, which is why military plutonium is made from uranium that is given a relatively short burnup in special production reactors. The formation of ^{240}Pu occurs only after a lot of ^{239}Pu has been formed.)

In a fast reactor, there is much less radiative capture. Fast (unmoderated) neutrons cause the fissions in such reactors, hence their name. The number of neutrons produced per fissile atom destroyed is about 2.4–2.5 for ^{239}Pu and about 2.2 for ^{235}U. Here, then, plutonium is the superior fuel. The situation is further improved in that ^{240}Pu, which only *absorbs* neutrons in a thermal reactor, *fissions* to some extent in a fast reactor. The number of neutrons produced per plutonium atom destroyed is large enough so that, after the neutron needed to maintain the chain reaction is subtracted, and after other unavoidable losses of neutrons to other reactor materials are accounted for, there remains, on average, slightly more than one neutron to form fresh plutonium from ^{238}U. Not only is the plutonium a fuel material that can make a fast reactor critical in a comparatively low concentration, but in this way it permits good breeding: the formation of more fissile material than is used up to produce power.

A fast reactor could nevertheless be started up on enriched ^{235}U. The scheme would be to load with that material, reprocess after burnup, reload the uranium and plutonium, burn again, and so on. It is therefore possible to calculate an indifference value for plutonium in fast reactors as well as in LWRs. The assumptions going into the calculation are much the same as those made for the LWR case, except that it is assumed that the required ^{235}U enrichment would be about 20%, the equivalent plutonium concentration about 16%, and the cost of fabricating fast reactor fuel containing plutonium would be $1000/kg fuel. The resulting indifference value of plutonium is about $50/g—just about twice its value in a thermal reactor. It has a further value, of course, because if a breeder could be started on plutonium, its first few charges would have fissile material to sell, from the breeding, whereas breeding is very minor with ^{235}U. Because of this fact, we can confidently assess the value of plutonium in a breeder reactor as about $60/g.

5. RECYCLE OR STORE?

Other things being equal, it therefore seems that plutonium should be used preferentially as a fuel for fast breeders; it is simply more valuable in that use. But other things are not equal. We have LWRs now, and we will not have fast breeders, at least in quantity, until after the year 2000. We *could* have had them much sooner, as it appears France will have, had it not been for a series of extremely bad management decisions made by Congress and the AEC in the 1960s, compounded by the misunderstanding of the proliferation problem in recent administrations; but that is still another story. The choice is therefore between using plutonium now in a way that is less than optimal, or storing it for later use in more optimal fashion.

Economically, there is always an incentive to make productive use of a resource now rather than later. Future economic benefits are discounted relative to present benefits. The analogy is that money I put into savings now will draw compound interest starting from now so that I will have more money later; reasoning in reverse, future money is therefore not as valuable to me as present money, even disregarding inflation. The measure of this value difference is the delay time and the discount rate. This latter tells the annual decrease in the value of a deferred economic benefit.

Unfortunately, there is little unanimity as to what the discount rate should be. In private business, discount rates well above 10% per year are used for planning purposes. These, however, are rates for "current dollars"—dollars that are losing their value at a rate of over 7% each year because of inflation. If we calculate in "constant dollars"—dollars of constant purchasing power in terms of what they will buy—we get business rates between about 2.5 and 10%. The lower value is classical "true interest" while the higher values include an "opportunity value"—the value of being able to take advantage of a business opportunity to make more money than simply by depositing in a savings account.

For public accounting we have an analogous spread in constant-value dollars. The lower limit is around 2.5% again and is set by the GNP growth rate, while an upper limit of around 10% is used by analogy with industry. There are ferocious arguments between those preferring the low discount rate and those preferring the high one. In qualitative support of the lower rate is the observation that only then does government support of basic research make sense, for the practical value of basic research is usually long deferred. Support for the higher discount rates usually comes from those who are demanding the maximum of relevance to present problems in the present spending pattern. All that we can conclude is that the range of discount rates covers a range of attitudes about the value of the future. Those who argue for low ones value the future highly, and conversely.

What is the relative value of using plutonium valued at $24/g in 1980, or storing it to the year 2000 when it will be used at a value of $60/g? We can calculate the indifference value of the discount rate for that choice. At a rate of $(60/24)^{0.05} - 1$, or 4.59% per year, the two actions will have equal value. At a higher discount rate, we do better by burning it now, and at a lower rate we do better by saving it. These numbers serve only to sharpen the dilemma, for 4.59% is within the range of possible discount rates we might use.

We must look therefore to other arguments. One that used to be popular was that the world faced a shortage of fissile material for a durable nuclear era. The standard scenario was one of very rapid buildup of thermal reactors, using up most of the world's supplies of easily accessible uranium. Under these circumstances, the need for breeders would arise abruptly, and continued demand for electric power could only be satisfied by deploying them at the maximum rate possible. This scenario argued strongly for storing the plutonium. However, events have made this scenario highly implausible. The entire rate of buildup of nuclear power—burners and breeders alike—has been much slower than anticipated. The result of an extension of the time scale is that the expected rate of deployment of breeders will be less abrupt than the scenario had suggested. In turn, this means that the production of plutonium from LWRs that are operating at the time when breeders are deployed will be sufficient to provide the plutonium the breeders need. This is not to say that it would be better to delay the breeders. The sort of scenario now envisaged would still be easier to accommodate if breeders were available early: less mining of increasingly dilute uranium ore would be needed.

Today's nuclear deployment scenarios argue, albeit somewhat mildly, for early recycle of plutonium. A more decisive argument comes from the need for technological continuity. The optimal fast breeder reactor cycle is one in which the out-of-reactor process steps, reprocessing and fuel fabrication, are fast, efficient, and economical. There are no general technological reasons for considering that these goals cannot be met. However, to achieve them, there is no substitute for experience. Moreover, such experience cannot be merely at laboratory or pilot-plant scale. There is no substitute for full-scale industrial experience. And it is this experience that we can expect to develop by recycling plutonium in thermal reactors. To summarize, recycling of plutonium in LWRs gives us the know-how that we will need later, regardless.

There is another advantage to recycling plutonium as soon as we can. Doing so would also have advantages with regard to nonproliferation of nuclear weapons. The U.S. would then have experience that is relevant to developing institutional measures that are needed to make a success of this enterprise: the process designs for safeguarding the fuel against theft; security measures and managerial techniques toward the same end; international institutions to permit the benefits of U.S. plutonium operations to be shared also by foreign countries, which might otherwise be tempted to proliferate plutonium operations unwisely; and the selection of sites that might serve in the future as nuclear fuel-cycle centers.

6. SAFEGUARDS AND WEAPONS PROLIFERATION

Plutonium is a material that is useful—indeed, extremely so—as a fuel for the nuclear industry. However, it is necessary to safeguard it—to take whatever steps are necessary to prevent it from being stolen or "diverted" by people who would want to make it into bombs. It is also a very toxic material that should be kept out of the biosphere. To put these remarks in perspective, we must immediately also observe that plutonium is by no means a problem of the nuclear power industry alone. The bigger problem of plutonium is the depressingly large stockpile of bombs that major governments have seen fit to make. We just have to keep hoping that some day the world will realize the folly of nuclear war and demand that this stockpile be used constructively as nuclear fuel. The weapons business has also put several *tons* of plutonium into the biosphere, as fallout from weapons tests. This is a very, very bad way of disposing of plutonium; yet the fact that *no* effects of this material have been observed puts the hazard of plutonium toxicity into perspective.

The use of plutonium as a fuel serves to minimize the world's plutonium inventory and keep it out of mischief, so to speak. This reduction of inventory is most marked if plutonium is recycled rapidly. Furthermore, recycling is most compatible with safeguarding if it is quantitative, automated, and automatically analyzed and inspected. The development of these techniques on an industrial scale is an important contribution the U.S. can make to nonproliferation of weapons; only practice on that large scale can do the job properly, and this takes time. Our delay simply would mean that other countries will be doing the job; if we mistrust other countries' motives with regard to diversion, it is strange that we should trust them to develop antidiversion recycle techniques.

It goes without saying that plutonium recycling would also call into play a variety of security measures. These include elaborations on the type of security already in force at plants where explosives and toxic chemicals are manufactured, or at repositories of gold and other precious materials. As an aside, it may also be necessary, in these troubled times, to consider increasing the guard forces at such vulnerable energy facilities as dams, oil refineries, and natural gas storage depots. The collocation of nuclear fuel fabrication plants and of some—but not, by any means, all—reactors with reprocessing plants is a basic security measure for the nuclear industry. It reduces the amount of guarding needed for the transport of fissile material, and permits the guard forces of several facilities to be mutually self-supporting.

All these measures are primarily effective in assuring domestic security against theft or other antisocietal diversion. One must also consider the problem of national nuclear weapons proliferation. Nuclear opponents allege that worldwide application of nuclear power carries with it grave dangers of contributing to such proliferation. The logical conclusion is, however, quite different: that nuclear power is of very marginal significance in this respect.

To understand the reason for this, it is necessary to ask what the reasons are that so few countries in fact have nuclear weapons. After all, it is relatively cheap and straightforward to make plutonium in production reactors. Their design is far simpler than that of power reactors; indeed, the technology used for that purpose by the U.S. in 1944 is by now within the capability of not merely industrial countries, but of any country that has some minimum cadre of physicists, chemists, and engineers. The plutonium turned out by such reactors makes much better bombs than does "reactor-grade" plutonium, as has already been noted.

And, indeed, this is the route that has been taken by India, the most recent member of the "nuclear club." India made its nuclear explosive in a reactor that had been donated by Canada for research. It was a heavy-water research reactor, easy to build, and also easy to turn into a plutonium production reactor. India has power reactors, but did not get its plutonium from them.

Among other forces, the most powerful force against more countries developing nuclear weapons has been world opinion. The Indian example is again instructive. One reason for not using power reactor fuel for bomb plutonium was that India had signed a solemn agreement not to use that fuel for that purpose. This is essentially what a country adhering to IAEA safeguards does. Additionally, India was sufficiently aware of the force of world opinion that it took great pains to advertise that its bomb was not a bomb at all, but a "peaceful nuclear explosive."

And, of course, as there is no way (short of war) to force a sovereign country to do, or not to do, anything it pleases in its own territory, the reason that other countries that could have atomic bombs don't is that world opinion is a strong enough deterrent. When backed up by even nominal safeguards, it is a very strong deterrent. When those safeguards apply to known facilities—as is the case with IAEA safeguards—and these are subject to inspection, world opinion is an extremely strong deterrent. This, and no technical reason, is why so few countries have decided to "join the nuclear weapons club."

It is not wrong, of course, to be concerned about safeguarding the material that comes through the plutonium recycling technology. Not to be concerned would be reckless. But this is a case of using appropriate means to counteract the concern: security and technological efficiency to protect the material and keep it in legitimate hands; and international agreements, such as the Non-Proliferation Treaty and IAEA safeguards, to keep it channeled in legitimate uses. With these measures, we can stop worrying about the connection between nuclear power and nuclear weapons, and devote ourselves to the far more important task of building a world in which nuclear weapons are worthless.

10 Alternative Fuels, Fuel Cycles, and Reactors

Are They Useful? Are They Necessary?

Bernard I. Spinrad

1. INTRODUCTION

The nuclear industry of the U.S., and most of the world, is based on light-water reactors (LWRs). These reactors were developed in America, where we had the capacity to enrich natural uranium slightly. Using slightly enriched uranium as fuel made it possible to use ordinary water as reactor coolant and moderator for power reactors. This, in turn, permitted us to use well-known steam technology in designing the reactor system. The resulting reactors turned out to be cheaper than their main competitors and quite reliable.

The fuel cycle contemplated for these reactors was also a straightforward one. After reactor irradiation, spent fuel was to be chemically reprocessed, the remaining uranium returned to the power reactor cycle as an alternative feed material for enrichment plants, and the plutonium (formed from the principal isotope of uranium, ^{238}U) itself used as a nuclear fuel. The plutonium could be used to substitute for fissile ^{235}U in LWRs, reducing further the requirements for mining natural uranium, or held as fuel for more advanced reactors, and particularly fast breeder reactors.

The main drawback of LWRs is that, with present designs, they use a great deal of natural uranium. About 6000 tons of uranium must be procured from ore to provide the lifetime fuel for a present-generation LWR. This isn't much— of the order of one-millionth of the amount of coal or oil to fuel an equivalent power plant. As long as uranium is plentiful and cheap, it is a minor cost factor. However, rapid growth of the nuclear industry will use up this ore and force us

---

BERNARD I. SPINRAD • Department of Nuclear Engineering, Iowa State University, Ames, Iowa 50011.

to mine leaner, more costly uranium deposits. Indeed, even such growth as occurred in the early 1970s, when the world installed about 100 GWe of nuclear power, forced the price of uranium up considerably. So the question remains whether different fuel cycles or different reactors would be better.

Coupled with this question is the matter of safeguards. When the fuel is reprocessed, plutonium is isolated from the radioactivity that comes from fission products in spent fuel. This makes it possible to work with the plutonium, even though it is still so radioactive that it takes special equipment and special precautions.

And, of course, reactor-grade plutonium can be used to make nuclear weapons—not very good weapons, and not the sort that can readily be designed by an amateur, but nuclear weapons nevertheless.

For these reasons, the nuclear community early took the position that the reprocessing of spent nuclear fuel should be a safeguarded operation. In this safeguarding, there are three components. Of overriding importance is the acceptance of safeguarding as an institution. To this end, a scheme of international safeguard agreements was set up involving the International Atomic Energy Agency (IAEA), and in addition nuclear supplier countries imposed bilateral safeguards on their customers. To implement safeguards, the three pillars of security, accountability, and inspection are brought into play. The safeguard schemes were expected to evolve, and did evolve, under the general understanding that it would be a major breach of international trust to violate or circumvent them. Thus, the overwhelming fraction of safeguard measures can always be expected to serve the purposes of assuring the international community that nuclear facilities are entirely dedicated to peaceful purposes. Only a small fraction of safeguarded facilities could ever be considered suspect cases, and these are given priority attention in inspections.

A further function of safeguards is to assure that plutonium, or other fissile material, is secured against theft. Again, the security function has been a part of nuclear operations from the beginning. After all, the entire nuclear industry emerged from a top-secret project of World War II. Security operations, based on military practices, were standard at all facilities, military or civilian, where sensitive material was handled.

For reasons that can only be political, reservations about the efficacy of nuclear fuel or plutonium security measures have been focused particularly on the reprocessing operation. That the reasons must be political is clear from the fact that the reprocessing operation is, at its head end (where reprocessing begins), safeguarded intrinsically by the enormously high radiation level to be handled. Only remote operations, behind thick shielded walls, can be undertaken. This means that the physical structure of a reprocessing plant makes it, for the most part, impenetrable. It is only at the product end that it becomes safe to withdraw or divert plutonium and at this point, the product is concentrated, still radioactive,

10 • Alternative Fuels, Fuel Cycles, and Reactors

and subject to the same type of guarding as the A-bombs at nuclear weapon sites. It therefore seems that the concern about safeguards in reprocessing, and particularly domestic reprocessing, is in reality motivated by a desire to stop nuclear power, for reprocessing is the key to long-range nuclear utilization.

Against this background, the nuclear community has nevertheless responded to the political necessity that it look into ways of making reprocessing (and the subsequent step of nuclear fuel refabrication) even more foolproof against diversion, as well as looking into ways of generating nuclear power that reduce the demands for natural uranium.

2. BREEDING

Let us first examine the easiest way to free ourselves from the constraint of mining a great deal of uranium to make nuclear power. This is by taking advantage of the ability to make new fissile isotopes by neutron reactions with a group of nuclides that are called "fertile." The two important ones are ^{238}U, the principal constituent of natural uranium (about 99.3% of the total), and ^{232}Th, the only isotope of natural thorium. From the former, one can make ^{239}Pu; from the latter, ^{233}U. ^{239}Pu is the best fuel material for reactors whose neutrons are unmoderated—which are known as fast reactors. ^{233}U is the best fuel material for thermal reactors. With both these fuels, it is possible to make a reactor that produces more new fissile material than it consumes. The margin is large enough in fast reactors so that it is not even very difficult to design such a "breeder." For thermal reactors using ^{233}U, the margin for breeding is smaller, and a neutronically more careful design is required. However, both ^{239}Pu and ^{233}U can be made into nuclear weapons. Indeed, it is hard to say which is the nastier material when used for weapons. As we shall see, this complicates life considerably.

The significance of breeding becomes clear when we observe that the uranium we mine from nature contains only about 0.7% of nuclear fuel, in the form of the fissile isotope ^{235}U, and that we only get about that fraction of fissions from it in LWRs. Even by recycling the plutonium in spent fuel in LWRs, we could only expect to get about double that many fissions. In other words, we get less than 2% of all the uranium to fission. With breeding, however, we can continue recycling indefinitely, while at the same time we produce extra fuel for use in other reactors. We can thus use as fuel *all* the uranium that is mined, rather than less than 2%. With the excess fissile material, we can even get a breeding, or near-breeding, thorium cycle started, and use thorium as a fuel. Not only does breeding multiply the energy available from mining heavy elements by a factor of the order of 100, but it also virtually eliminates the cost of uranium as a factor in nuclear power. As a result, very small amounts of uranium, available at high cost from ordinary shales, become usable nuclear fuels, and there is so much of this sort of uranium (and thorium) in the world that the nuclear fuel supply becomes virtually infinite.

Because breeding opens the door to an era of "infinite fuel," all the countries of the world that have had the money and the technical background to do so have mounted research and development programs in nuclear breeding. Because the easiest type of breeder to construct with assurance that it will in fact breed is the fast reactor, this is the type they have all concentrated on. The logical and best fuel for a fast breeder is plutonium.

A fuel loading in a fast breeder will not last forever. Periodically, it is necessary to remove it and separate out the fission products, because these latter both cause physical deterioration of the fuel and absorb neutrons that could be used for breeding. This requires chemical processing, in the course of which it is also possible to separate out the excess fuel that has been bred, to make it available for use in another reactor. In short, reprocessing of fast breeder fuel at intervals is both necessary and desirable. In fact, the annual throughput of plutonium from a fast breeder through a reprocessing plant is three to ten times as much as the throughput of plutonium from an LWR. Consequently, if there are concerns about safeguarding the plutonium that comes from LWRs while the nuclear fuel is being reprocessed, there are greater concerns about safeguarding the breeder fuel during reprocessing (and subsequent manufacture of new fuel). Yet, for conventional breeders, recycling the fuel is virtually mandatory.

3. ALTERNATE BREEDER CYCLES

There are only a few ways that one could conceptually deal with this problem *if* it is assumed that fast breeders cannot be safeguarded. Of course, we could cut the argument short by noting that the breeder *can* be safeguarded. This is exactly what the French, British, and Russians have maintained as they proceed to develop and commercialize fast breeders. But, to appease an obviously committed, doctrinaire group in the U.S. administration, these countries have gone along with a U.S. insistence that alternatives be looked for in an "International Fuel Cycle Evaluation" program.

First of all, what are the possibilities for reducing the throughput of plutonium in the reprocessing plant, or for making it more easily safeguarded? For the former purpose, there are a few possibilities. Although these do not eliminate the necessity of reprocessing plutonium, they reduce the quantity handled.

One possibility depends on the fact that a significant part of the new plutonium is produced not in the reactor core, where most of the power is generated, but in a "blanket" around the core where this plutonium is made in rods of ^{238}U, loaded to catch neutrons that leak from the core. In principle, there is no need to reprocess this blanket material. Instead, one might wash and rearrange these blanket rods after sufficient plutonium has been formed in them and use them directly as fuel for thermal reactors. The method has not been developed, but it is technologically possible and has some economic advantages besides (because it eliminates a fuel fabrication step for the thermal reactors). Or, since ^{233}U is a superior fuel for thermal reactors, the same technique could be performed with a thorium blanket.

One can also design the fast reactor so that the maximum amount of breeding occurs in blanket elements. In effect, one could deplete the plutonium in the core and build it up in the blanket. Now we are talking about a hybrid reactor cycle: plutonium made in LWRs is loaded into fast breeders, where it is consumed but makes fuel for thermal reactors.

These schemes, intriguing as they are to the reactor developer, are not panaceas, however. All they do is reduce the amount of plutonium to be reprocessed, and not by a large factor—less than a factor of 2. The schemes might still be implemented, but only if they simplify operations, reduce costs, or exhibit other technical advantages.

A final proposal has been to go to the Th–^{233}U cycle in fast breeders. The breeding would not be as good as with plutonium, but such a system could breed a little. Unfortunately, this simply replaces the plutonium safeguards problem with an entirely equivalent ^{233}U safeguards problem. However, there is a way out. This involves the so-called "denatured uranium" fuel cycle. In this cycle, the fuel is ^{233}U diluted not into thorium but into ^{238}U. Now the weapons material can only be recovered by performing isotopic separation—a step that necessarily requires a major national or industrial facility. (An amateur would not have the slightest chance!) On irradiation, plutonium is formed from the ^{238}U, and this must be dealt with in the reprocessing operation; but, at least for the first decade of operation, the plutonium throughput would be equivalent to that from an LWR, rather than several times larger. And if there is a source of ^{233}U—which would have to come from an operation analogous to plutonium reprocessing—the reactor could run indefinitely with low plutonium throughput.

I think the problems of this scheme are obvious. First, if the reactor is to be fueled permanently with ^{233}U, this implies a reprocessing operation for that material that is entirely equivalent to a plutonium operation. The material added to the breeder fuel would have to be concentrated, rather than denatured, as its purpose is to bring the spent, denatured uranium back up to a fairly high concentration. The ability of the reactor to breed also suffers. And, of course, all this complexity merely serves to reduce the quantity of plutonium to be handled, and that reduction is replaced by the need to handle the undenatured ^{233}U.

4. ALTERNATIVE REPROCESSING SCHEMES

If the fuel cycle cannot be altered to the point that the putative danger of reprocessing is strongly reduced, can the reprocessing scheme itself be changed to make safeguarding easier? Remember, there is really no reason to do this except to appease fearmongers; but such are the ways of politics that the nuclear profession must take them into account. In other words, what we are being asked to do is unnecessary to achieve safeguards and safety, and probably uneconomic; but it is necessary to overcome the reservations that people hostile to nuclear power have foisted on the public.

Returning to the question, the answer is "yes": reprocessing can be turned into an operation that produces "intrinsically safeguarded" plutonium. The key to doing this is to keep enough fission products in the recovered plutonium so that the radiation from it remains murderous. This is not so much to foil the would-be diverter's ability to make weapons from it as to make diversion almost impossible. The high radiation level of such material would make counters scream all over the place if anyone tried to remove the material in an unauthorized fashion.

In fact, the very first breeder pilot plant used such a scheme. The Experimental Breeder Reactor No. 2 (EBR-2) was placed in operation in 1963, at the National Reactor Testing Station in Idaho, complete with a fuel cycle facility that reprocessed and refabricated the fuel on the spot. The process used, known as pyrometallurgy, was different from that favored now. One of its key features was that it retained in the fuel many highly radioactive fission products that are useful alloying additions to nuclear fuel. The fuel came out of the reactor, went through processing and refabrication, and returned to the reactor by way of a highly shielded circular process line. The only materials shipped out of the plant were radioactive waste, maintenance waste from the facility, and a small quantity of the fuel, mixed with waste, which represented the bred plutonium.

EBR-2 is an interesting reactor in many ways. It has been running for 15 years, producing power and demonstrating that a breeder is operable with complete safety; yet the AEC, and later the ERDA and DOE, never were willing to admit that the breeder had in fact been "demonstrated." Somehow, when it came to planning the future of breeding, this reactor was passed over quickly. It was turned into a fuel test facility, in which function it has been extremely useful. Its design was, in fact, the point of departure for the successful French and British programs. The student of politics within the nuclear community can learn many lessons from it, but particularly the lesson that technical decisions made by bureaucracies are usually wrong.

For a variety of reasons, the EBR-2 technology was not followed up. The decision was made that breeder reactor reprocessing was to be done in the same way as the reprocessing of LWR fuel. The high contents of plutonium and fission

10 • Alternative Fuels, Fuel Cycles, and Reactors

product of breeder reactor fuel were to be circumvented during reprocessing by diluting the fuel in the core with uranium from the blanket. There remained technical problems arising from the fact that this dilution must be done after the fuel has been dissolved, and that the dissolving therefore is under different process conditions than is the case with LWR fuel. These problems have caused difficulties in laboratory operations but seem to have been resolved in France, where this scheme is in successful operation.

To safeguard the handling of fissile material using this scheme, the CIVEX process has been designed. In CIVEX, both the separation of plutonium from uranium and the removal of fission products from the uranium–plutonium mixture are only partially done. The plutonium is always associated with about five times its weight in uranium, and the resulting mixture is intentionally made to contain some highly radioactive fission products. The CIVEX process has the virtue that it is actually a simplification of the standard process (known as PUREX) from which it is derived. The main difference is that several of the PUREX steps are completely omitted. The product of the CIVEX process is highly radioactive.

CIVEX has not been demonstrated, although it appears to have no technological problems that cannot be overcome. The conception of CIVEX was not, however, greeted with joy by the antinuclear community: their goal in raising the diversionary red herring was not to stimulate the nuclear community into constructive action but to destroy nuclear power. And CIVEX has some considerable drawbacks that make the nuclear community unwilling to embrace it unless it solves the political problem. Chief among these drawbacks is the fact that, in producing highly radioactive plutonium, it requires the fabrication of fuel from this material to be a process that *must* be at least remote and very heavily shielded, and should ultimately be automatic. Such automatic processes have not yet been commercially developed. Again, they are developable, and probably desirable in the long run in any event. The question is, therefore, whether preferably to develop them carefully, logically, and economically, or to be forced into a crash program that would be less economic and might miss some important opportunities that a more deliberate program could accept.

5. ALTERNATIVE REACTORS

The two reactors that are real breeders are known as the liquid-metal-cooled fast breeder reactor (LMFBR), which is cooled with liquid sodium, and the gas-cooled fast breeder reactor (GCFBR), which is cooled with pressurized helium. Of these, LMFBR has been demonstrated in several countries and is the mainline breeder. LMFBR and GCFBR have nearly identical fuel cycles, and there are no safeguard reasons for preferring GCFBR.

As previously noted, the ^{233}U–Th cycle provides the best neutron economy in thermal reactors. At least four types of reactors are capable in principle of showing very slight breeding of this cycle. One of them is a very ingenious concept that has been carried through to the reactor experiment stage at Oak Ridge. This is the molten salt reactor (MSR), which is fueled with a mixture of uranium and thorium fluorides dissolved in other fluoride salts. The fluoride salts are extremely corrosive to most metals, and Oak Ridge has developed a special steel, Hastelloy-N, to contain them. There is essentially no industrial experience in making components of this material, and no long-term experience with its behavior. Therefore, this reactor must be counted only as an interesting possibility. The reactor fuel cycle operations are performed at the reactor: a sidestream of molten fluoride is processed by a salt–salt extraction process, which purges fission products and extracts the parent nuclide of ^{233}U—^{233}Pa—for storage until it decays into uranium fuel. (The half-life for this decay is about 30 days.) With the help of this on-site processing, the reactor could breed slightly.

The heavy-water reactors, developed in Canada under the name CANDU, could also breed slightly. For this, they would have to be loaded with thorium fuel containing a fairly high ^{233}U content—about 6%. They could not be run to high fuel exposure, because the buildup of fission products would jeopardize breeding; in other words, they would require frequent reprocessing to breed, and the throughput of undenatured ^{233}U through the reprocessing plant would be as high as that of fast breeder plutonium. This is true also of the graphite-moderated, helium-cooled high-temperature gas-cooled reactor (HTGR), developed to the point of commercialization in the U.S., but represented only by a demonstration unit, Fort St. Vrain, in Colorado. It too could breed if it were loaded heavily with ^{233}U and Th and reprocessed frequently.

To complete the list, there is one reactor under active development that could breed, or come close to breeding, with light water as coolant and moderator. This is the light-water breeder reactor, being developed by the Naval Reactors development organization. A core to demonstrate the principle has been loaded into the Naval Reactors team's PWR at Shippingport. Although light water, as a moderator, absorbs more neutrons than does either graphite or heavy water, an ingenious arrangement of the fuel minimizes the bad effect of water on breeding. The fissile material is segregated into heavily loaded zones, in which the absorption of the water is not important, and the fertile material is placed around it in blankets that have only the necessary minimum of water in them for cooling. The reactor is calculated to breed—barely—and is usually considered as more nearly a 1 : 1 converter. By this, one means a reactor that, once loaded, needs no further input of fissile material to continue operating, but one that likewise produces no excess of fissile material for use in other reactors.

All of these reactors appear to be less economic than LWRs, however. If the price of uranium rises sufficiently in the future, they may be competitive, but not as alternative breeders. In all of these cases it can be shown that even with very-high-priced uranium, the most economical fuel cycle is achieved when the reactor is less heavily loaded with fuel than is required for breeding, and is run to high burnup. The reason is that the breeding cycles for all these reactors require high fuel inventory charges, frequent reprocessing with its attendant costs, and frequent fuel fabrication. Thus, we look to these reactors for the future for an entirely different function: as reactors that require only small continuing fuel inputs to run economically. The fuel would be furnished by breeders in a mutually supporting or "symbiotic" system. The assumption is that one or more of these reactors might turn out to be cheaper to build than a fast reactor that is run on a nonbreeding fuel cycle (also possible).

6. SYMBIOTIC REACTOR SYSTEMS

The concept of a system of LWRs making plutonium for use in fast breeders, with the breeders in turn supplying ^{233}U for use in the LWRs, is a simple example of a symbiotic system. In fact, the LWRs with ^{233}U fuel instead of ^{235}U would run a little better than they do now. This improvement in operation would permit a system with approximately equal numbers of breeders and nonbreeders to operate without any further input of natural uranium. For such a system to expand, though, there would have to be a large excess of breeders or a further supply of fresh national uranium.

Faced with the political reality that unusually carefully guarded reprocessing and refabrication facilities are now being demanded, there is a natural inclination to lump these together in large "fuel cycle parks" that could serve many reactors. The advantages of such parks for security are fairly obvious: a single perimeter requires much less guarding than the perimeters of several individual plants; the guard forces would be mutually supportive; the internal transportation links would be short and easy to secure; and so forth. Since breeders require a lot of reprocessing capability, there is also an advantage in locating breeders within these parks. But we are not really talking about energy centers; it would still be very desirable to have most reactors reasonably freely locatable, to serve demand centers efficiently and to improve the reliability of the electrical network. Too much power, transmitted over too long a distance, leads to power loss in transmission and to system instability if the transmission lines fail.

Therefore, over the long term, we might find it attractive to look for symbiotic systems in which one breeder serves to provide fuel for several nonbreeder reactors that just slightly miss being 1 : 1 converters. For this long-range possibility, the reactors just described could be very attractive. For different reasons, the HTGR and CANDU types are most often mentioned: HTGR because it has already been carried through the "demonstration" stage in the U.S. and would be less difficult to bring to commercial operation than other reactors; and CANDU because it already is commercial, and has accumulated considerable operating experience in Canada.

It must be stressed, however, that it will probably be a long time before these reactors are competitive with LWRs. They are more expensive to build, and since it is assumed that they will be attractive on a ^{233}U–Th cycle, the fuel cycle facilities will have to be developed virtually from scratch. An optimum nuclear power strategy for the U.S. would thus include a broad spectrum of reactor and fuel cycle development: breeders, the advanced converters being discussed here, and their fuel cycle facilities. Unfortunately, all of these programs compete for the limited development funds that are available. Under these circumstances, the breeder and its fuel cycle command priority, simply because they ensure against nuclear fuel shortages. The advanced converters, while desirable, would just serve to improve the system.

7. AN INTERIM STRATEGY

There is one way that a start can be made on the long-range symbiotic system that has so many attractive features. This is to begin obtaining experience with thorium fuel cycles in LWRs. The U.S. is not entirely lacking in experience in such systems. One of the first commercial reactors, the PWR Indian Point No. 1, was originally fueled with a mixture of uranium and thorium fuels. The reactor ran successfully, but in the absence of commercial reprocessing for the thorium fraction, it became more economic to switch to the more standard slightly enriched uranium fuel. However, reactor loadings of this type could be repeated and the necessary reprocessing facilities developed. At the same time, breeder development could continue, with the standard blanket being made of thorium rather than depleted uranium. This would require further development of breeder core reprocessing, which is currently based on the concept of dilution with a low-plutonium-content uranium blanket. Nevertheless, all of these developments are feasible, and they have the advantage that only one reactor type (the breeder) would be under development. The development of fuel cycle capabilities and facilities does not require the same type of personnel, nor the same type of corporate involvement, as reactor development does.

8. OTHER SOURCES OF NUCLEAR FUEL

Except for mining natural uranium for its ^{235}U content, or deploying breeder reactors, are there any other ways of getting fissile material to fuel a nuclear industry? There are two ways, in principle. One is by no means technologically feasible with present knowledge, and the other is not economic. The first is a scheme known as the "fusion–fission hybrid;" the other is called "accelerator breeding." In spite of what must only be rated as a speculative possibility that either will be successful, they have been used by certain antinuclear circles as reasons for delaying or discarding broader development.

The fusion–fission hybrid is a system in which the thermonuclear reaction of a fusion system is used to produce energy, neutrons, and fissile material in a blanket around the fusion system. Its basic advantage is that the pure fusion reaction produces a neutron every time about 18 MeV of energy is generated, whereas the fission reaction, even with plutonium, produces one neutron for every 70 MeV of energy generated. The fusion reaction, in other words, is energy-poor and neutron-rich, while the reverse is true of the fission reaction. The hybrid combination would "multiply" the energy of fusion and permit the fusion neutrons to power a multiplying fission blanket that does not need to be a critical reactor. It is hoped that under these circumstances the fission system would require very little in the way of add-on safety devices and would thus not add much expense to the fusion system. A variant of this scheme would have a nonmultiplying (or almost so) thorium blanket in which ^{233}U would be generated, with less development of power. About two new fissile atoms could be made for each fusion reaction, while still retaining the ability (required in most fusion concepts) of regenerating the fusion fuel. That fuel is tritium, made by absorbing neutrons in lithium, which then splits into tritium and helium.

Both of these concepts, of course, rely on achieving workable fusion in comparatively inexpensive fusion systems. Neither the feasibility of fusion, nor the form a feasible reactor might take, nor its cost can be predicted with any confidence now, and realistic estimates place the deployment of fusion power on a large scale as 50 years into the future at best. It may be that fusion power will not turn out to be economic at all, but if it does achieve economy it will be the result of simultaneous improvements in technologies that are now only complex laboratory constructs. Thus, we must consider both of these schemes as proper justification for continuing R & D on fusion; they do not vitiate the case for breeders, however.

Related to these fission–fusion concepts is the idea of making fissile material with neutrons generated in accelerators. Again, this concept has been considered in conjunction with pure fertile material blankets ("nonmultiplying") or with subcritical assemblies as blankets ("multiplying"). The accelerator breeder is buildable with technology available today. However, the cost of new fissile material, even assuming considerable advances in the economics of accelerator construction, is a few times greater than that that would pertain even to a not-very-advanced breeder reactor. Indeed, the multiplying concept has been criticized as doing nothing that could not be done with a well-designed fission reactor, except that the control rods are replaced by an extraordinarily expensive substitute: the accelerator.

Both the fusion–fission breeder and the accelerator–breeder postulate a situation similar to that foreseen by the symbiotic breeder reactor—advanced converter reactor system. This is one in which, for reasons of siting and economics, most of our electric energy is generated in dispersed converter reactors. We may take some heart from both ideas in at least one respect: if we are foolish enough to burn down the world's stock of natural fissile material without replacing it by breeding, these ideas permit a comeback. The comeback would be slow and expensive, but possible. Of course, both time and money spent can be equated with subtractions from human happiness.

Another important point with both of these schemes, fission–fusion and accelerator breeding, is that they require reprocessing for fissile material recovery. In this sense, their only advantage over the fast breeder is that the throughput of fissile material in the reprocessing plant is somewhat reduced. The reduction ranges from a factor of about 5 in the nonmultiplying version to factors of less than 2 in the multiplying version.

We can summarize the situation by observing that both ideas are interesting ones and worthy of further exploration. In the case of fusion systems, the basic program is still at the research stage: trying to make a fusion reactor work. After that step is accomplished, we will undoubtedly find all sorts of engineering problems that must be solved before we can start designing "demonstration" reactors, and yet more before commercial system can be installed. These may turn out to be such excellent power producers that fission energy cannot compete. More power to them if that is the case. Conversely, it may turn out that—as we believe—fission energy will be permanently cheaper and more manageable than fusion energy. The situation may be summed up as "Let the best win out, but let the contest be a fair one." Compared with this competition, the fusion–fission scheme can at present be seen as an interesting sidelight.

In the case of accelerator systems, the situation is different. They are not economic sources of fissile material. There are still enough possibilities for economies in accelerator design that the idea does not deserve simply to be discarded, however. Rather, it is a proper concept to keep in mind as new accelerators and accelerator hardware systems are developed. If this improves the prospects of accelerator breeding to the point where the economics have a chance of being attractive, then and only then is it worth taking the concept seriously enough to begin a development program. We are of the opinion that this time will not come soon, and may not come ever.

But opinions are not facts. Serious fusion researchers and accelerator designers do not promote their ideas by playing fission energy down, but by pointing out the potentials of their own ideas. This we respect. R & D in these areas is very welcome; the world can only gain by competition among different technologies.

10 • Alternative Fuels, Fuel Cycles, and Reactors

9. DISCUSSION

The search for reactors, fuel cycles, and fuel production concepts other than those considered conventional in the nuclear community is motivated by two separate interests. One is to look for improvements with the aim of providing cheaper and more durable energy systems. The other is to contribute toward solution of the threat of weapons material diversion and weapons proliferation problems.

Toward the former aim, it would be desirable, but not pressing, to carry through development on high-gain converter reactors. On the other hand, it is a pressing matter to begin to gain experience in the recycle of nuclear fuels, achieved by chemical reprocessing. This is also a necessary step in the development of breeder reactors, which in the long run are the best insurance the world has that energy can be had abundantly and at a reasonable price.

Toward the latter aim, technological alternatives can only make a minor contribution. Of overwhelming importance to safeguarding nuclear fuels that could be made into usable, if inefficient, nuclear weapons are diplomatic, political, and administrative measures. These include international agreements, security practices, and good fuel management practices including accounting, inspecting, and record-keeping. Technological measures can only serve as adjuncts to these basic methods. Nevertheless, searches for such measures have been undertaken in good faith.

In reviewing the results of these searches, the fundamental result has been that reprocessing of spent fuel is an important safeguard measure in itself. This is because this reprocessing operation gathers in the huge quantity of potential weapons material contained in spent fuel and, first, reduces its volume so that guarding it becomes easier, and second, prepares it for reuse in reactors, at which point it becomes inaccessible for diversion.

Thus, the most important technological measures against diversion are those that concern improvements or simplifications in the measures that can be taken to guard recycled fuel. There are essentially three such possibilities:

1. Diluting the product with an isotope that jeopardizes its value for weapons. The best known of these concepts is the "denatured" ^{233}U cycle in which ^{233}U is mixed with ^{238}U. Unfortunately, there are few reactors that can be efficiently run on this fuel without forming significant quantities of plutonium on recycle (since plutonium is made from the ^{238}U added to safeguard the ^{233}U). Alternatively, plutonium may be "spiked" with the isotope ^{238}Pu, which generates so much heat that weapons parts would lose their shape on storage; this, of course, enormously complicates fuel fabrication operations.

2. Adding enough radioactivity to the product that diversion becomes more difficult and, particularly, much more easily detectable. The CIVEX process does this for standard solvent-extraction reprocessing, while the specialized methods used at EBR-2 and MSR have shown that such techniques can also be invented for reprocessing at the reactor site. All of these methods require fabrication to be a remote, shielded operation.
3. Rapid turnaround of product into new fuel assemblies. This is the main purpose of fuel-cycle parks. If further protection of the freshly manufactured fuel is desired, this can be achieved by a variety of measures, including preirradiation of the assemblies, incorporation of marker isotopes in the cladding, and so on.

Of these measures, the one that presents the fewest problems to the industry is the third. It is also, incidentally, the one most compatible with the inclusion of breeder reactors in fuel-cycle centers. The second measure has long-range attractiveness, in that it may turn out that fabrication of highly radioactive fuel is not as difficult as is thought on the basis of experience to date; but this awaits development. If automatic fabrication is developable, the attractive feature of this option is that it is probably easier to leave radioactivity in reprocessed fuel than to "decontaminate" the fuel. The first set of measures, and particularly the use of denatured uranium, has received the most publicity, however.

Denatured fuel can only be used to reduce the throughput of chemically separable weapons material in a reprocessing plant, not to eliminate such throughput. It turns out that only a few reactor types, and particularly the CANDU reactor, permit an appreciable reduction of plutonium throughput. The LWBR replaces plutonium throughput with ^{233}U throughput that cannot be denatured within the LWBR fuel cycle. The reason is that the modular cores of the LWBR require enriched fuel. When denatured fuel cycles are completely analyzed for CANDU, one finds that significant production of plutonium cannot be avoided if high burnup is contemplated, and high burnup is needed in order to achieve good fuel-cycle economy. The conclusion is that the denatured cycle can achieve some reduction in weapons-usable material output, but at a significant economic penalty. As the diversion and proliferation problems are not eliminated, but simply reduced in terms of quantity of material to be safeguarded, it doesn't seem to be worth the expense.

Other technological measures that have been examined include fusion–fission and accelerator breeding, to see whether they compete with fission breeding. They do not compete now and probably never will. Nor do these schemes eliminate the need for reprocessing. The best that can be said for them is that, if ^{233}U is the desired fuel, it can be denatured during reprocessing. As noted, this simply delays dealing with plutonium until after the next irradiation cycle.

10 • Alternative Fuels, Fuel Cycles, and Reactors

The most robust and workable option that we have found is a simple variant of what has been considered all along to be the "standard" nuclear fuel cycle of the world. This is one in which breeder reactors are deployed within fuel-cycle parks. Their fuel is reprocessed in a quick turnaround, so that the quantity of plutonium in storage is small. Their blankets can be thorium, from which denatured or undenatured fuel elements (containing ^{233}U diluted in ^{238}U, or thorium, respectively) can be derived. These can be shipped to dispersed reactor sites. The reactors at dispersed sites will be LWRs at first, and they will produce more than enough plutonium to keep the breeders going. Until more sophisticated advanced converter reactors are deployed, it will therefore be desirable to recycle much of the plutonium in the LWRs. This fuel, too, could be manufactured in the fuel-cycle park. If technological measures are desired to improve safeguarding this fuel in storage or in transit, preirradiation or "spiking" the cladding would be completely workable.

11 The Nuclear Weapons Proliferation Issue

E. L. Zebroski and Bernard I. Spinrad

1. SOME DEFINITIONS

There are four issues that various people include under the term "proliferation." These are:

1. *Vertical proliferation:* Increased numbers, types, and methods of deployment of nuclear weapons by countries already having such weapons, and particularly by the two superpowers.
2. *Horizontal proliferation:* Additional countries with deployable or deployed nuclear weapons.
3. *Potential capability:* Still other countries with a capability to rapidly manufacture nuclear weapons, should a decision be made to do so.
4. *Subnational potential:* The possibility of manufacture of nuclear weapons from material diverted or stolen, as by terrorist or criminal groups intent on blackmail.

Many books and articles on the subject, and many of the popular media presentations, mix several of these issues. This is a disturbing fact, because the problems of proliferation are serious. In discussing them, we must know precisely what we are talking about. For that reason, students of proliferation problems have devised a nomenclature to identify clearly which of the issues is being addressed. The terms are defined above, and we will use them as so defined.

E. L. ZEBROSKI • Electric Power Research Institute, Palo Alto, California 94303. BERNARD I. SPINRAD • Department of Nuclear Engineering, Iowa State University, Ames, Iowa 50011.

2. VERTICAL AND HORIZONTAL PROLIFERATION AND NUCLEAR POWER

The two most vexing proliferation problems are those of vertical and horizontal proliferation. These are also the two proliferation problems to which civilian nuclear power is largely irrelevant.

For vertical proliferation, the reasons are particularly transparent. The sort of fissile material—the stuff of which bombs can be made—that is turned out by nuclear power plants is low quality, compared to that which dedicated processes and facilities can make. All of the countries with nuclear arsenals have these dedicated processes and facilities. There is thus no incentive to use nuclear power facilities to make bigger or better or more bombs. If any incentive exists, it can well be a reverse one: the demand for fuel, to use in nuclear power plants, will ultimately lead to pressure to convert weapons material to fuel for energy production.

For horizontal proliferation, the situation is not so obvious. The history of nuclear weapons countries is that nuclear power is a by-product of a weapons program, rather than the reverse. Five out of five (or, counting India, six out of six) nuclear weapons states have not used civilian power reactors to produce plutonium for weapons [1]. We must also understand why civilian nuclear power was not used.

One obvious reason is that, except for India and France, all of the weapons powers built bombs *before* they built power plants. Another compelling reason is the same as for vertical proliferation: better bombs come from other routes. But the final compelling reason involves time and cost. To understand this, we can refer to Table I, which presents eight ways* of procuring nuclear weapons material, three by reactor production and five by isotope separation [2]. For reference, production reactors and isotope separation by gaseous diffusion are the methods that were used historically. Of the three reactor routes, using power reactors is the slowest, by far the most expensive and most complicated, and the least flexible in terms of scale of operation and ability to expand. While production reactors have been used to make by-product power (in England, France, Soviet Union, and America), power reactors have never been used to make by-product weapons plutonium. The isotope separation routes are capable of feeding enriched fuel to nuclear power plants, but are not concerned with nuclear power-plant products. The historical connection of isotope separation with nuclear power is that a nuclear power industry can be a customer for the output of an isotope separation plant which was originally built for weapons production. In recent years, the technology for isotope separation has advanced rapidly in many countries. Relatively small facilities are becoming practical for

---

* Calling this "the eightfold way" is a physicist's irony. The term "eightfold way" has long been used as a shorthand and mnemonic for certain symmetry properties of elementary particles.

TABLE I[a]
The Eightfold Way to Nuclear Weapons Materials Production[2]

	Cost	Support technology	Support industry required
Research reactor			
natural U-D$_2$O	Very low	Low	Minor
Enriched U	Low	Medium	Large
Production reactor (natural U)	Medium	Medium	Medium
Power reactor			
Natural U	High	Medium	Medium
Enriched U	High	High	Large
Diffusion cascade	High	High	Large
Centrifuge cascade	Medium	Medium	Medium
Aerodynamic jet cascade	High	Medium	Large
Electromagnetic separation	Medium	High	Medium
Neutron generator (accelerator)	Medium	Medium	Medium

[a] Since this table was published two additional methods have been developed, for pilot-scale use; namely, Laser Isotope Separation and Chemical Ion Exchange Separation.

production. Several nonweapons countries have isotope separation facilities which were justified on the basis of providing assured domestic supply of low-enrichment fuel for their power reactors. In principle, such facilities could be expanded, or operated to produce high-enrichment, that is, weapons-usable material.

Dealing with Horizontal and Vertical Proliferation

Under these circumstances, it has long been recognized that the problems of both vertical and horizontal proliferation are primarily to be dealt with by political and diplomatic means. Self-interest restraints, reinforced by the communities of nations (or strong bilateral relationships), have the continuing ability to persuade individual countries not to develop and produce nuclear weapons. Diplomacy and economic relationships backed selectively with sanctions or embargoes, are the tools of persuasion. The visible signs of successful diplomacy are formal proposals, treaties, trade agreements, and cooperative institutions. Table II presents a number of these; one list contains the various steps toward demilitarizing nuclear energy, while another lists various arms control agreements, all but one of which implicitly or specifically are concerned with nuclear weapons.

This situation has been stable for more than a decade. The concern arises from the evident *potential* weapons capability in at least a dozen more countries, and the periodic rumors or allegations that the potential capabilities have been carried to actual construction of explosives—for example, such allegations have

TABLE II
Steps toward Control of Nuclear Weapons[a]

International agreements on arms control	Demilitarizing nuclear energy
1. Antarctic Treaty 1959	1945 Baruch plan for international control (Truman–Atlee–King proposals)
2. Limited Test Ban Treaty 1963	1946 Russia rejects—will consider only on basis of equality
3. Outer Space Treaty 1967	
	1946 America withdraws cooperative agreements with England (France)
4. Treaty for the Prohibition of Nuclear Weapons in Latin America 1967	
	1953 Eisenhower–Strauss plan Atoms for Peace
5. Treaty on the Non-Proliferation of Nuclear Weapons 1968	1954 Atomic Energy Act
6. Sea-Bed Arms Control Treaty 1971	1955 Geneva Conference
7. Biological Weapons Convention 1972	1956 International Atomic Energy Agency
8. SALT-ABM Treaty (SALT-I) and SALT Interim Agreement 1972	Worldwide safeguards plan IAEA-CIR-66
9. SALT-II 1979	Civilian/military distinctions Euratom safeguards system

[a] From reference 1.

occurred, especially in election years, for Argentina, Brazil, India, Israel, and Pakistan, among others.

Vertical proliferation is a significant matter primarily for the U.S. and USSR. Ultimately, China will also have to be considered, if China comes to seek nuclear parity with the USSR. Of the other nuclear weapons states, England and France, although their political and military strategies are different, have stabilized their nuclear weapons capabilities at levels that are a fraction of those of the two superpowers. Finally, India has made no moves toward vertical proliferation after experimenting with one test explosion: The Indian position with regard to the test (i.e., that it was a test of "peaceful nuclear explosive" designs) indicates that India has not yet made the decision to establish or proclaim itself a nuclear weapons power, and may not do so.

Thus, the problem of vertical proliferation is uniquely a U.S.–USSR problem, now and for a considerable time to come. That the SALT-II treaty actually lowered some levels of nuclear armaments is a hopeful sign, but not yet reassuring, given that the treaty is not ratified, and may not be observed. What is important, however, is that the pressures on the superpowers originate from the

Non-Proliferation Treaty (the NPT), and that these pressures may eventually be taken seriously for mutual advantage rather than by coercion.

Horizontal proliferation is easier to measure. By 1955, there were three nuclear weapons powers: the United States, Soviet Union, and Britain. As of now, 1984, only two other countries, France and China, have joined the circle, and one other country, India, demonstrated its potential capability. Actually, that's a rather remarkable record. If horizontal proliferation proceeds at a rate of less than one country per decade, there should be time for the world to work out institutions to make nuclear war impractical. The problem is pace only: the capability of war to be fearfully destructive constantly increases, and in the long run can only be counteracted by eliminating the motivations for war and terrorism as human activities.

In addition to the restraints from prudent self-interest, two institutions have contributed to the slowness of horizontal proliferation so far: the nuclear umbrellas of the superpowers, and the NPT (and related bilateral treaties). The nuclear–military umbrellas have made it unnecessary for close allies of the superpowers to have their own nuclear arsenals. This becomes even more relevant when we realize that France and China, the two more recent nuclear weapons powers, decided to build up their own capabilities *after* deciding to distance themselves from their respective superpower allies. The absence of new club members recently can be attributed partly, still, to the superpower umbrellas; but the NPT must get some of the credit, too, in view of the technical capabilities of many of the nonaligned countries.

3. PROLIFERATION POTENTIAL AND NUCLEAR POWER

To this point we have been talking history and facts. When we get to proliferation potential, we enter the realm of scenarios and speculation about what a country could do—or might be doing—secretly. Proliferation potential is, of course, related to horizontal rather than vertical proliferation. It describes a condition preceding horizontal proliferation.

Proliferation potential is a technical capability. If a particular less industrialized country is inclined to acquire its own nuclear weapons, can it do so? For more industrialized countries there is no real uncertainty about *whether* it could do so, but only *how rapidly* and what *the possible motivations* are which could trigger a decision to proceed.

What are the likely routes? What is the relation to nuclear power? Are there certain facilities that make this relation suspect? Can these facilities be organized to decouple nuclear power from proliferation? These are some of the questions to be addressed.

3.1. Potential Technical Capability

It is necessary first to define and describe what the requirements are for being able to acquire nuclear weapons. Whether or not there is civilian nuclear power, the necessary conditions are:

- Availability of uranium
- Availability of people to design and build weapons production facilities
- Availability of special equipment and materials, or industrial infrastructure to produce them
- A national motivation to become nuclear armed, and to allocate major resources, and to accept the consequent risks.

3.1.1. Availability of Uranium

To the first point, availability of uranium: access to uranium is necessary in order to implement any of the different routes to nuclear weapons. Uranium is widely distributed in the earth's crust, being associated with many types of sedimentary rock. The global distribution of known reserves is indicated by Table III. For the U.S., the total recoverable resources are estimated to be between 3 and 7 times larger than the known reserves. Almost every country that has mounted a widespread exploration program has found significant tonnages of uranium. The smaller tonnages in Table III—say, under 20,000 tons—are correlated with relatively small efforts at exploration to date. *Even the smallest reserve shown in Table III has enough uranium to produce several nuclear weapons, if such a decision were to be made.* Each 1000 tons of uranium can yield over 4 tons of separated ^{235}U by isotope separation (over 200 weapons). It can also be converted to about 800 kg of plutonium (over 100 weapons) using natural uranium reactors. Sixteen of the countries listed (plus the USSR and China) have reserves large enough to mount very large-scale military or civilian nuclear power programs, and still export large tonnages of uranium.

The mining and milling of uranium is relatively conventional technology within the same range of mining and recovery practices used with common metals such as copper, zinc, or lead. On both a U.S. and a worldwide basis, the main limitation in the rate of growth of uranium supply is the rate of investment in conventional mining and milling equipment. The equipment and chemicals for mining and milling are all easy to manufacture and are freely available worldwide in normal trade channels; and any attempts to restrict their production or transfer in commerce would fail. There is already a large stock of such equipment worldwide and all the reagents can be prepared from the most common chemicals.

Thus, many countries have, or can get, enough uranium to make, not just one, but many nuclear weapons. Restrictions on uranium trade tend to be more

TABLE III
World Distribution of Uranium[a,b]

	Reasonably assured resources to $30/lb.	Total assured plus estimated
Algeria	28	28+
Argentina	21	60
Australia	213	293
Brazil	10	19
Canada	166	585
Central Africa Republic	8	16
Denmark (Greenland)	6	16
Finland	2	2+
France	55	95
Gabon	20	30
Germany	1	5
India	29	52
Italy	1	2
Japan	7	7+
Korea	6	6+
Mexico	6	6+
Niger	50	80
Portugal	7	7+
South Africa	276	350
Spain	104	211
Sweden	300	300+
Turkey	3	3+
United Kingdom	2	6
United States	454	1266+
Yugoslavia	7	22
Zaire	2	3+
Total tons (thousands; rounded)	1810	3470+

[a] From reference 3.
[b] Excludes Russia and China.

limiting to civilian nuclear power, which is thirsty for uranium as fuel, than they are on potential nuclear armaments, which can become threatening with much smaller amounts.

3.1.2. Availability of Know-How

After years of struggling with problems that eventually turned out to be solvable, some of the scientists of the Manhattan Project were alarmed to discover that most of the theoretically possible routes to the atomic bomb eventually proved feasible. Given talent, resources, and facilities, every route could be made to work: several types of isotope separation plants, production reactors, chemical separation plants, and, alas, the bombs themselves. Moreover, this was *with the technology* largely available in the 1940s and 1950s in advanced countries, and now in many more countries as well. The conclusion that there were no secrets that could not be independently surmounted was dramatized by the atomic explosions by Russia, then England, then France, China, and finally India. For most of these, only 3 or 4 years' work as a national effort produced the initial explosive device.

Since World War II, nuclear technology has become codified and published. Students can find textbooks explaining what the theory and practice are, for nuclear fuel fabrication, reactor construction and operation, nuclear fuel reprocessing, isotope separation by gaseous diffusion, isotope separation by centrifuge, and other methods. (Actual manufacturing technology and know-how are generally not published, but can be developed—given the motivation, talent, and resources.) Students using hand-held calculators can work out the behavior of reactors better and faster than Enrico Fermi could. Undergraduate chemists can work with isotope separation stages in student laboratories. The theory and practice of uranium recovery from chemical mixtures is taught; there are standard radiation protection techniques that can be used to make all operations safe.

This knowledge is, by now, well known, not only in America, but worldwide. Any country that has a modern science or engineering university is likely to have the talent to undertake its own "Manhattan Project." The projects that most countries would be able to set up would be cheaper and faster than the Manhattan Project in America in World War II. Science and technology have progressed, one human generation and at least two technical generations worth.

To get an approximate idea of the required talent, Table IV might be helpful. Every country on the list has the manpower to start a nuclear weapons program,

TABLE IV
Research Reactor and Power Reactor Countries (1977)

	Research reactors (MWt)		Power reactors (No. of units, 1975–82)[a]
	Natural uranium	Enriched uranium	
Argentina		1	2
Australia	10		
Austria		12	1
Belgium	4	100	8
Brazil		5	3
Bulgaria		1	4
Canada	200+	2	15
Chile		5	

TABLE IV
(Continued)

	Research reactors (MWt)		Power reactors (No. of units, 1975–82)[a]
	Natural uranium	Enriched uranium	
China		(many)	
Czechoslovakia	4		4
Denmark	10	5	
Egypt	2		
Finland			4
France		(many)	39
Germany	69	30	40
E. Germany	(several)		8
Greece		5	
Hungary		2	2
India	40	1	5
Iran		5	4
Iraq		2	
Israel	26	5	2
Italy	45	13	7
Japan	75	25	27
Korea	3		4
Kuwait			1 (1983)
Luxembourg			1
Mexico		1	2
Netherlands		46	2
Norway	22		
Pakistan		5	1
Phillipines		1	2
Poland	10	30	2
Portugal		1	
Romania	3		
South Africa	20		2
Spain		3	17
Sweden	50	1	14
Switzerland	20	5	7
Taiwan			3
Thailand		1	6
Turkey		1	1 (1984)
United Kingdom		(many)	40
United States		(many)	143
Uruguay		1	
USSR		(many)	31
Venezuela		3	
Yugoslavia	10	3	1
Zaire		1	
Total	623[b]	322[b]	455

[a] Cumulative commitments to January 1977 total 455 power reactors in 49 countries with a capacity of 343 GW (excluding America).
[b] Excluding countries with many units.

with one or two possible exceptions. There are also countries not on the list that should be included: Columbia, Egypt, Indonesia, among others. The ability to operate a research reactor implies the presence of the *technical personnel* needed to produce at least small quantities of weapons-usable materials.

3.1.3. Weapons Design and Manufacture

The relatively high availability of design data and operational experience for power reactors does not apply to weapons design or manufacture. The *general* scientific principles of producing nuclear explosives are widely known (or can be reinvented using generally available textbook information). The specific designs, however, involve a number of specialized technologies and kinds of know-how and data which are not available in textbooks. Acquiring such know-how and data involves a great deal of trial and error and testing—which takes time, skilled manpower, and considerable resources. Such an effort—after one to three years (and with sufficient weapons material)—might be able to produce an initial "nuclear device." This would be bulky and heavy, and have rather low yield and poor reliability—but it would be dangerous, nevertheless.

In a military sense, it would not be a very practical "weapon," but the symbolic and balance-of-power impact of even a primitive nuclear device can be very significant.

However, having one or a few primitive nuclear devices can be very dangerous to the country owning them. Even the rumors of such a capability are highly provocative to neighboring countries, who are impelled to mount similar efforts for deterrence. If the country which feels threatened is associated with one of the known weapons countries, it also may be able to react to a nuclear threat swiftly—with much more potent weapons than an initial effort could produce.

More potent capabilities—true military weapons—involve much larger efforts than the minimum which might make initial devices. Literally thousands of man-years and people skilled in several dozen scientific, engineering, and manufacturing disciplines are required to make devices that have desirable military attributes such as:

- Reliability
- Reasonable shelf-life
- Known yield
- Compactness
- Packaging for various environments
- Adaptation to versatile delivery systems
- Requirement of only moderate amounts of materials

Such attributes are not attainable from theory or laboratory work alone. Actual testing is essential to be able to develop specialized designs and processes, and to find and correct the thousands of things that can go wrong in a complex new design and with inexperienced fabrication and assembly operations.

No actual proliferation of weapons-in-hand has been evident since mainland China in 1965 [1]. This is in contrast to the dozens of countries which have visible and evident potential technical and industrial capability if they wish to apply it [1]. Possibly one of the main practical factors which has led to this restraint is the recognition that a primitive nuclear capability is useless in a military sense. Brandishing it or using it would very likely be suicidal—since enormously potent or even preemptive responses are possible. A further factor which restrains conversion of potential capability to actual production is the very large effort and infrastructure and testing which must be committed to produce a true military capability.

For all but the largest countries, such an effort is large enough to drain the civilian economy noticeably—and to degrade conventional military capabilities. The scale of such an effort—and the testing required—could not be concealed. The political and economic costs, including likely widespread sanctions (or even preemptive strikes) by some countries, are further deterrents to the actual proliferation of weapons.

These controlling factors are important to give perspective to the perennial save-the-world proposals based on ever more intensified export controls. The export of parts, materials, or equipment directly useful for the production of weapons is obviously imprudent—especially to hostile countries. The extension of such controls to civilian materials or equipment already in world commerce—that have only a tenuous, indirect, or alleged connection with some supporting infrastructure—cannot be effective. More often such controls are directly counterproductive to their stated intent since they stimulate and provoke the development of more widespread capacity for production in more countries. This has clearly been the case with isotope separation. Erratic policies and procedures on export controls have helped to stimulate such developments in many countries.

3.1.4. Availability of Equipment or Infrastructure

A list of countries that possess major engineering schools, with laboratories that have been equipped since 1950, would have more than 50 entries. The ability to maintain these laboratories and to build and repair machinery is a considerable part of the infrastructure that is needed to build up at least a primitive nuclear weapons program.

The upshot of these considerations is that many, many countries can, if they want to, build one or a few atomic bombs of the Hiroshima or Nagasaki type, and it is not only possible but rather likely that several have already done so, secretly. Technical capability, in most cases, is not the limiting factor in whether potential technical capability is converted to a few actual devices.

3.2. What Are the Routes?

We refer again to Table I, the "eightfold ways." There are two routes on it that are moderate in cost, sophistication ("Support Technology"), and required infrastructure ("Support Industry Required"). These are the production reactor route and the centrifuge cascade route. There is also a low-cost, incidental route to a small capacity for making material for a few weapons. This is through the use of natural-uranium, heavy-water research and test reactors which have significant power levels (say, above 30 MW). This latter route requires only a laboratory-scale reprocessing line in addition to the reactor to extract enough plutonium for a bomb every year or two. This is the route that India used in making its bomb.

3.2.1. Production Reactor Route to Weapons Material

Production reactors can be built using graphite moderator, low-temperature light-water cooling, uranium metal fuel, and aluminum for structural material. Such reactors were used originally at Hanford. The technology is similar to that of research reactors except that they must be much larger in size and dissipate greater amounts of heat. The technology remains that of a low-temperature device with simple materials requirements and little or no additional technical know-how over that required for a research reactor (except the ability to organize construction projects on a much larger scale). Natural-uranium production reactors using either graphite or heavy water as moderator have produced essentially all of the weapons-grade plutonium existing in the world today.

Associated with the production reactor must be a fuel fabrication facility and a reprocessing plant. Both technologies are now within the range of conventional metallurgical and chemical process industry.

3.2.2. Gas Centrifuge Route

The last six options (in Table I) for producing fissionable material involve a separation of ^{235}U isotope from natural uranium. The bulk of the enriched ^{235}U in the world has been produced by diffusion cascades. Future enrichment capacity is likely to use mainly centrifuge cascades in the U.S. and Western Europe, because of the lower energy input requirements as compared with a diffusion plant. Laser isotope separation is now technically proven, and is the method of choice for some applications. The remaining options appear to be more expensive, complicated, or speculative in respect to the costs attainable.

11 • Proliferation

The gas centrifuge is likely to be attractive to a country embarking on a weapons program. The main reasons are: a relatively small number of centrifuges are needed in order to achieve the "few bombs a year" production rate; the process can be built up stepwise to an increasing production rate; and usable but low-performance centrifuges (for example, of the Zippe type) are commercially available. They would need rework, but not complicated rework.

The choice of ^{235}U for weapons is likely for those countries that either have doubts about their weapons design capabilities, fear of detectability of production reactors due to their size, heat emissions, and a desire to proceed to at least initial capability secretly.* Uranium weapons are simpler to design and to make than plutonium weapons. The enrichment of uranium does not involve dealing with the high levels of radioactivity of a production reactor. A production reactor complex might be discovered by surveillance overflights. a centrifuge cascade could be more readily concealed.

Summing up these two "preferred" routes, the advantages of a production reactor are that the technology is simple and a capability for one bomb a year can be built up for less than $100 million, and start of production in 3 years or less. High-quality plutonium ("weapons grade") is produced and capacities can be increased, stepwise. The advantages of a centrifuge cascade are that a relatively small-scale installation can produce a bomb a year, the ^{235}U is easily handled, and the operation can be more easily kept secret. The cost of the first modules of the cascade plant are likely to be somewhat more than $100 million, but as experience is gained the cost would drop.

4. WHAT IS THE RELATION TO NUCLEAR POWER?

Most of the power reactors in operation, or under construction today and for the next decade, are light-water-moderated reactors fueled with slightly enriched uranium (^{235}U content is between 3 and 5%). Such reactors involved much more sophisticated and exotic materials, sophisticated heat transfer design, and much heavier and more advanced fabrication and construction technology than that required for production reactors. As power reactors must compete economically in costs of the power produced with alternate sources of generation of electricity, it is not sufficient simply to build a machine that works, as in the case of a production reactor, but it is also necessary that that machine work reasonably well (which means operation for 65% of the time or better each year), and to be buildable at acceptable capital costs. Because of the economic constraints, several competing types of reactors have not attained widespread use

---

* A small centrifuge plant was built in Argentina starting in the late 1970s, but became known to the world only in 1983.

due to either higher costs, relative to light-water reactors, or less assurance of operability.

An intermediate type of power reactor derived from the production reactor has also been built in considerable numbers. Such "dual-purpose reactors" use a fuel cycle (low burnup, natural-uranium or very slightly enriched uranium fuel) for maximum plutonium production but operate at somewhat higher temperatures so as to permit production of some steam for power purposes. Examples are the Russian light-water-cooled, graphite-moderated reactors and the N-reactors at Hanford, Washington. Worldwide, there are no further plans to build additional reactors of this type except in the Soviet Union, which is building very large reactors (1500 MWe) of this type. Apparently the by-product plutonium production is valued highly, either for weapons purposes or to produce fuel for future breeder reactors.

Heavy-water power reactors have also been developed and built in at least eight countries, but only in Canada, India, and Argentina are there plans for continued construction of this type. With very large uranium reserves and resources, relative to foreseeable domestic requirements, the Canadians use a "once-through" fuel cycle. As the cost of natural uranium increases in the future, there is some expectation that the spent fuel from these reactors could be reprocessed to recover its plutonium content. The stated purpose of the French, Russian, Japanese, and Indian programs has been to produce enough plutonium to be able to fuel fast-breeder reactors, starting in the 1990s.

Of the power reactors, the enriched-uranium light-water reactors have not been used to produce weapons material (although this mode of operation is technically possible—at much higher cost of operation, and considerable loss of electrical energy production). The natural-uranium-based power reactors, as with the natural-uranium production reactors, can somewhat more readily be used to produce weapons-grade plutonium although with large economic penalties relative to power-only operation. Table V shows the relative strategic cost and timing characteristics of production reactors, power reactors, and the main supporting fuel cycle services (uranium enrichment and fuel reprocessing).

4.1. The Plutonium Connection

Power reactors of all types produce some plutonium as a by-product. This plutonium, typically about one part in 200, is intimately mixed with uranium oxide and fission products in the used fuel which is discharged after about 3 years' exposure in a (light-water-type) power reactor. This fuel, initially, is intensely radioactive so it can only be handled with remote handling devices behind 3 to 5 feet of dense concrete shielding, or under 12 to 20 feet of water.

The plutonium in the fuel consists of a mixture of four different isotopes, ^{239}Pu and small amounts of ^{240}Pu, ^{241}Pu, and ^{242}Pu. In contrast, the weapons-

TABLE V
Strategic Characteristics of Power Reactors versus Production Reactors and Fuel Cycle Support of Each[a]

	Production reactors			Power reactors				
	Low temperature, natural uranium			High temperature, enriched uranium				
	Research reactor	Small production reactor	Minimum plutonium recovery plant	Reactor only	Fuel cycle reprocessing[b]	Low-enrichment plant[c]	High-enrichment plant[d]	
Cost ($ millions)	5–20	50–100	10–20	500–1000	150–500	200–400	20–100	
Time (years)	2–3	3–4	2–3	5–8	5–8	4–6	3–5	
Technology construction	Low and simple	Low and simple	Low and moderate	High and complex	High and complex	High and complex	High and complex	
Organization	Small	Medium	Medium	Large	Large	Medium	Medium	
Potential explosive devices/year	1–4	10–20	10–20	None, but diversion possible	30	None	2–20	
Time to double capacity (years)	2	3	2	5–8	4–6	4–6	3–5	
Detectability of clandestine units	Low	Low +	Low	High	High	Low to moderate	Low	
Attractiveness to get nuclear weapons option	High	Very high	Very high	Low; (moderate for D$_2$O reactors)	Medium (national), low (with international) safeguards	Low	High	

[a] From reference 2. [b] Most countries would prefer to buy reprocessing services rather than build costly small-capacity plants.
[c] Most countries still plan to buy enrichment services to supply power reactors rather than build, unless supply becomes uncertain.
[d] Not required for LWRs; required for military reactor fuel, HTGR fuel, or weapons-grade ^{235}U.

grade plutonium from a production reactor contains mainly ^{239}Pu, with only a small amount of ^{240}Pu.

The "reactor-grade" plutonium from power reactors can, in principle, be used to make nuclear explosives (although it has some disadvantages relative to weapons-grade plutonium from production reactors). This gives rise to the concern that the spent fuel—now stored at hundreds of reactors in over 20 countries—could be stolen, diverted, or seized by a government intent on making nuclear weapons.

In order to extract the plutonium from spent fuel, the fuel must be dissolved in nitric acid, and extracted with an organic solvent in a process with many stages. This is necessary to separate the small amount of plutonium from the much larger amounts of uranium and structural materials, and also to eliminate the intense radioactivity from the fission products in the fuel. After this reprocessing, the plutonium would undergo a series of metallurgical and mechanical manufacturing steps, and finally incorporated in a complex device involving sophisticated types and forms of special high explosives. Blanket or reflector shells are used to reduce the amount of material needed. Triggering devices and their associated electronics and safety systems complete a working "laboratory" device.

To convert the laboratory device to a practical weapon involves protective packaging to maintain operability under various environments, including the shock and vibration of transportation. Much greater engineering and manufacturing sophistication and facilities are required to miniaturize such devices, and to get the level of reliability desirable for military use.

Reactor-grade plutonium is not used by any of the present weapons countries because it has several disadvantages relative to weapons-grade plutonium or ^{235}U. For example:

- It has a high level of residual radiation (γ rays and neutrons) which makes handling more difficult.
- It produces a lower and less predictable energy yield, including the possibility of a "fizzle"—or very-low-yield explosion.
- The shelf life of devices made with reactor-grade material is relatively short, and leads to the possibility of inadvertent explosions resulting from radiation-induced deterioration of the chemical explosive triggers.

Despite these disadvantages, there is obvious need to provide the best attainable safeguards against diversion, theft, or seizure of stored spent fuel—especially abroad. Domestically, fuel storage facilities have extensive physical security measures, guard forces, and provision for detection and forceful reaction to attempted thefts.

While similar measures are applied in nonweapons countries, they do not necessarily deter the possibility of a national decision to seize spent fuel. The

Soviet Union, which has supplied reactors to most of its satellite countries including Cuba, minimizes this issue by requiring the return of spent fuel to the Soviet Union. There it may be stored, or reprocessed to provide fuel for breeder reactors, or some might be used for blending with weapons-grade material to extend the supply of weapons.

The U.S. has sidestepped this issue. It does impose requirements for prior approval by the U.S. for transfer or reprocessing of U.S.-supplied fuels. Approval is rarely granted, and slow in coming. However, the proportion of U.S.-supplied fuel overseas has dropped from nearly 100% in the early 1970s, to 35% currently, and declining, so transfer restrictions have less and less effect as time passes. Several countries are making agreements with China to accept their spent fuel. Return of U.S.-origin spent fuel to U.S. for storage at protected facilities would manifestly be much safer—in respect to future proliferation potential—than the present situation. Multinational or binational storage facilities have been recommended by several international studies, and are encouraged in principle by the U.S. However, such schemes have made little progress after several decades of discussions.

4.2. Attractive Nuisance Theories

The fact that some natural-uranium reactors (with graphite or heavy-water moderator) have been used incidentally for power generation, while having been intended primarily as weapons material production facilities, is the link between this type of nuclear power plant and nuclear weapons.

The salient matter is a feature of the NPT that amounts to a bargain struck between the existing nuclear weapons states and nonnuclear weapons (NNW) states. In return for foregoing the option of developing or acquiring nuclear weapons, the NNWs interpreted the NPT as a firm assurance that *all operations* connected with civilian nuclear power would be available or made available to them. And what the debate made clear was that this meant that NNWs would not be barred from the options of access to, or else providing themselves with any or all of the associated fuel cycle facilities. The reservation was that such facilities would come under the safeguards inspection of the International Atomic Energy Agency. Among the facilities specifically contemplated were the two that are also usable to produce weapons-grade materials, namely: uranium enrichment, and reactor fuel processing.

4.2.1. Theory 1: Nuclear Power as a Blind

Among the countries that were insistent on getting the right to operate fuel cycle facilities as a quid pro quo for the NPT pledge were West Germany and Japan. When the NPT was being negotiated, 25 years after the end of World

War II, these two defeated countries were finally reemerging as major industrial powers and exporters. Both were anticipating being able to compete in what was expected to be an expanding world market for nuclear power. And because America, England, and France were offering fuel cycle services package deals with reactors, they wanted to be able to do the same.

Suspicions that more than that was intended arose. Germany and Japan were among the countries that argued for minimum information to and minimum access by the inspectors of the International Atomic Energy Agency to the operations of their nuclear facilities. The reason given was their concern about protecting proprietary industrial process developments. This is commonly a way of life in many foreign countries. Business is conducted with a minimum of disclosure or publication of anything—financial or technical.

All of this led to intense negotiations about the allowable procedures for the IAEA safeguards that were to be applied under NPT agreements. The possibility was that nuclear power commerce was the cover for formerly militaristic-dominated countries to develop nuclear weapons secretly. Against this, the Russians concentrated on prohibiting enrichment or reprocessing capabilities in satellite countries (except perhaps in China, which initially had Russian support). These situations led some people, including some of those who were the earliest supporters of the NPT, to believe that the NPT was a mistake. This ultimately led to U.S. resistance to the sale of technology (by other countries as well as by the U.S.) which could lead to enrichment or isotope separation pilot plants being placed in several countries. This action, as was noted by the IAEA, was substantively a unilateral American abrogation of a major element of the quid pro quo of the NPT.

More than 10 years has passed since the NPT was originally negotiated. During that time, Germany and Japan have not only prospered, but have come to dominate many high-technology markets. They are full members of the community of nations, and their alliances with nuclear weapons states are so extensive that it may seem moot (to Western countries) whether or not they have their own weapons.

The chief targets of suspicions now are a number of industrializing countries that have not signed the NPT: Argentina, Brazil, India, Israel, South Africa, and Pakistan. The concern remains that under the guise of developing a complete nuclear power capability, countries can develop the capacity to produce nuclear weapons fairly quickly—not one at a time, but in the tens per year.

Argentina and South Africa have announced they have built and operated uranium enrichment facilities—nominally for low-enrichment fuel only. Pakistan is apparently developing such facilities. India has reprocessing capability in connection with its breeder projects. Israel has a secret, high-security site, Dimona, with suspected capability to assemble nuclear weapons.

Against these contingencies, safeguards can be effective for civilian power facilities—but not for secret military facilities. If a country has civilian nuclear power, it has a choice between establishing an entirely separate military production complex, or attempting to divert material from the nuclear industry. But, with a safeguarded industry, even an occasional surveillance would raise alarms if bulk fuel were seized, and would be likely to detect and raise alarms for any diversion that might be occurring on the scale of several bomb-equivalents per year. The relatively small cost of separate production facilities (compared with a modern large power plant) will generally be the preferred route. It also is much more quickly and cheaply expanded—to get higher production, and the resulting weapons materials are of higher quality.

Thus, although the scenario of diversion or seizure of civilian reactors has provided the scenario for many papers, books, and at least one movie, it is a limited and implausible threat. The real threat remains clandestine production, which is feasible with or without civilian nuclear reactors.

4.2.2. Theory 2: Nuclear Power as a Lure

In this scenario, a country originally has no intention of building nuclear weapons. It does install civilian nuclear power, however. It signs the NPT, agrees to IAEA inspections, and proceeds to build up a sizable nuclear power capacity. Such a country would initially contract with a major nuclear supplier country for its enriched fuel and for storage or reprocessing services on spent fuel. Up to this point, there is little opportunity and incentive for proliferation.

As the country's investment in nuclear power grows, however, a legitimate interest in having a nationally owned and operated enrichment or reprocessing facility could develop to ensure continuity of supply of fuel. To this end, pilot plants could be set up to gain experience in these operations. Such a plant might be a small bank of centrifuges, or a pilot-size reprocessing plant with a capacity for 50 to 100 tons of fuel per year.

It is now argued that the existence of such facilities would operate as a lure to changing the original policy against developing nuclear weapons. Previously, the country would be discouraged against taking this step not merely by the political factors that induced it to adhere to the NPT, but by the realization that any steps toward producing weapons could take a long time, during which time they could be put under the pressure of severe sanctions, including the threat of forceful international intervention. However, even a 50 ton/year reprocessing plant could extract enough plutonium from spent fuel to manufacture several dozens of bombs a year. Thus, a policy change could be quickly implemented (provided that the weapons technology and facilities were developed secretly). A *fait accompli* could be presented to the world. This, it is argued, makes such a *fait accompli* more likely.

This scenario is ingenious and might be a factor for one or two countries.

However, these are the same countries, in general, for which the "nuclear power as a blind" argument could also be operative for countries that are already among the technical or industrial leaders of the world, or are moving in that direction. For such countries, several clandestine routes are possible on an accelerated time scale, too. For the same reasons that make an entirely clandestine route preferable under the "blind" theory, it is also preferable under the "lure" theory.

Some of the circumstances which might lead to a decision to exploit a potential capability are worth considering. The most obvious one is to be able to react to a perceived or evident threat to national survival. Until multilateral sanctions against incipient weapons developers become generally applied and visibly effective, the need to be able to deter possible nuclear blackmail is a powerful motivation for development and maintaining a weapons-production option. Limited sanctions and diplomatic pressures have been applied in several cases—with apparent success to the extent that overt production and testing of nuclear explosives by added countries has not occurred in the last two decades.

4.2.3. Theory 3: Nuclear Power as a Cover

Civilian nuclear power or research has been viewed as providing some countries with a "cover" for acquiring at least *potential* technical capability for weapons. India used a "peaceful nuclear explosion" as the rationale for a test which could equally well be a preliminary weapons test. The U.S. and USSR both have carried on programs for "peaceful uses of nuclear explosions." Other countries have been suspected of using nuclear "research and development" facilities, including those ostensibly connected with nuclear power but not opened for inspection, for developing facilities that could be useful support facilities if the country were to embark on production of weapons.

Many of the scenarios linking civilian uses and weapons potential are far-fetched in practical reality because they neglect the greater availability and security advantages of the secret military routes, not dependent on any civilian-related capacities. There is, however, a political linkage which has been overlooked in most public discussions of the subject. That is, if civilian nuclear power becomes a significant fraction of a country's electric energy supply, then security of continuing fuel supply becomes important. To the extent to which the U.S. (or other suppliers such as France or Russia) is perceived to be unreliable, or excessively demanding in terms and conditions of supply, an incentive to build domestic fuel production capability is provided—at least as a backup. This has been used to justify R & D facilities, and pilot facilities for enrichment and/or reprocessing in several countries. Such efforts build up substantial national laboratories, their budgets, and large concentrations of highly trained people—scientists, engineers, technicians, and support staffs. Once the habit of allocating large national resources to such activities is established, the possibility arises of diverting small portions of those efforts to "option-keeping," i.e., the various technologies unrelated to civilian nuclear power which are required to make a nuclear explosive work.

The indiscriminate, slow, or excessively onerous terms of export controls have apparently provided some of the justification for establishing large-scale R & D facilities—and some pilot projects. Many international studies and conferences (notably the INFCE study involving more than 50 countries) have called out the vital importance of highly reliable, preferably multilateral supply arrangements. These should be pursued by the supplier countries—and especially the U.S.—to limit the motivation of bolstering the security of fuel supply, which can also lead to increased potential capabilities for making weapons.

5. SUBNATIONAL DIVERSION

The fourth issue under the general heading of "proliferation" is the potential for access of terrorists or subnational groups to nuclear weapons materials or to nuclear weapons themselves. The hazards of concern here include: the demented idealist; the gangster; the politicized gangster—such as the various nationally sponsored and international terrorist "armies." Further concerns include guerilla attacks of substantial force which could arise either from terrorist groups, by subnational coups, civil wars, or covert acts of terrorism or war sponsored by a foreign power. Finally, there is the possibility of "unauthorized use" which is the military euphemism for potential diversion and use of nuclear weapons or weapons materials by people normally authorized to have control of such materials, but who have become persuaded that national command and control apparatus is secondary to their personal convictions.

Various aspects of the terrorist threat have been a favored theme of the entertainment media, and are used as part of the scare tactics of antitechnology and antienergy groups. "Fear sells"—papers, books, lectures, and theatre. The various elements of the terrorist threat have been analyzed and managed for a very large tonnage of weapons materials and weapons in the U.S. for over 30 years, in the USSR for 28 years, and in British and France for over 20 years each. The safeguards measures which have been so effective have been combinations of: physical security, alarms and instruments for detecting attempted diversion, and an extensive command and control system which can provide indefinitely large and timely response to attempted diversion. As all of these operations are kept secret in each of the countries possessing nuclear weapons, the functioning of the safeguards systems has been largely invisible to the public. However, the very large scale and effectiveness of this effort can be discerned on a circumstantial basis. There are several hundred locations of establishments which handle nuclear weapons or nuclear weapons materials in substantial quantities. Several such locations are present in all but two or three states of the U.S. From the published statements on the number of tactical and strategic weapons deployed in the U.S., one may conclude that on the *average* each of these locations contains over 100 nuclear weapons. In several locations, much larger quantities of weapons materials are stored or processed.

The safeguarding apparatus is not limited to the U.S. Weapons are deployed by the U.S. in various regions of the world, in several hundred locations, averaging somewhat less than 100 per location.

The bases, airfields, storage and deployment areas impose a set of safeguards tasks overseas which are in some respects much greater than the task of safeguarding materials located with the U.S. There is a continuing debate within Congress and the military on the wisdom of stationing large numbers of land-based weapons in other countries.

As a basic element of national defense policy, the U.S. has accepted the necessity of providing extremely good protection and safeguards for very large amounts of nuclear weapons material and nuclear weapons. There is also a very large flow of such materials across the U.S., and the face of the earth generally—which is also continuously safeguarded. There is no growth of civilian nuclear power in the foreseeable future which would require adding even 1% to the flows of "weapons-usable" materials which already exist.

The obligation of safeguarding very large numbers of nuclear weapons and large flows of weapons-usable materials has been accepted politically for two reasons. First, the hazard of not having such forces appears to bring with it risks of both survival and/or domination which appear to be far greater than the unavoidable risks of safeguarding the weapons. A risk-versus-risk judgment is clearly operative here. Second, despite a large number of seeming "close calls" in which nuclear weapons were accidentally dropped, or seemingly lost, the safeguards enterprise has been successful in each of the weapons countries. There is no evidence that any of the agencies involved, the military, the Department of Energy, the Nuclear Regulatory Commission, the Environmental Protection Agency, the U.S. Congress, or the Judiciary, has pressed serious concerns in recent years that the safeguards risks associated with nuclear weapons are unacceptable or that military safeguards need general upgrading. (There have been some concerns expressed over the security level at some test sites and security measures were upgraded generally after the embassy bombing in Lebanon.)

Safeguarding against Diversion in a Civilian Industry

The question at issue is clearly not whether effective security can prevent diversion of nuclear weapons material. This has been accomplished for over three decades. The question is whether we will have the will and the capability to implement appropriate levels of security on the sensitive parts of a civilian industrial enterprise.

The structure of the British and French nuclear industries is such that all steps of the nuclear fuel cycle in which pure or concentrated fissile material is handled are run by the government, under security equivalent to military protection. The U.S. also has retained federal control of enrichment technology production and security. The operations performed at U.S. enrichment plants are done by industrial contractors much as they would be if they were private industry, but on federal reservations with military protection. They are secure, and there has been no invasion of civil liberties in achieving security. Thus, the issue of safeguarding the nuclear fuel cycle industry in the U.S. is reduced to questions of safeguards for reprocessing, which was for a time assigned to the private sector but which is now uneconomic for civilian use. (Large-scale military reprocessing continues, however, in the U.S. and other weapons countries, and commercial reprocessing is available in France and Britain.)

The various measures that are available to protect fissile material at the reprocessing plant are all incremental to, or supportive of, ordinary measures of physical security. Diversion-resistant processes are discussed in Chapter 12. Such processes are not in use now, since only government-owned reprocessing facilities are in operation, but they are developable if needed eventually.

6. DISCUSSION

The limitation of nuclear weapons is universally desired. However, the wide variety of means available for producing weapons materials makes it impossible, by technical measures alone, to prevent additional countries from building them if they want to do so. Therefore, the slowing or prevention of horizontal proliferation must involve reducing the incentives and increasing the penalties for countries to have their own nuclear arsenals. The absence of added nuclear weapons states in the last 20 years—despite evident widespread technical and material capabilities—appears to be due ot several main factors:

1. The "nuclear umbrellas" of the superpowers provision to many countries of the equivalent of a nuclear deterrent force against the principal evident threats so that they do not need their own nuclear bombs.
2. A growing recognition that regions harboring nuclear weapons are more likely to become targets.
3. The evident resolve of most present supplier countries to apply economic sanctions, as well as to cut off the supply of nuclear fuel materials, from countries in which weapons production activities are detected or strongly suspected.

4. The work of institutions such as IAEA Safeguards Inspections, NPT, and Pugwash conferences to spotlight weapons tendencies and to provide institutional and moral pressure against proliferation.
5. The explicit tradeoff of foreswearing nuclear weapons production—in return for technical, material, and financial support of civilian nuclear power. This tradeoff has been advantageous for the large majority of countries that have chosen to accept this bargain—as long as it is perceived to be effective in helping to meet their energy needs.

These restraint factors have been effective in limiting national decisions to embark on weapons production—even with the temptations of evident technical capability and availability of materials. None of these restraints would necessarily be very influential for subnational factions or for terrorist groups, who might seek to acquire nuclear materials for use for blackmail as a means to gain power or wealth, or who might hope to benefit from the chaos and fear following an explosion. Preventing subnational factions or terrorists from getting materials for nuclear weapons, or actual bombs, clearly depends on continued very strong physical security— backed up by adequate forces with the capability for timely detection and reaction to attempts at overt theft, seizure, or covert diversion of material.

A continuing world problem is the vertical proliferation of the nuclear weapons and missiles of the superpowers. The Damoclean Sword of the existing weapons arsenals threatens all civilization. Yet, mutual nuclear deterrence has been effective for over 35 years. As yet, no alternative policies have been proposed which appear likely to be equally effective—or to be adopted. A nuclear "freeze," while popular, would not change the basic balance-of-forces aspect of deterrence.

The very evident potential for added countries to acquire self-produced nuclear weapons raises the obvious concern that regional nuclear arms races will develop. These would further complicate the strategic balance of power and make the problems of a verifiable freeze or a testing moratorium even more difficult.

The limitation and reversal of vertical, as well as horizontal, proliferation commends the most strenuous possible efforts of political leaders of the world. The NPT signed and ratified by the U.S. and USSR requires this, and our global future makes it imperative. As with other proliferation issues, this matter is being addressed—however slowly—by existing institutions, and we can only support the acceleration of those constructive efforts.

The blurring of distinctions between nuclear weapons and civilian reactors—which is evident in many antinuclear statements—is counterproductive for weapons control. It diverts attention from the real problems. It lessens the usefulness of civilian nuclear power domestically, and it has led to policies that increase the risks of horizontal nuclear proliferation.

REFERENCES

1. Joint Committee Print, U.S. Senate, Congressional Research Service, Library of Congress, *Nuclear Proliferation Fact Book,* U.S. Government Printing Office, Washington (Sept. 1977).
2. C. Starr and E. Zebroski, Nuclear power in weapons proliferation: The thin link, *Proc. Am. Power Conf. 39,* 26 (1977).
3. Joint Report on Uranium Resources, OECD and IAEA (Dec. 1975).

12 Paths to a World with More Reliable Nuclear Safeguards

E. L. Zebroski

1. INTRODUCTION

The possibility of further growth in the number of countries having overt nuclear weapons capability calls for reexamination of the role of security measures—often called "nuclear materials safeguards." The effectiveness of safeguards in reducing proliferation risks depends on timely application of safeguards as well as the specific attributes of the national and international safeguards systems and procedures. Effectiveness also depends on which specific aspects of "proliferation" are actually amenable to effective international safeguards measures. The issue of proliferation of nuclear weapons and its relationship to nuclear energy use involves several basic topics:

- In the context that "proliferation" encompasses at least four different aspects that may require distinct approaches and policies (see Chapter 11)
- In the context of the regulatory process, by which the perceived risks to society are managed
- In the context of the realistic options and objectives for a realistically attainable world nuclear control structure

The unreliability of the U.S. as a supplier of nuclear fuel and equipment has evidently stimulated in many countries an increased rate of educational activities, developmental efforts, and the installation of added basic scientific and engineering capabilities—as well as equipment fabrication facilities involved in nuclear power technologies [1, 2]. This effect is evident in about 20 of the

---

E. L. ZEBROSKI • Electric Power Research Institute, Palo Alto, California 94303.

44 nonnuclear weapons countries that have set up substantial nuclear research and development establishments. There is clear circumstantial evidence that many countries have acquired or developed the material, technical, and industrial skills required to have a contingency capability to make nuclear weapons. When and if a threat arises that appears to make this course of action necessary, the main question is how rapidly the potential capability is transformed into actual production.

2. FUTURE WORLD OPTIONS WITH RESPECT TO SAFEGUARDS

The third basic topic is the perception of the range of world options, and the extent and means for possible U.S. influence, to help achieve a less dangerous future. Three basic options for the future are definable:

- Option I: The Structured Nuclear World. A politically and technically well-structured nuclear world implies widespread (but not necessarily universal) acceptance and implementation of "full-scope" safeguards, reinforced by incentives of secure and economic energy supplies; with a minimum number of imminent nuclear weapons capabilities in evidence; a minimum number of nuclear weapons states; and a decreased rate (and eventually extent) of deployment of nuclear weapons by the weapons states. While not "perfectly safe," this is the safest credibly attainable option in this century. It will more likely involve several regional safeguards blocs, rather than a worldwide system, although the latter can be a useful "umbrella" for regional blocs.
- Option II: The Unstructured Nuclear World. This option implies increasing rates, types, and numbers of indigenous nuclear developments; an increasing number of states with overt, covert, or imminent capability of producing nuclear weapons. Increasing deployment of nuclear weapons and delivery systems by the weapons states also continues. Effective international safeguards systems are "full-scope" only in client states and even then only to the extent possible to enforce by control of enriched uranium supply. Such control becomes increasingly subject to being bypassed by the further development of improved production technologies, and by more supplier countries for enriched uranium, and by new, relatively small-scale methods of isotope separation. Also evident is the increasing commitment of natural-uranium reactors (heavy-water or graphite reactors). Even if these are nominally used for electricity production, they also bring the capability for production of weapons-grade plutonium.
- Option III: The Alternative Energy World. This is a hypothetical nonnuclear world in which the numbers of both civilian electric power and

military production reactors diminish. The amounts of nuclear fuel materials in use and production diminish. The scientific principles for nuclear weapons and nuclear weapons production technology are somehow forgotten and banished worldwide. Sources of alternative, low-cost energy are rapidly developed and installed worldwide. The increasing supply of alternative energy permits reduction in the use of nuclear power without increasing world tensions or local or major conflicts over limited energy supplies.

The Drift toward Option II

As seen from overseas, U.S. nuclear policy seems uncertain whether to work to implement Option I, hope for Option III without any actual plan or evidence of feasibility, or to continue the drift toward Option II. This uncertainty in approach, if continued, contributes directly to the most dangerous future, namely Option II.

Even those who adhere to anti-energy and anti-nuclear ideologies rarely advocate Option III explicitly, since for this century it has evident utopian elements. The *timely* availability of alternative, low-cost energy sources is highly problematical both technically and in terms of investment required. The continued pursuit of Option III as the implicit goal of unilateral U.S. policy, however unlikely it seems when stated explicitly, has appeal to some. Such a course in the near-term appears rather risky even for the U.S.—which is richer in fossil fuel resources than most other countries. Attempting to mandate Option III for other countries—friendly, neutral, or hostile—appears to parallel the historical difficulties of using unilateral disarmament as a means of getting all others to disarm.

The strategy of denial of practical steps toward Option I, by advocating delay—limited or indefinite—is often proposed with the implicit but unconsidered assumption that this will lead to Option III. Lack of assured "perfect safety" in available safeguards measures is one of the reasons often used for denial or delay of steps toward Option I. Such delays, however sincerely motivated, must carry the burden of responsibility leading with increasing speed to Option II.

The policy of inhibiting the full and effective use of nuclear energy by the U.S. and its allies contributes to maintaining the degree of dependence on imported oil, and the continued dependence of both Western and Third World economies on such oil. The increased risks of massive economic disruption and conflicts over limited supplies of oil are acutely evident to most Western countries and Japan. The slow pace of action in the U.S. in deploying its available strengths in energy production, both nuclear and fossil, is puzzling to non-Americans. Increased margins of domestic energy capacity could reduce such risks not only for the U.S., but also for other countries that are more completely dependent on oil. The temporary decline in U.S. oil consumption and imports in 1980–1983 is due partly to improved efficiency of the most wasteful uses (transportation and domestic heating) but also to recession in many kinds of productive activities. Neither of these "solutions" can be legitimately offered to most countries, which are *heavily* oil-dependent and have limited domestic energy resources.

3. PATHS TO IMPROVED SAFEGUARDS SYSTEMS

The evident limitations of existing international systems for control of nuclear materials in nonweapons countries has led to a needed rethinking of measures to improve such systems towards "full-scope" safeguards. The Ford and Carter announcements (September 1976 and April 1977) were attempts of "consciousness raising" for the world nuclear and diplomatic communities.

The Nuclear Non-Proliferation Act of 1978 (NNPA) intensified U.S. efforts on the control of exports, ostensibly as a further step toward Option I. The NNPA was paralleled by a two-year international review of nonproliferation options. Fifty-six countries and five international agencies paticipated in the International Fuel Cycle Evaluation Study. This study politely but decisively rejected the U.S. policy of discouraging and attempting to deny other countries the full use of civilian nuclear energy. The final report essentially ignores the thrust of most U.S. proposals, concepts, and procedures as embodied in the NNPA [2].

The NNPA and other positions of the Carter Administration were perceived by many countries as the unilateral abrogation of essential parts of the Non-Proliferation Treaty (NPT) which the U.S. had promoted and signed a decade earlier. The NPT calls for the nuclear weapons states to assist the nonweapons states in developing civilian nuclear power in return for their pledge not to develop nuclear armaments, and to accept international inspections of their facilities to verify that the pledge is being observed.

The NNPA, in contrast, calls for discouraging nuclear energy development abroad wherever possible, providing assistance in developing solar energy and other alternate energy sources, and enforcing this view mainly by restricting the export of civilian nuclear power equipment, components, and fuel, and requiring prior U.S. approval for reprocessing or retransfers of U.S.-supplied fuel or technology to third world countries. These requirements have also been used to delay or block the sale of nuclear power equipment or engineering abroad. The effect of NNPA on the steps needed for increased international cooperation on actual safeguards implementation is clearly negative. It has spurred the growth of developmental capabilities in many countries, investment in and construction of pilot-scale production facilities, and in a few cases, production-scale facilities for enrichment or for reprocessing. It remains a major obstacle to the types of cooperation which can minimize the spread of "option keeping" and which can best deter covert production of weapons.

The effect of the denial-by-delay approach of indefinitely restudying and improving safeguards with the aim of near-pefection amounts to indefinite delay in actual improvements in safeguards, and thus increased risk of continued proliferation of weapons capabilities.

There is an evident constituency for a low-energy, deindustrialized, low-productivity society. The path to deindustrialization has been idealistic but has used the tactics of denial-by-delay through litigation, regulation, and legislation. The method of denial-by-delay has been made highly effective by legislative and regulatory tradition of neglecting the time–cost effects of delays even if the issues raised are found to be without merit. This amounts to condoning of the "unlimited growth of due process." The handbooks of the anti-energy and anti-nuclear lobbies extol the methods of denial-by-delay that are described as the most general and basic tactics available. This approach has developed a large constituency in legislative, regulatory, and legal communities as well as among "activists." This tactic avoids confrontation—and responsibility—for the large and yet dimly perceived consequences that are being forced on society by policies and practices that lead to low-energy, and in some respects, low-productivity future. These lead there, not necessarily by stated political will and intent of a majority, but merely as a by-product of the new tradition of condoning increasing delays of many kinds of productive activities.

The tactic of denial-by-delay has become a normal and accepted part of the American scene. Its ultimate impact will be determined by the rate and extent to which public perception develops that inadvertent deindustrialization by delay may have unacceptable personal and social costs.

In an international context, the tactic of denial-by-delay is dangerous to the prospects for improving safeguards and slowing down weapons proliferation. The denial-by-delay of productive activities is seen by much of the world as the basic cause of the American decline, both economically in an increasing number of industries and also as a stabilizing force in world nuclear affairs.

The extension of the domestic pattern of denial-by-delay to international energy matters is especially dangerous to the improvement of international safeguards for two reasons. First, regardless of the quality of judgments involved and sincerity of intent, it will be subject to test. For example, the inevitable test of history will be whether international discussions such as INFCE lead to better and more widespread safeguards, or whether they simply were a continuation of indefinite denial-by-delay for constructive national and international actions by the U.S. Second, even if the sincerity of constructive intent prevails and is fully credible, the pace of tangible constructive actions is determined mainly by developments that are not subject to unilateral control by the U.S. The habit of delay in domestic matters tends to preclude timely U.S. contributions to the development and acceptance of many desirable features of international safeguards systems.

The recent pauses in international developments of effective safeguards are partly due to U.S. passivity, and probably also to the greater drama of superpower negotiations or posturing on missiles and weapons limitations and nuclear freeze. Meanwhile, some added procedural safeguards are developing—largely unilaterally—by added requirements and controls on enriched uranium supplied by the U.S. However, the rapid growth of alternate production capacity overseas is such that *over two-thirds of the supply of enriched uranium abroad is already not subject to unilateral U.S. controls* [4]. Such materials will be subject to a diversity of bilateral control managements by the U.S., France, the USSR, and by some general umbrella of controls under IAEA auspices. Small but increasing amounts will be produced in nonweapons countries, not subject to external agreements.

Many of the arrangements which have led to a minor role for the U.S. in the future supply of enriched uranium fuels were already in discussion and contract stage in the early and mid-70s—because of the lengthy lead-times for, and start up or expansion of, uranium enrichment capacity. The pause in new supply arrangements by the U.S. associated with NNPA and INFCE have not contributed to the implementation of better safeguards. Procedurally, the U.S. seems to be sulking because allied, neutral, Third World, and Soviet-bloc countries have continued to develop and explore added nuclear activities despite official displeasure of the U.S. Practical actions, to implement better systems with evident mutual benefits to buyer countries, are possible and needed. However, such steps are unlikely until the U.S. accepts the reality that further enrichment, reprocessing, and breeder developments are proceeding despite U.S. objections.

This would require a considerable shifting in approach (by ACDA, NRC, State, DOE, and Congress) from the preoccupation with denials or delays and with unilateral controls enforced by the U.S. These have resulted in simple bypassing of the U.S. by advanced allied countries, and by clients in Third World countries, who switch to alternative sources of supply. The now entrenched domestic tradition of denial-by-delay of any nuclear activities resists even constructive changes in approach. This situation, if not corrected, is one of the most dangerous obstacles in the path toward a safer nuclear future.

4. MODIFIED FUEL CYCLES—INTERIM SAFEGUARDS

The near-term technical focus of U.S. safeguards development in the fuel cycle has been to initiate expansion of spent-fuel storage capacity and to provide for sufficient enrichment capacity to help avoid the need for near-term fuel reprocessing in the U.S. The mid- to long-term focus has been on alternate fuel management schemes, and other measures which cut off most future prospects for growth of U.S. nuclear capacity. There are continuing discussions but no progress on alternate reactor types. These circumstances also rationalize a delay in the timing of need for breeders and reprocessing in the U.S., and a delay in the domestic need to develop more "proliferation-resistant" cycles which include reprocessing.

While the U.S. can reasonably delay the deployment of breeders and reprocessing due to abundant uranium and fossil fuel reserves, many other countries do not have such reserves. Breeder development, including construction projects, is proceeding in seven industrialized countries, paced by France and the USSR. Pilot scale reprocessing is also under way in at least three countries, and industrial-scale reprocessing continues to expand in the USSR, France, and Britain. The U.S. meanwhile has largely withdrawn from breeder development, and continues reprocessing only for military purposes. In doing so, it has retreated from the mainstream of world nuclear energy developments with corresponding loss of credibility and influence on the course of such developments. There continues some academic interest in alternate reactor designs and alternate "more resistant" fuel cycles, but no prospect of practical uses.

The alternate cycles all share the defect that the beneficial impacts on cumulative resource requirements or on diversion resistance are virtually negligible in this century—and limited in the early part of the next century. The combination of some breeders in the 1990s together with continued growth of water reactors appears to be a prudent pattern for many countries, as well as for conservation of U.S. resources. This holds over a wide range of resource assumptions and growth assumptions [5].

There is a theoretical possibility of some contribution from advanced burner-type reactors, which can be designed to have high conversion ratios. The practical obstacles of large early investment required as well as many remaining risks in technical performance and licensing make this route interesting for R & D, but with significant use highly improbable for the next two decades [5].

The "proliferation resistance" aspect of alternate cycles (more precisely the diversion resistance) can be measured in relative terms by several methods including probabilistic estimates, and by taking inventory of the resources and skill required to produce significant diversion. One of the main results of such studies is that some "inherent" differences between different reactor types and fuel types can be analyzed in theory. These differences appear to be minor in terms of the differences in the demand on total national resources committed if there is a national decision to acquire nuclear weapons capability. For example, if there is a national decision to abrogate an NPT pledge,—by seizing spent fuel as a starting material—the overall national effort required is many times larger than the *differences* in technical effort required for different fuel cycles [6].

The inherent differences, furthermore, are not generic to different countries, nor are they stable in time. They are obviously different in each country, depending on the facilities and supporting technology available.

Possible Role of Thorium—Uranium Cycles

Thorium-using cycles have been proposed for over 30 years, but have not been used for compelling technical and economic reasons. One evident reason is the very minor market penetration which such cycles could have for several decades—relative to the conventional uranium cycles—even if the handicaps of higher user costs and added industrial investments were overcome. Still to be considered are possible special purpose uses in regions of the world which lack adequate indigenous energy reserves, but which are subject to frequent conflicts or terrorism. For such regions, reactor fuel cycles using thorium and low-enriched uranium have a theoretical advantage of relatively low rates of production of plutonium, but significant production of uranium-233. In principle, such cycles could be used in developing nonweapons countries to limit the effects of possible diversion or seizure of spent fuel. These cycles are not without merit for newly-starting civilian power projects in some countries. However, they provide only marginal benefit and they are highly unlikely to see any significant practical applications in this century for the following reasons:

- Higher costs, at least initially, will discourage use in competion with conventional fuel cycles while uranium costs are low.
- Thorium cycles would have negligible proliferation reduction benefit in countries which already have large stocks of spent fuel containing plutonium.
- Initial uses could not occur earlier than the 1990s. By that time, the advances in isotope separation methods which are occurring will make practical the recovery of uranium-233 from the excess of uranium-238. (U-233 is a product of thorium fuel cycles, and a potential weapons-usable material.)
- The political-diplomatic problem of "selling" or imposing an unusual fuel cycle appears insurmountable (unless volunteered by a developing country).
- The proposed use of thorium cycles in unstable regions that have substantial exposures to political turmoil, or war, and/or terrorism—in some cases might be acceptable to a client state (e.g., in the mid-East). However, the more general objections to any civilian nuclear projects in such regions—together with the evident options, motivations, and military environment which facilitates the bypassing of civilian safeguards—make this an extremely unlikely application.

5. NEED FOR ENHANCED DIVERSION RESISTANCE AND DETECTION FOR REPROCESSING

Developments of plutonium recycle and breeder cycles are mainly strategic at present—either to provide a hedge against possible interruptions in the supply of uranium, or as a hedge against the eventual depletion of low-cost uranium reserves.

Reprocessing of spent fuel makes possible more complete use of uranium reserves as an energy resource. Recycle in thermal reactors improves utilization of uranium in water-cooled reactors by 50–100%. Recycle in breeder reactors can theoretically improve utilization by *factors* of 10–50, greatly exceeding the energy worth of all known and potential fossil fuel reserves. Neither recycle nor breeder reactors are as yet economically competitive with conventional reactors and fuels, while uranium supply exceeds demand and prices are low.

Reprocessing plants, to be eventually economic, require a substantial scale of operation—serving 5000–10,000 megawatts of capacity as a minimum, preferably 30,000 or more megawatts. This makes the justification for installing reprocessing capactiy valid (eventually) only in countries with very large nuclear power programs. The reprocessing for countries with smaller nuclear power programs (say, less than 10,000–20,000 megawatts in the next 20 years) is thus likely to be done most economically using facilities in countries with larger nuclear power programs.

Reprocessing continues on a large scale in the five weapons countries. France, Britain, and the USSR employ reprocessing for civilian power, while China and the U.S. use it for military purposes only. Germany, Japan, India, Belgium, and Italy have built and operated large pilot projects. Many more countries have demonstrated laboratory-scale capability for reprocessing.

France, Britain, and the USSR currently provide reprocessing services to other countries. Eventually, regional reprocessing centers are likely—in which two or more countries join to provide a sufficient load to make a reprocessing plant economic. The large national centers—so far—are all in weapons countries, with established military security and physical protection adapted from preexisting military reprocessing programs.

For the five current reprocessing capabilities in nonweapons countries, international safeguards systems apply. Enhanced diversion detection and resistance can be provided in several ways for such facilities and for future national and regional facilities.

Large differences between alternate approaches to reprocessing can be found using different physical and procedural security and safeguards designs. The main "inherent" differences between reprocessing systems arise from whether or to what extent highly purified, weapons-divertable, fissile material is produced; the ease of detection of diversion; whether appreciable amounts of readily accessible fissile materials are in continuing inventory; and the degree of difficulty of added process steps required to derive directly usable weapons material.

Criteria for Enhanced Diversion Resistance for Reprocessing

The perceptions of needed enhancements have led to a set of criteria or attributes desired for diversion-resistant fuel cycles involving reprocessing and the associated process and physical plant designs [7] (Table 1). Such criteria in turn can be met by a variety of processes and plant designs of which the proposed CIVEX process is one tangible example [10] (Figs. 1, 2). A pyroprocessing system for breeder fuel operated for five years at the experimental breeder reactor in Idaho (EBR-2). This is another example of a process that meets such criteria. The main novel aspect of such processes are:

- The limited removal of fission products in reprocessing—about 99%—leading to the requirement that fuel fabrication and handling be done by remote handling equipment.
- The avoidance of separation of uranium and plutonium.

The EBR-2 facility produced—by remote handling—four full-core loads of fuel. The incremental cost of remote fabrication is due to the added capital cost and maintenance cost of the remotely operated fabrication equipment. Typically this would require more than a 100% increase over conventional fuel fabrication costs, and an increase of 10–20% in overall costs of the fuel cycle. For this reason, this approach is not at present attractive for the reprocessing being done in the weapons states, or in Germany, Japan, and India. The higher costs are partly offset by the smaller size and lower cost of the extraction equipment and facility. For breeder fuel, some degree of remote handling is required in any event. As the costs associated with occupational exposure allowances are in-

TABLE I
Criteria for Terrorist Proofing

1. No pure plutonium in storage.
2. No pure plutonium at any intermediate point.
3. No way to produce pure plutonium by simple process adjustment.
4. No way to produce pure plutonium without equipment modifications.
5. No way to carry out equipment modifications with facilities and components normally on site.
6. No way to carry out the required equipment modifications without plant decontamination or entry into extremely high radiation fields.
7. Length of time required for successful diversion should be such that adequate time is available for national and/or international responses to occur.
8. Any alternative proposed must be technically credible.

12 • Nuclear Safeguards

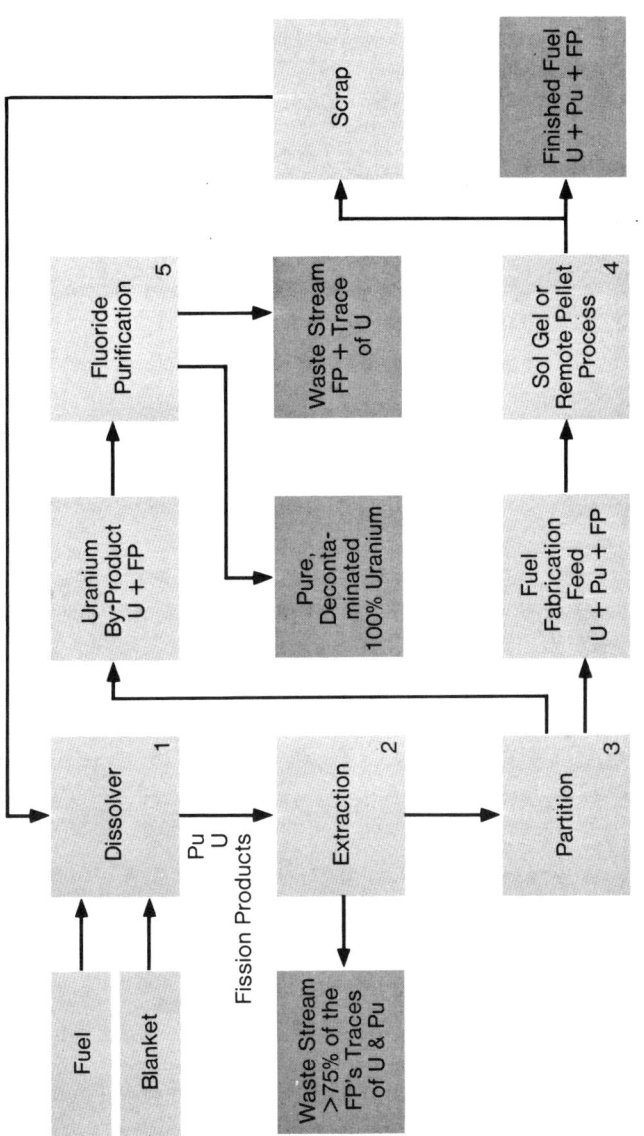

Figure 1. The CIVEX process for diversion-resistant reprocessing.

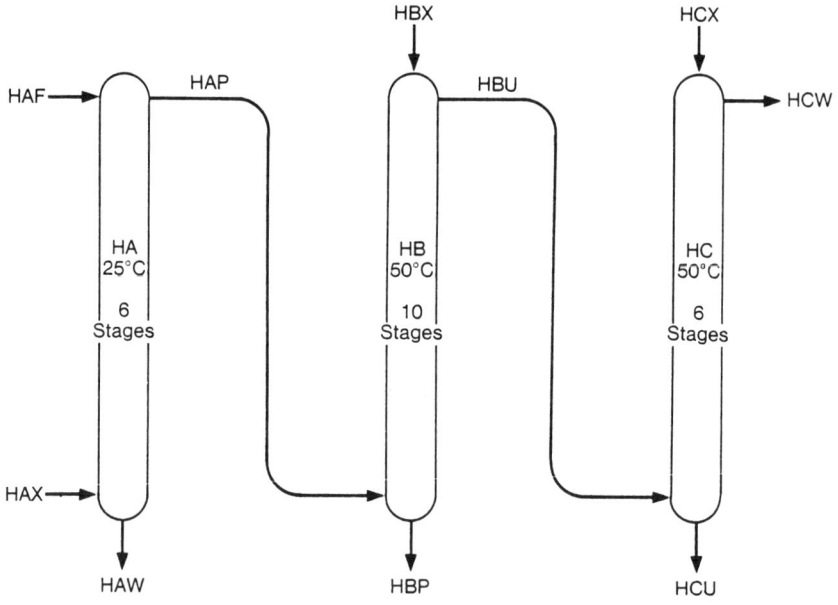

STREAM	RATE	UM	PuM	HNO$_3$ M	TBP
HAF	100	1.100	0.1590	2.000	—
HAX	350	—	—	—	30%
HAW	84	10^{-4}	10^{-4}	1.120	—
HAP	371	0.296	0.0429	0.286	30%
HBP	211	0.302	0.0754	0.584	—
HBX*	200	—	—	0.100	—
HBU	360	0.128	10^{-7}	0.025	30%
HCU	489	0.094	—	0.120	—
HCX	465	—	—	0.100	—
HCW	350	10^{-3}	10^{-4}	5×10^{-3}	30%

*This stream also contains $N_2NOH + H_2N_2$ reductant.

Figure 2. A typical reference flowsheet for CIVEX.

creasingly recognized, eventually there may be little or no cost penalty for remote operation. This can eventually become the cost-effective option, when credit for improved security in storage, shipping, and use is evaluated. From a utility viewpoint—and a national security viewpoint—whatever difference in cost exists may be acceptable if it is perceived to be a vital part of a system for assuring the security of the fuel cycle internationally and for unlocking the large increases in domestic and world energy resources that are available only with reprocessing.

The application of CIVEX-type approaches (coprocessing of uranium and plutonium, with fission product retention or radioactive "spiking," and embedded safeguards) to the reprocessing of light-water fuels, may have other added values. A desirable sequence of application was suggested by Starr and Marshall [7–9] (Fig. 3). During the period in which only the weapons countries, plus Germany and Japan, have sufficient numbers of reactors to warrant economic-size reprocessing plants, it is practical and desirable for these countries to provide reliable reprocessing services to other countries. CIVEX-type or radioactively-spiked fuel may be evaluated for the added security in international shipping of recycle fuel for conventional reactors or breeder startup fuel. When there is upwards of 30 GW of reactor capacity operating in a given region (outside the weapons countries, plus Germany and Japan), then regional reprocessing centers may become practical and desirable under international auspices, either regional or worldwide. Such auspices may advantageously include one or more of the weapons countries to provide the experience in the use of the most effective security and safeguards systems. The CIVEX process approach is not a panacea. It simply makes diversion more difficult—and much more readily detectable. It still requires a "structured nuclear world" to be widely applied and to be effective. (An obvious alternative is the continuation of government-operated reprocessing with unilateral military safeguards, but this is not fully satisfactory when exchanges of separated fissile material between several countries are involved.) A substantial degree of international concensus is needed to agree to the adoption of realistic diversion criteria for reprocessing and serious development of "embedded safeguards" for the back-end of the fuel cycle. These are analagous to the embedded safeguards being designed into the new enrichment processes [11]. Positive steps of this kind are essential if the credibility of the U.S. as a partner to world security and to energy-productive actions is to be sustained.

The ideological posture of expecting that reprocessing and breeder development will somehow disappear from the face of the earth was at odds with reality in the late 1970s and is even more so in the 1980s. The remnants of this posture are now viewed as merely quixotic, obstructionist, and ineffective by most of the civilized world. Most seriously, the remnants of this posture deny the possbility of constructive U.S. contributions to more effective international safeguards, which will require a substantial degree of international consensus to be adopted. In this sense, *the denial of reality reflected by some aspects of U.S. policy is a serious contribution to increased proliferation risks.*

The immediate result of denying or delaying a given export is clear, and is often politically attractive. However, we must also find a way to recognize responsibility for the mid-term and long-term effects of such denials, which tend to stimulate more rapid developments off-shore. Such trends may become inescapably evident only after current or recent decision-makers are gone from the scene. Timely steps are needed to help avoid continuation of a "Pontius Pilate

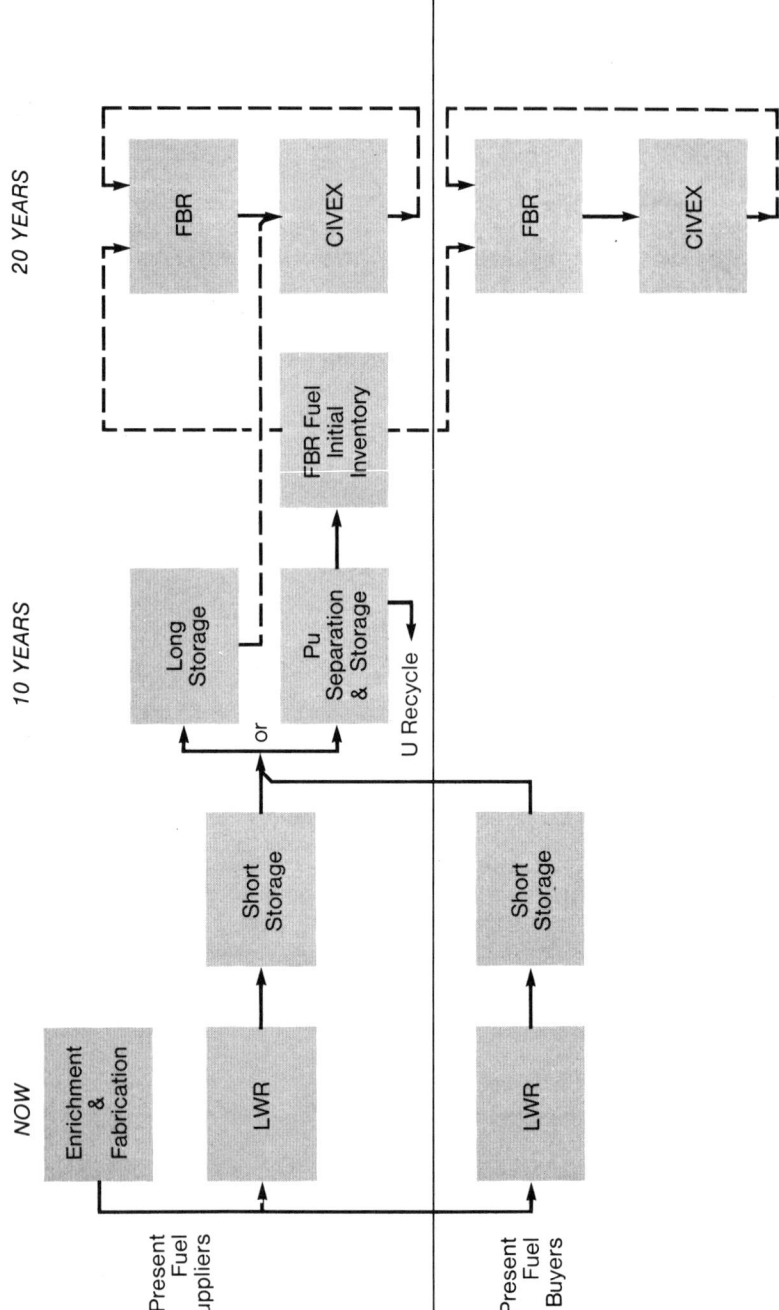

Figure 3. Proposed international nuclear power sequence. (LWR) Light-water reactors, (FBR) fast breeder reactors.

Policy, producing clean hands but disastrous overall effects" [11]. Such steps are needed if we are to find ways to a safer nuclear future.

REFERENCES

1. C. Starr and E. Zebroski, Nuclear power and weapons proliferation: The thin link, *Proc. Am. Power Conf. 39*, 77 (1977).
2. M. Levinson and E. Zebroski, Overview of the fuel cycle, *Proc. Am. Power Conf. 39*, 67 (1977).
3. International Fuel Cycle Evaluation Summary Report INFCE/PC/2/9, IAEA, Vienna (1980).
4. S. T. Brewer, Testimony on U.S. Uranium Enrichment Enterprise; for the subcommittee on Energy Conservation and Power; Committee on Energy and Commerce, U.S. House of Representatives (Oct. 21, 1983).
5. E. Zebroski and B. Sehgal, Goals: Near and mid-term opportunities for advanced reactor development, *Trans. Am. Nucl. Soc. 25*, 169–173 (1977).
6. T. Albert and E. Straker, Analysis of Proliferation Resistance of Alternative Fuel Cycles, SAI-77-872-LJ/F, Science Applications Inc., La Jolla, California (Dec. 1977).
7. C. Starr, Proliferation resistant technology, *Proceedings of the Fifth Energy Technology Conference,* Government Institutes Inc., Washington, D.C., p. 103 (1978).
8. W. Marshall, Nuclear Power and the Proliferation Issue, Graham Young Memorial Lecture, University of Glasgow, *Atom (London)* no. 258:78–104 (April 1978); also in *Phys. Technol.* 9:3, 115 (May 1978), also *Combustion 49:* 12 (June 1978).
9. R. H. Flowers, K. D. B. Johnson, J. H. Miles, and R. K. Webster, Possible long term options for the fast reactor plutonium fuel cycle, in: *Proceedings of the Fifth Energy Conference,* Government Institutes Inc., Washington, D.C., p. 256 (1978).
10. M. Levinson and E. Zebroski, A fast breeder system concept, in: *Proceedings of the Fifth Energy Technology Conference,* Government Institutes Inc., Washington, D.C., p. 230 (1978).
11. J. Nye, Nonproliferation: A long term strategy, *Foreign Affairs 62*(no.4), 618 (1978).

13 The Homemade Bomb Issue

Bernard I. Spinrad and E. L. Zebroski

If a terrorist gang or an equivalent group of malevolent psychopaths could get hold of, say, ten kilos of reactor-grade plutonium, could they mount a credible threat to society in the form of a hidden nuclear bomb?

This question brings to the hearth, so to speak, the issue of nuclear fuel reprocessing. Nuclear opponents have suggested that the theft of plutonium from a reprocessing plant would be a strong probability, and the manufacture of a several kiloton bomb from this material a straightforward matter. The nuclear community has maintained that both of these steps, material theft and bomb making, would be so different that the scenario becomes a virtual impossibility. What facts can we bring out so that we can intelligently understand this debate?

The scenario is a two-step one, involving both theft and bomb making. Of these two steps, theft from a reprocessing plant has been discussed in other papers here, and we need only summarize some points:

1. Plutonium as produced in a reprocessing plant is relatively easy to keep secure by a system of sealed vaults, guard stations, and well-enforced operating rules. Relative even to the commercial value of this plutonium, security is cheap.
2. Maintaining the security of plutonium is technically easier than keeping gems or precious metals secure because plutonium is radioactive. This both makes it riskier for a thief to handle and permits automatic surveillance of portals by radiation detectors, as a supplement to normal control practices.
3. Stealing plutonium from earlier stages in the reprocessing operation

---

BERNARD I. SPINRAD • Department of Nuclear Engineering, Iowa State University, Ames, Iowa 50011.   E. L. ZEBROSKI • Electric Power Research Institute, Palo Alto, California 94303.

means dealing with a dilute material that must be removed in bulk. For example, even after uranium and plutonium have been separated from fission products, light-water reactor (LWR) plutonium, i.e., the plutonium produced in power reactors of today, contains about 200 times as much uranium as plutonium. A 1 M solution of this material contains about 1.5 g of plutonium per liter, and it would take 7000 liters, about 1800 gallons, of solution to accumulate 10 kg of plutonium. Only at the "end of the line" is the plutonium concentrated enough to be attractive to a thief.

When all these factors are taken into account, military weapons makers use, not just any old plutonium-containing material, but a very special form, plutonium metal whose plutonium consists almost exclusively of the single fissile isotope ^{239}Pu. The metal is preferred for two reasons:

1. It is denser by almost a factor of 2 than any other plutonium compound, and the denser the material, the smaller the amount that is needed for a bomb.
2. Plutonium emits high-energy α particles that will react with the nuclei of elements such as oxygen to produce neutrons; plutonium compounds— or even impurities on plutonium metal—make "premature disassembly" likely, with consequent small explosive yield.

High isotopic purity is likewise called for for two reasons:

1. The presence of nuclei such as ^{240}Pu dilutes the primary fissile material and again increases bomb size for a given explosive yield.
2. The chief impurity isotope, ^{240}Pu, is a source of neutrons arising from its "spontaneous fission," and again could initiate premature disassembly.

Because of this second requirement, military plutonium is made in special reactors that are designed for only short-term irradiation of uranium and fast charge and discharge.

Commercial power reactors, on the other hand, are designed for long-term irradiation of their fuel and shutdowns for refueling are required only at intervals of 12–18 months. Unless the maximum energy is obtained from fuel that is loaded into a power reactor, the associated expenses of the fuel cycle (uranium purchase, enrichment, fabrication, conversion, and reprocessing charges) will add to the fuel cost. In addition, short fuel exposures call for frequent shutdowns for fuel unloading and replacement, and no power is generated during such periods. Therefore, the commercial incentives have all been in the one direction of even longer irradiations, or, in the vocabulary of the industry, "high burnup" of fuel.

A typical LWR will produce plutonium in its spent fuel that has the composition listed in Table I.

TABLE I
Isotopic Composition of LWR Plutonium[a]

Isotope	Percent of plutonium
^{239}Pu	58
^{240}Pu	23
^{241}Pu	14
^{242}Pu	5

[a] Source: Uranium Resource Utilization Improvements in the Once-Through PWR Fuel Cycle, Report CEND-380/COO-2426-199, Combustion Engineering, Inc., Windsor, Conn. (1980).

The higher isotopes build up by neutron capture in the plutonium that is formed initially. If plutonium were to be recycled, the higher isotope content would be even higher.

This sort of plutonium is quite undesirable for military use. The 23% content of ^{240}Pu provides a significant neutron source that increases the likelihood of predetonation. Finally, the 14% content of ^{241}Pu, a fissile isotope, decays with about a 13-year half-life into a nonfissile isotope, ^{241}Am (americium). This short half-life means that the bomb-maker must deal with a much more radioactive material than "clean" plutonium, and one whose radiation is much more penetrating besides. Additionally, the bomb would not have a long shelf-life.

The most obvious targets for would-be nuclear theft, with the purpose of acquiring bombs, are nuclear weapons themselves. U.S. weapons are stored or deployed in over two hundred locations in the U.S. and in several hundred more locations in Europe and Asia. Several hundred more locations are known or suspected in the Soviet Union. Theft—or even seriously threatening attempts—have not occurred so far as is known. Three elements account for this record: (1) physical barriers—such as vaults, bunkers, or fences; (2) sophisticated surveillance means—including guards and electronic detection and alarm systems; and (3) the availability of response forces, and secure, timely communications to alert and direct them.

The experience of bank vaults and fortifications (e.g., the Maginot Line) is that passive barriers and guards can be circumvented, given enough time and resources for attack. However, the prospect of timely detection and indefinitely large response and recovery forces poses an intractable problem for successful theft. Sensitive radiation detectors as well as conventional sensors can trigger alarms if there is unauthorized movement or attempted removal of nuclear material. The combination of these conditions accounts for the rarity of attempts at theft of nuclear weapons.

The favorable experience with preventing theft of actual weapons also can be brought to bear on the question of preventing theft from reprocessing facilities. Several reprocessing facilities have operated for over 40 years in the U.S., and some continue to operate for military production and for navy fuel. large-scale facilities also operate in the Soviet Union, France, and England, and smaller-scale facilities operate, or have operated, in Germany, Japan, Italy, Belgium, India, and China. Laboratory-scale facilities have been demonstrated in at least six more countries. What are the obstacles to theft from such facilities?

The same obstacles to theft are operative for reprocessing as for nuclear weapons. In each country, reprocessing has been done under the control of the central government—typically in a well guarded and protected facility, with response forces available. In addition, there is a formidable set of technical problems in converting a dilute and highly radioactive raw material to workable form.

The alleged simplicity of a nuclear bomb is true in theory only. Naive reporters periodically write "College Student Designs Nuclear Bomb"—for a naive public. The college student might well also design, in general principle and largely correct in *outline,* a high-power jet engine, a manned moon rocket, or a supercomputer. But he would pose very little threat to Pratt and Whitney, NASA, or IBM. Even with full access to the raw materials, he would never produce working devices, unless a much larger range of resources, knowledge, skills, and facilities were available.

A "crude" nuclear explosive requires several times more material than a highly developed and tested military weapon. It might be assembled with the level of knowledge, skills, and facilities available in a large modern technical university. However, it would be bulky (weighing a ton or more), unreliable (might not work), and uncertain in yield (anything from a blockbuster or town-buster to a fizzle). The process of making such a device would be very hazardous to its makers for several reasons. For one thing, specially shaped high explosives are required, and (unless very specific nuclear data, instruments and tests were available) nuclear criticality accidents would, be likely—not true explosions but lethal locally. Unless considerable resources and time were available, many elements of the systems involved would be inadequately tested, so failures would be very likely.

If only smaller amounts of bomb material are available, typical of the amounts used in military weapons, much higher levels of resources; facilities; data; and scientific, engineering, and technical or craft skills are needed. For the weapons countries this step has commonly taken a national-level effort of about three years duration by large teams of people with essentially unlimited access to resources, special materials, and equipment. Even then initial trials have sometimes failed or misfired.

13 • The Homemade Bomb Issue

At this point the reader seeking assurance of zero risk might say, "So a homemade bomb is not impossible after all." Few things in life are more unprovable than zero or infinitesimal risks. But there is some comfort in gaining a perspective on how small a risk is, relative to other risks. The following observations are offered, not as "proof," but as common-sense remarks based on reasonably good knowledge of the relevant information:

1. The successful diversion or theft of raw nuclear weapons material from protected national reprocessing facilities is much less likely than, say, theft of major stores of gold or large diamonds from vaults.
2. Crude explosive devices require fairly high levels of technical sophistication, resources, and facilities and a wide range of specialized materials and skills, both scientific and technological.
3. Crude devices are possible, but are bulky, heavy, unreliable, and very dangerous to make or to attempt to make.
4. True weapons (compact, light, reliable, with minimum amounts of bomb material), require several years of large-scale national levels of effort and resources, even for highly industrialized countries, and more extended time frames for less industrialized countries.
5. Clandestine subnational efforts are highly susceptible to detection and interception.
6. Subnational groups—whether criminals, terrorists, or seekers of political power—have many other, more accessible and swifter means, requiring much smaller resources, for mounting threats or causing mischief, sabotage, damage, or casualties.
7. Two scenarios remain possible and plausible for subnational or terrorist threats (neither of which actually involves a homemade bomb):
   - An apparently subnational group could be provided with nuclear explosives as a cat's-paw, that is, as a covert act of war or terrorism by a nuclear weapons state.
   - A terrorist or criminal individual or group could allege possession of a bomb (or enough material for a bomb), hoping for naivete and panic by officials to produce a payoff.

Some limits on the scenarios in (7) above are evident, if not entirely comforting. The source of a "leaked" device from a weapons state can be reliably identified—if it is exploded. This can tend to deter "leakage" as a tactic. In addition, nuclear bomb threats have occurred a number of times. Special teams, organized under the military applications side of the Department of Energy, are trained and equipped to evaluate and react appropriately to such threats.

IV Risk Assessment

Part IV of this volume is concerned with risks of nuclear power to people's health and safety. Probably no other topic of modern times has been subject to such gross misrepresentation, and the sin is at least 99% due to misinformation circulated by nuclear opponents. That we must continually point this out is distressing as well as distasteful, for all of us would rather spend our time discovering new facts and inventing improved devices, rather than struggling with distortions of information as popularized by skilled propagandists.

The first paper is "LWR Risk Assessment" by G. S. Lellouche. It is an expository paper; its purpose is to make it clear that the risk assessment process is a valid and unbiased one and that in particular the famous Rasmussen report produced robust results. By "robust" is meant that the conclusions are relatively insensitive to changes resulting from detailed criticism. There are improvements that could be made in the input data, but as many of them would tend to decrease the risk estimate as to increase it. It must therefore be considered to be a meaningful report.

The theme is amplified by T. G. Theofanous and R. Wilson in "Accident Analysis and Risk Assessment." Some of the more specific issues in reactor safety are addressed from the point of view of the analyst and designer. How do we improve our data; how do we apply it? The answer is that we must use engineering judgment, but based on knowledge of fundamental phenomena, tests on components and subsystems, and experience. When uncertainty exists, conservatism is used to cover it, and as our knowledge improves, we can provide even greater assurance of safety. Both uneventful operating experience and accidents add to such knowledge of safety.

The next paper is B. L. Cohen's "The Waste Disposal Risk." The relatively simple process of drilling deep holes and dropping nuclear waste cannisters into them has a completely trivial risk, either to us or to our remote descendants, even if we make the unlikely assumption that our descendants will be neither wiser nor more knowledgeable than ourselves. The paper explains why this is the case. In a follow-up paper, "Radon Problems," Cohen then examines a radiation problem that is actually more severe: the exposure of people to radon gas. It turns out that the use of nuclear power will diminish this risk, while the burning of coal and (particularly) the overzealous insulation and caulking of buildings very much increase it.

The final paper, also by B. L. Cohen, is "Risks in Our Society." This paper, abridged from lengthier essays by the author, demonstrates that risk reduction priorities must be concentrated on other aspects of life than industry, other industries than the energy industry, and other energy technologies than nuclear. It is in this sense that we can characterize nuclear energy as a benign technology.

Our message is not that nuclear energy is free of risks, but rather that its risks to health and safety have been grossly overstated, both descriptively, e.g., sensationalized headlines and unjustified adjectives, and in relation to risks that we accept routinely. Far from "poisoning the earth" as so many critics declaim, nuclear power adds to our own health and safety and leaves our children a safer and cleaner environment, as it displaces other energy alternatives.

B.I.S.

14 LWR Risk Assessment

Gerald S. Lellouche

1. INTRODUCTION

Reality is not free of risk. The world around us continually places individuals, groups, and nations in the position of making decisions, and decisions always entail a change in risk, either up or down. It is clear that reality does not impose high-risk (of death say) consequences on much of humanity with any particularly large probability since the human race would die out. In fact, as the birth rates are seldom greater than a few percent, the risk of death to the population cannot exceed that few percent for very long without the population disappearing. This implies, and it is backed up by simple observation, that very-large-consequence occurrences cannot, and indeed do not, have very large probabilities, while high-frequency events (those with large probabilities) must have, relatively speaking, small consequences.

Nuclear power plants, dams, and LNG terminals have been perceived by some to be particularly sensitive in the category of low-probability but large-consequence risks to the public. Over the years the design philosophy of nuclear power systems has matured to the rather elaborate concept of "defense-in-depth" based upon layered engineered safeguards (including both prevention and containment of postulated events). As each layer is added, portions of the overall risk are reduced, making certain events less likely. With this procedure two things follow: the likelihood of the events become smaller but the prediction of the reduced likelihood becomes more difficult.

The Reactor Safety Study (sometimes referred to as the Rasmussen report) attempted to establish the likelihood of the various events that could cause

---

GERALD S. LELLOUCHE • Nuclear Power Division, Electric Power Research Institute, Palo Alto, California 94303.

significant radioactive releases to the ex-plant environment and thereby establish a risk relation for nuclear power plants.

1.1. Overview

This essay will attempt to present a brief but reasonable description of what a risk assessment is and how it is prepared with emphasis on light-water reactor (LWR) analysis. We first assay a definition of risk. This is not simple, for the word is used in many ways. The next step is to describe what a risk assessment is and how it is performed. Along the way definitions and descriptions of fault and event trees will appear (almost in a natural way). The consequence of an event is clearly part of the risk, and its position in the calculation is described. The additional calculations necessary to quantify the consequences are briefly discussed; a subsequent essay by Theofanous and Wilson goes into the details of the calculations of the consequences; the final sections discuss the questions of adequacy, presentation, completeness, and believability.

1.2. Risk

What is risk? How is it determined? People will say things such as "that's pretty risky" or "I'll risk it." What they mean is they perceive that the consequence of an action is dangerous (pretty risky), or worth the reward (I'll risk it). Such statements usually relate to a single situation, or at least a relatively rare one (gamblers excluded). Even police and firemen know that while the work is "risky," certain precautions can be taken that may minimize the consequences (which can be deadly). In business, risks are also taken; the payoff is balanced against the possible loss and decisions are made to proceed or not. These types of risk are generally felt to be understood by those involved, but how about risks that may or may not come to pass? Aircraft wings are designed to allow for types of events that may never occur. This is done by designing to wing loads that are not expected to occur (although statistically loads exceeding the design level do have a very small probability of occurring). This kind of design is performed because the wings are desired to be very reliable. We also want the aircraft control systems to be reliable and the design is carried out (very often) by means of a probabilistic analysis. Such an analysis tells us what would happen if some particular portion (a transistor for example) should fail. In warplanes the systems must be able to take shock loads, but one cannot design to an arbitrary load (the plane cannot be made arbitrarily heavy), the loads are determined probabilistically, and the behavior of the system given that load is calculated (and often experimentally determined for comparison and "prove out"). The result of such a design is that one can determine that there is one chance in 100 (say) that given that load the system will fail. The consequences of such failure are varied (death, loss of mission, etc.). Such a consequence is the risk associated with that portion of the design.

There is a risk associated with every action, even not doing something carries its share of risk. How does one determine what the varied risks are for an enterprise or for a technology? It is clear that what is needed is a predictive analysis for risk. That is to say a probabilistic risk assessment. It has to be probabilistic because it is only those portions of the risk that have not happened yet that we need to analyze for and things that have not happened yet only have probabilities attached to them (not certainties).

1.3. Risk Assessment

"Risk assessment" may mean different things to different people. In our context we shall use the term to mean the frequency that an event may occur with consequences greater than some specified amount. Note that risk and risk assessment are not the same: a risk assessment determines or quantifies the frequencies of the various events and their consequences; the risk is the product of the frequency and the consequence.

The consequences may be stated in terms of: monetary cost, health effects, early deaths, delayed deaths; or any of a large number of other more neutral indicators such as: number of curies released, 30-day whole body dose, 50-year dose, etc. Normally, several consequence indicators will be used in order to provide further insight into the meaning of the risk.

Finally, the risk can itself be broken into two parts: the frequency of an event occurring times the consequences of that event. It is informative to consider the risk of events with a consequence greater than C_0:

Risk of events with consequences greater than C_0 = Sum of the product of the frequency of event E_i times consequences of event E_i for those events having consequences greater than C_0

Mathematically $= \Sigma_{E_i} f(E_i) C(E_i) \quad C > C_0$

From this relation we see that the "total" risk is the sum over all events having any consequences at all. Such a number describes the total impact of an LWR. Generally speaking, for any industry or life itself, events with small or no consequences are tolerable with a high frequency while events with high consequences must have very small frequency.

Risk assessment has several possible uses. It can guide design to a very small risk consistent with affordable costs (one can drive the risk of a product toward zero at very high cost but then no one would buy the product) and it can be used in licensing or technology assessment to help evaluate acceptability questions. The important thing to remember is that no risk assessment has meaning by itself as it is only in comparision that one can say whether a risk is large or small (i.e., if the total risk is 0.01 death per year, one cannot say that 0.01 is large or small in absolute terms; it is larger than 0.001 and smaller than 0.1 and much smaller than the 50,000 per year due to traffic accidents). The acceptability of a risk depends not only on the number but on how one perceives the risk; some people are averse to risks that others readily accept; some see a distributed risk of a specific size as allowable while they would not accept the same size risk at one time (1 death a year for 10 years may be acceptable but 10 deaths in a single event once in 10 years may not be). Such considerations are part of the social process of determining acceptability and are outside of our purview. Nonetheless the number gives an experienced assessor a "feel" to the risk and by isolating different portions of the risk one can intercompare them and make decisions based on such intercomparison. In a typical application of this, one would conclude that it does not make sense to try to reduce the consequences of events of type X if they comprise only 1% of the total risk; one should first wish to address the larger contributors.

Key characteristics of risk assessment are the following:

- Risk assessment is based on a probabilistic modeling of a system. The modeling is undertaken with a set of prespecified guidelines in mind.
- The consequence portion of the assessment is based on a model of reality that of necessity contains uncertainties.
- The data base from which failure rates, etc., are taken is incomplete.

At first glance it would seem that these three points should invalidate the usefulness of risk assessments, but such a conclusion would be inappropriate. Any decision process is *always* started and completed in the absence of "sufficient" data. If "sufficient" data existed, there would be no need for a prolonged decision process, for the risks and rewards would be documented; it is *precisely* because of the absence of such sufficiency that we need a risk assessment.

A risk assessment of an LWR, for example, uses as its source material the blueprints, bills of material, Q/A programs, etc., in fact, all information necessary to build (or rebuild) the system including any site-specific information. For each system one must determine the nature of the potential impact (dam bursting, chemical releases, natural gas explosions, sulfate emission, radiation release). Then one has to identify the potential initiating events that might ultimately lead to an impact (i.e., have untoward consequences). This step is an important one, for potentially some initiating event might not be identified. In fact, one can clearly state at the outset that some (perhaps many) initiators *will* be missed. We shall address why this should not be a serious problem below.

1.4. Event Trees

Having the initiating events, one must determine, from a knowledge of *how* the system works, which systems are *needed* to work (and in what order) in order to *eliminate* or *mitigate* the consequences of the initiator. This step permits the construction of the "event tree" for the progress of the possibly hazardous event. Thus, if following the occurrence of an event E we must have systems A, B, and C work,* in that order, to mitigate the consequences, then the event tree would look like

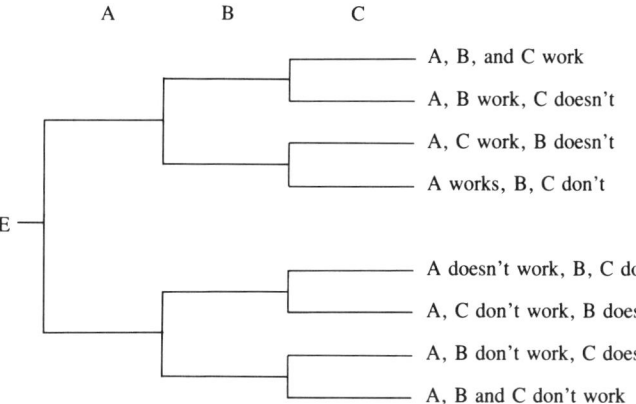

Several things should be noted here. If it is correct that A, B, and C, in that order, are the only systems needed to fully mitigate the consequences of the event, then the above diagram is a complete one, exhausting all possibilities. The probability of each of the eight branches depends on the three probabilities P_A, P_B, and P_C, which are the failure probabilities of A, B, and C, respectively. Thus, for example, the probability of the bottom branch (A, B, and C don't work) is $P_A P_B P_C$, while that of the top line is $(1 - P_A)(1 - P_B)(1 - P_C)$, for the probability of a system either not working or working must add to unity. It is important to note that the probability numbers that enter the calculation may depend on the state of the preceding events. Thus, the calculation for the failure

---

* For many initiators, only a single system (i.e., the scram system) need work to fully mitigate the progression.

probability of system C may depend on whether we assume that systems A and B are successful or not. This means that the probability of C failing is a conditional one, and if a model is constructed to synthesize the C system failure probability (by a fault tree for example), a different model may be needed depending on whether A and/or B have failed or not. In general, the numbers for the probability of each branch will be different and the overall system response will be different as well, going from full mitigation at the top line to least mitigation on the way to the bottom. In order to determine the degree of mitigation, one has to understand the general response of the system as it proceeds along which branch.

Based on the event trees, one uses deterministic dynamic models of system behavior (pressure/temperature response for example) to determine under what, if any, circumstances an initiator could lead to, for example, an overpressure condition. These questions are addressed in a subsequent essay. Such models are used to lead to an understanding of which engineering safeguard systems are important for mitigating the effects of, for example, the release of radiation. It is clearly important in determining any impact to know that the, in this case, radiation is released slowly or quickly (if it is slow enough, there may be no impact from a health viewpoint but there may still be land contamination). At this stage in the risk assessment we should have a comprehension of the following:

- The plant and site design and description
- The impacts one needs to consider
- The initiators that could lead to consequences of significance
- The systems necessary to mitigate any impact given an initiator occurs
- The event trees leading to less than perfect mitigation

1.5. Fault Trees

Assuming that we have completed the construction of the event trees, we now must determine the values for the failure probabilities of the various systems of importance. If there are data on the failure rates of these systems, then the determination of such probabilities is fairly straightforward. If not, one must synthesize the failure probability using probabilistic methods.* One general method for doing this is referred to as fault tree methodology. A fault tree is a logical diagram created from the blueprints, wiring schematics, operating and maintenance procedures, qualification procedures, piping layout, etc. of a system. It describes the connections between the parts of the system using, generally, the

---

* Clearly, for rare events that have not occurred yet, we cannot have actuarial data but most systems are made up of components for which failure data are available. The rare-event frequency is then determined by modeling the system behavior (by means of a logical diagram) in terms of the component behavior and using the component failure data for quantification.

logical concepts AND and OR. An AND means *all* of two or more events must occur for the next higher level event to occur. An OR means *any* of two or more events must occur for the next higher event to occur. Taking a simple case of a radio we have:

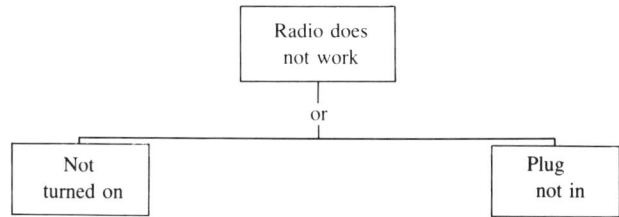

This is *not* a complete fault tree for the failure of a radio: there are many other OR possibilities (failure of diode 1, diode 2, etc.). Similarly, if a system has redundant circuits so that only one of several need be available to allow success, then we could have, for a car with two spare tires:

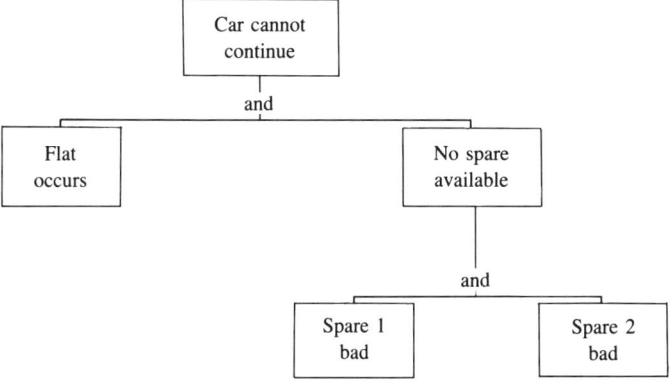

Basically, a fault tree requires deductive logic. The fault tree modeler asks himself numerous questions all of the form "what could cause X to occur?" The answers he gets are of the form "this *and* that," or "this *or* that." The modeler keeps moving to lower (more remote) levels of logic (events) until data are available for an event. The modeler could keep going beyond this point but no reason can generally be adduced for doing so. Building trees for the failure of systems A, B, and C permits the determination of P_A, P_B, and P_C. It must be pointed out that in some instances phenomena (such as an earthquake, or explosion) can occur that simultaneously fail multiple systems so that the failure of A, B, and C are no longer independent. In this case, the product $P_A P_B P_C$ is not the complete answer. This kind of problem is not intractable and can be handled by the same methods. If, for example, E is an earthquake, then we ask "given that E occurs, what are the chances that E itself (through shock waves) destroys the operability of A, B, or C?" More generally, we ask "given the failure of A, does the operating environment for B and/or C become so bad that they cannot function?" Such questions are answered in the normal process by considering what the initiators are, what conditions the systems (A, etc.) are qualified for, and what conditions occur during the transient.

At this point in the calculation we have all the information needed to define the time rate of release of radioactive material from the plant for a spectrum of initiating events*; we also have the probabilities of these releases. What remains now is to calculate the actual off-site impact.

2. CONSEQUENCE ANALYSIS

Calculating what happens outside the plant involves asking new questions. For example:

- What is the wind direction?
- How does it vary in time?
- Is there an inversion?
- Is it raining? Where?
- Where are the people?
- Do we need evacuation? In which direction?

If the radioactive material reaches populated areas, we have additional questions:

- How many people are at risk?
- How fast do they breathe?
- What are the somatic effects of radiation exposure?

To determine financial costs, we must consider property damage, land contamination, and cleanup models.

These questions are not trivial and there is controversy concerning the adequacy of a number of the assumptions used to answer them. Nonetheless, methods can be, and indeed have been, created to answer them and the final risk relationships can be calculated.

The Reactor Safety Study (WASH-1400) performed a risk analysis for two specific reactors and with some caveats estimated the risk for a population of 100 "similar" reactors. Figures 1 and 2 show the expected frequency of events

---

* The treatment of consequences is discussed in Chapter 15.

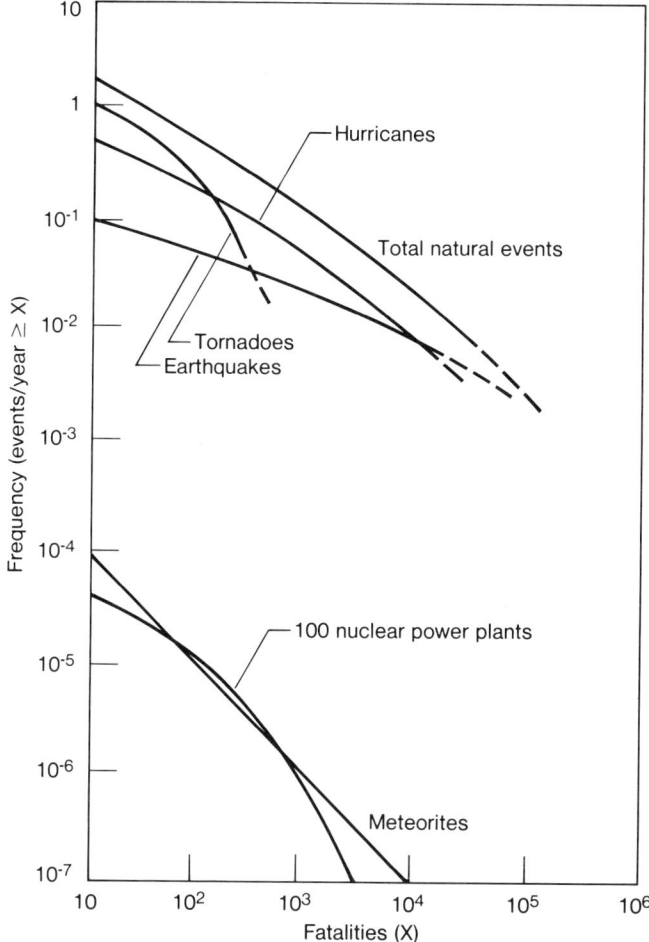

FIGURE 1. Frequency of natural events involving fatalities.

leading to consequences greater than some value X. Here we consider two indicators: off-site property damage expressed as dollars, and fatalities (here, the term means prompt, within 30 days, fatalities only). For comparison, there are also curves of property damage and fatalities due to natural and other human-caused events. We note here again that it is only by a comparison that a risk estimate assumes meaning. What is important here is that if the frequency were even 100 times larger, it would still not lead to fatalities comparable in size to

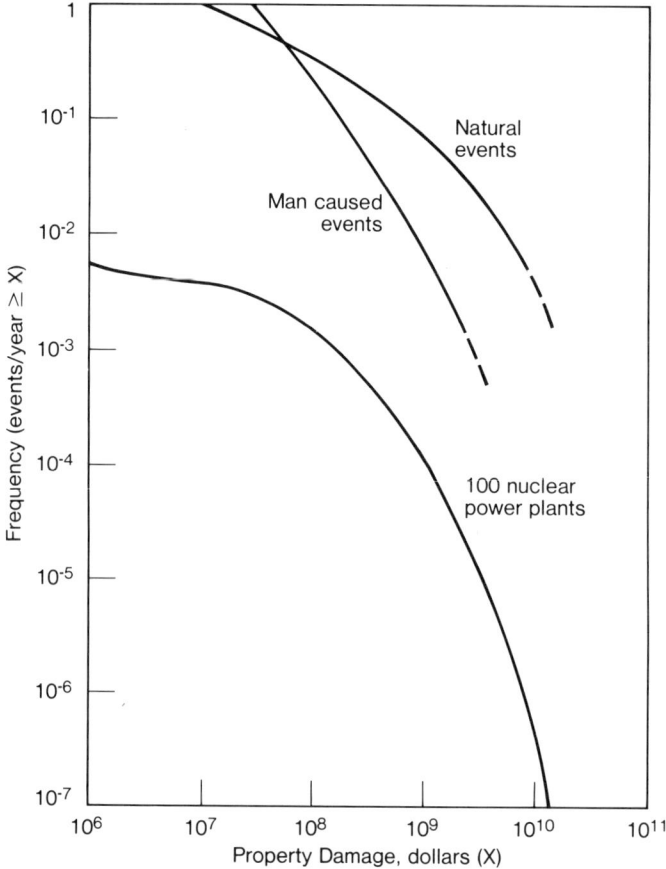

FIGURE 2. Frequency of accidents involving off-site property damage. Property damage due to auto accidents is not included because data are not available for low-probability events. Auto accidents cause about $15 billion damage each year.

natural phenomena. The property damage risk indicator shows that if the frequency were 100 times larger, LWR risk would start to become comparable with other human-caused events (excluding automobile events).* The question can then be asked "how far off may these assessments possibly be?"

* Figures 1 and 2 are of frequency versus consequence, but the frequency is of events with consequences *greater* than a given value. This frequency representation can be converted to a probability representation by normalizing by the total frequency of all events per year. The importance of this type of presentation is discussed in the section on completeness.

14 • LWR Risk Assessment

3. ADEQUACY OF THE RISK ANALYSIS

An analysis is performed for a purpose; the quality or adequacy of the answer can only be gauged with respect to that purpose. Thus, it is quite normal to ask for "a ballpark" figure, or "an order-of-magnitude" figure, or "an on-the-button" figure. When a great deal is known about a question, the answer will usually be quite accurate; as the specifics of the phenomena associated with the questions become less well known, the range of uncertainty for the answer becomes larger. It would appear from all of the comments made by qualified reviewers that the answers obtained in WASH-1400 are probably accurate to within an order of magnitude on the frequency and on the consequences. There is some indication that some of the consequence indicators may be off by somewhat more than an order of magnitude but this is not clear yet. What this means is that it is nearly as likely that the number could be higher or lower by as much as a factor of 10. A number of independent calculations made by EPRI, NRC, as well as the Lewis review and reviews by other generally qualified groups have shown that some aspects of WASH-1400 (such as the unavailability of the residual heat removal system of boiling water reactors) are considerably conservative, while on other aspects the various new studies have come to desperate conclusions: the breathing rate distribution of humans may have been underestimated (according to preliminary study). These newer studies do not tell everything because they do not answer the overall question of sufficiency or completeness.

Completeness*

It is important to discuss some of the limitations inherent in performing a risk assessment on a real system. We stress the word real as there is no question that probabilistic methods completely and correctly describe the expected behavior of things like dice or cards, at least of idealized models of dice and cards; idealized as the dice are not geometrically perfect and the card shuffling is not a perfect randomizer. But when the system is an LWR, a number of questions can arise that lead us to ask whether we really can model the system with all its significant interactions. Practically, we really cannot model all the interactions as has been mentioned above. Some interactions will be left out and the resulting model will be incomplete. The incompleteness will traditionally take two forms:

---

* There are groups who would only accept a risk analysis if it could be guaranteed to be complete in its consideration of all possible event sequences. Such a perception is no more reasonable or valid than that espoused by other groups who would accept nuclear power only if it could be shown to be completely free of risk. In the real world, no general theory is exactly true and no action or activity is free of risk.

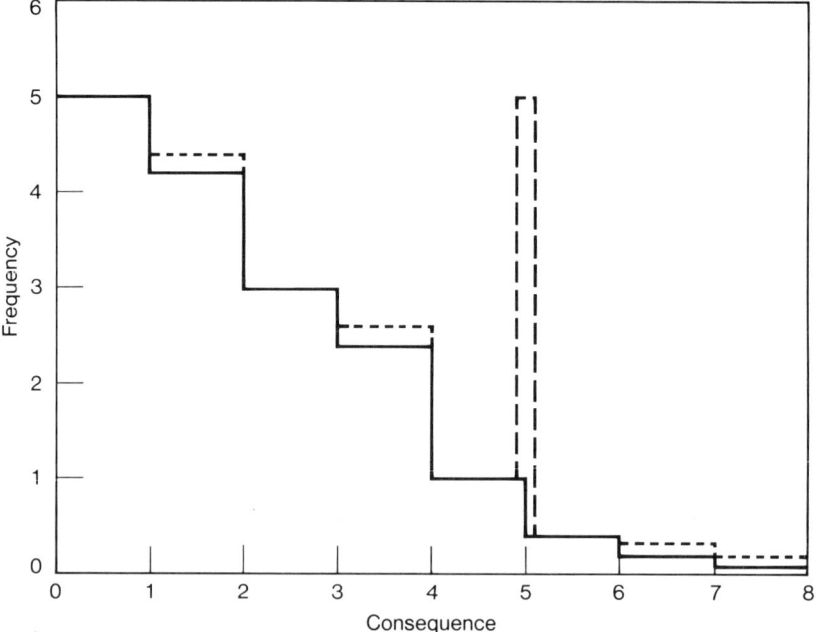

FIGURE 3. Frequency–consequence plot.

There will be event sequences that are overlooked, and there will be common-mode interactions that are unrecognized. Also, there is another type of event sequence that may be missed. This type can be considered to be a discrete event or class of events that does not fit on the same distribution as the others.

These missed event sequences affect portions of the risk probability distribution function (p.d.f., see Fig. 3) in that they will alter the calculated frequency at a given consequence. The results of the numerical calculations that lead to estimates of the p.d.f. may appear in the form of a histogram. The effects of both discrete and continuous uncertainties are illustrated in Fig. 3 where we use a solid line for the base case analysis, short dashed lines for the overlooked sequences and common-mode effects, and long dashed lines for a discrete high-frequency class of events. The scales are arbitrary. Rational arguments can be made that because the modeler does not know what the frequency of occurrence of the event sequence will be, he will leave out about the same number of event sequences in each consequence category. Unfortunately, this does not necessarily imply that the fractional change in frequency in each consequence category will be the same; if it did, then the p.d.f. implied by the curve in Fig. 3 would not change.

14 • LWR Risk Assessment

In fact, many sequences will be overlooked in each consequence category; but the frequency of such events must be small compared to the high-frequency events (at the low-consequence end of the spectrum) as otherwise they would be observed. As we move to higher-consequence, rarer events, the likelihood of missing a larger fraction of the total within a consequence category may increase. At the largest-consequence end of the spectrum, the likelihood of missing a large (even a multiple factor) fraction may well be very significant. Nonetheless, as is shown in Fig. 4, such situations do not seriously alter the cumulative distribution curve, and it is the cumulative curve that is important.

The cumulative distribution function (c.d.f.) tells us what the probability will be of having a consequence greater than any specified value. Clearly, this is the information necessary for decision-making, and not the actual probability related to any specific consequence. As the c.d.f. is an integral of the p.d.f., the detailed structure of the p.d.f. is not as important as it might appear to be at first glance. As an example (in Fig. 3), we consider frequency errors of 9% between $C = 1$ and 2, 17% between $C = 3$ and 4, 50% between 6 and 7, and 300% between 7 and 8. The short dashed line in Fig. 4 shows that such errors have little effect on cumulative distributions. Similar conclusions can be drawn

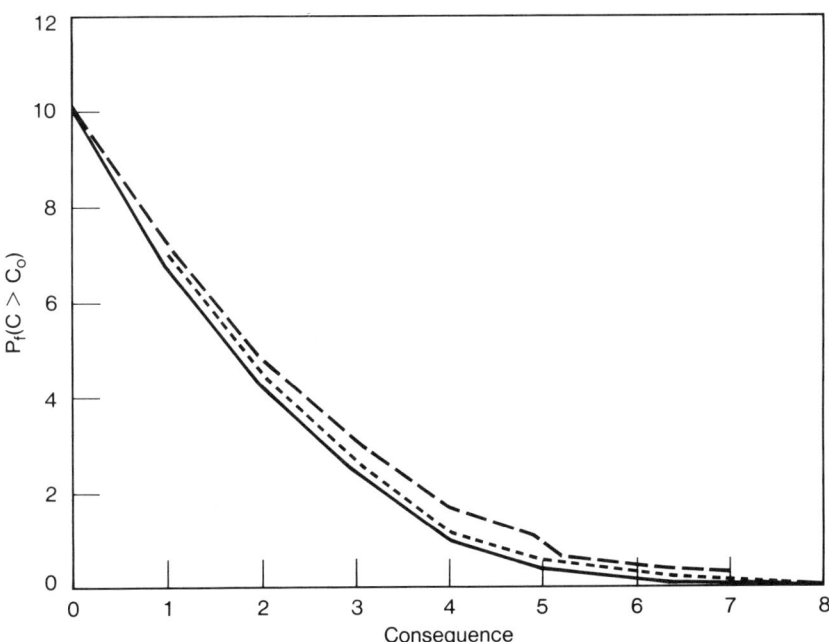

FIGURE 4. Cumulative probability plot.

for the discrete class. Again, very high frequencies cannot be ascribed because they would have been observed; here we choose a frequency equal to the highest and an area greater than that for all consequences following it. While it has some significant effects, it hardly trebles the tail risks. This is meant to be an example, not a proof; one could clearly put down other examples that showed very large deviations from the base case. But this would mean a poor base case study.

These perceptions and conclusions demonstrate that when the base case study is "properly" done, the question of completeness is not of significance. The assurance of achieving a "proper" base case study rests on taking proper account of error and uncertainty propagation effects and securing qualified peer review. A procedure to accomplish these ends should involve the following: The effects of data uncertainties (in failure rates, repair times, etc.) should be investigated, using sensitivity analysis, and documented as part of the base case; the effect of "loosening" the guidelines for the risk assessment should be investigated and documented as part of the base case; the technical peer review should always be substantive and searching. It may well occur that the first peer review produces changes; this may induce the decision-maker to seek a second review. When such a substantive iterative procedure is finished, the base case will be as complete as can be expected.

The above discussion when applied to WASH-1400 indicated that there really has not been much new discovered as far as the event initiators and pertinent event sequences are concerned. It is correct to say that a sufficiently detailed analysis of external event initiators (fire, earthquake, etc.) was not present in the first draft of WASH-1400, but these were reasonably well addressed in the final (it must be pointed out that a number of professional antinuclear advocates would disagree). It is also correct to say that a very few (one or two) new aspects of old initiators have been identified (e.g., the effects of asymmetric loads) but they have been shown not to impact probabilistically. Significant changes were made (and still are being made) in the consequence codes and in the deterministic dynamic models used at the event tree stage. Some of the changes can cause increases in the consequences (such as age and health variations) and some cause decreases (such as rain runoff).

The overall conclusion on the question of adequacy can be answered in the following way: If we were to say that the probabilities and consequences could be within a factor of 10 either way of the value set in WASH-1400, then it appears extremely unlikely that the true value would lie outside that band. There is always a chance that a scientific analysis is wrong; Newton's laws are approximate (as are Einstein's). Risk analysis is one more tool in the decision process; it has the advantages of producing a quantitative statement, and hence allows direct comparisons to be made. Its use will expand precisely because it is the only tool that can make such comparisons.

15 Accident Analysis and Risk Assessment

T. G. Theofanous and Richard Wilson

1. INTRODUCTION

Concerns about safety are part of life. Success depends on *anticipation:* From the all-too-well-known "drive defensively" in the "battleground" of the highways, to a good game of chess, to the intricate state of preparedness in the cold war, safety—and thus success of survival—depends upon our ability to *anticipate* future challenges and be *ready* with an adequate response.

Any particular sequence of a *set* of "exchanges," that follows an initiating event, represents a scenario. Clearly, a realistic safety (defense) approach must be based on priorities reflecting the relative likelihood of these scenarios. The whole defense process with these properly *ordered scenarios* yields a certain *level of safety*. Of equal importance is that this process also provides a mechanism for continuously *improving safety,* i.e., for guarding against an increasingly improbable set of circumstances. In this sense then, defense is an open-ended effort; it is never commmplete; we always strive to improve upon it.

Suppose an accident has been initiated; many outcomes are then possible, each one being affected by an interactive set of *automatic system actions, human decisions,* and *physical processes*. The laws that govern these three factors differ in essential ways:

Automatic system actions are subject to random failures; their likelihood can only be determined empirically (from experience) and taken into account probabilistically.

Human actions are even less quantifiable and bring in new degrees of freedom. Again the *range* of expected behavior must be based on actual experience taking into account completeness of instructions and potential for errors.

T. G. THEOFANOUS • School of Nuclear Engineering, Purdue University, West Lafayette, Indiana 47907.   RICHARD WILSON • Department of Physics, Harvard University, Cambridge, Massachusetts 02138.

It is easy to see that problems arise even in a probabilistic treatment of off-normal behavior and a most cautious attitude must be taken in that regard. Based on an incorrect analysis of the problem, people can perform a number of incorrect correlated actions. This happened at Three Mile Island.

Physical processes, on the other hand, are deterministic and obey well-known laws of nature. We take advantage of this determinism to significantly reduce the degrees of freedom in the interactive combination of these three factors. The whole progression of any particular accident scenario may be, therefore, viewed as a deterministic sequence of physical processes, *responding* to or being *redirected* by machine and human interference.

The study of physical phenomena and their mechanistic sequencing with *any particular* set of human and automatic system actions is called "accident analysis" (i.e., study of reactor accidents). The combination of such accident analysis with the probabilistic aspects of the human and automatic system factors, for the purpose of determining the likelihood of the various possible scenarios and in particular their outcomes, is called "risk assessment."

It is important to realize that it is the deterministic nature of "accident analysis" that renders the task of "risk assessment" manageable: deterministic accident-analysis results yield the magnitudes of accident consequences, within an uncertainty band.

The methodology for risk assessment is introduced in the preceding contribution to this book. Our purpose here is to focus on the study of reactor accidents (accident analysis) in the context outlined above. After an introduction on some basic features of the reactor systems and accidents of interest for public health and safety, we will discuss accident analysis tools and methodology. Current developments in both of these areas will be summarized next. We will conclude by speculating on future trends and directions. Throughout this essay, our aim is not to provide a comprehensive presentation of the treatment of reactor accidents but rather to illustrate the study process, to emphasize the breadth and high quality of engineering judgment (not a plug and chug approach) required and actually applied in the task, and to give an indication of further information needs. The approach taken is that physical events that cannot be excluded must be subjected to accident analysis. Evaluation of likelihood and hence impact on our perception of the level of safety is left for the risk assessment. It is crucial to emphasize that even if the probability for a particular sequence is prejudged to be small, this is not a reason, or an excuse, for sloppy accident analysis and inadequate definition of an uncertainty range in the outcomes.

2. REACTOR SYSTEMS AND ACCIDENT CONSIDERATIONS

In the discussion of accident analysis, we must consider individual designs. The two most important reactor types are the light-water reactor (LWR), which

15 • Accident Analysis

is extensively employed at present, and the liquid-metal fast breeder reactor (LMFBR or FBR), which appears to represent the choice for the future. The LWRs come in two versions: the pressurized-water reactor (PWR) and the boiling-water reactor (BWR).

The generic arrangement of reactors consists of fuel, contained in a cladding, surrounded by the coolant, placed in a reactor vessel, which is located in an airtight building ("containment"). Radioactive fission products are contained and their release is prevented by the successive barriers, represented by the sequence of fuel cladding, reactor vessel, and containment.

The normal way of controlling the nuclear reaction is by a bank of neutron-absorbing control rods distributed throughout the core region of the reactor. When the control rods are withdrawn, the nuclear reactions increase; when they are inserted further, the reaction ceases.

In particular, the control system is designed to provide *rapid* and *complete* insertion of all the control rods when the need for shutdown arises although ratchets prevent rapid removal. This action is automatic; it constitutes the heart of the reactor protection against accidents, and is known as "scram."

When the nuclear reaction has stopped, it is still necessary to take away the heat from the radioactivity of the core by pumps or natural circulation. Both of these heat removal modes demand, at least for a time, active devices.

In the LWRs, we convert nuclear heat to electricity by use of a steam cycle. The laws of thermodynamics tell us that the efficiency of this conversion increases with the temperature of the water. The part of the energy that is not converted to electricity is discharged as "waste heat." As we want to minimize the waste heat discharged to the environment as thermal pollution, we want to operate the power reactors at an as elevated coolant temperature as possible. We are limited by the fact that every liquid has a temperature at which it boils. This is for water at 100°C (212°F). From that point on, if we wish to increase the temperature and prevent the liquid from evaporating, we must increase the pressure. Thus, in order to achieve reasonable efficiencies with water as a working fluid, we are forced to operate with the coolant around 550°F. At this temperature the pressure must be at least 70 times that of the atmosphere and some power stations (PWR) operate at even higher pressures. The need for operating at high temperatures for achieving high efficiency is true for fossil-fuel power plants as well (up to 150 atm).

By contrast, the sodium in an LMFBR boils at 1600°F and temperatures as high as 1000°F are selected for operation. Accordingly, there is a considerable gain in efficiency compared to the LWR, and this without the development of any internal pressures. In the absence of forces to drive the coolant out, and with an appropriately routed piping and pumping system, the sodium will always remain within an LMFBR to cool the core even in the event a pipe should break.

Coping with a loss-of-coolant situation in an LWR is the major class of accidents considered in great detail. Such a situation may lead to impaired cooling of the core and it did so at TMI. Provisions must be made that (1) power is reduced and (2) the coolant inventory is replenished as soon as possible and certainly before the fuel reaches dangerously high temperatures.

Even if the control system would fail to *scram* an LWR promptly, the reactor has *intrinsic* safety features. As the reactor continues to heat up, the coolant water in the reactor core will expand and eventually boil and produce bubbles. The *density* of the water will go down. But in the LWR the water does more than act as a coolant: the hydrogen atoms also cause the neutrons to slow down ("to be moderated") to enable the nuclear fission reaction to proceed. Hence, any significant reduction in water density will reduce the slowing down of neutrons, and shut off the fission reactions before any major reduction in coolability of the core occurs. By the time water content of the core diminished to the point of practical cooling-capability breakdown, we can be *certain* that power shutdown has been spontaneously achieved. This occurs because the fuel and the water are normally close to the optimum configuration for maintaining the nuclear reaction, and about any change reduces the reaction level. We call this behavior inherently safe.* One of our tasks in safety is to make systems as inherently safe as possible, i.e., incorporating defense mechanisms we can count on in an absolute and unqualified way.

Provisions for quickly replenishing the coolant are needed, even although—as we showed above—we can count on a spontaneous shutdown. The reason is that the fuel rods will continue to be heated by decay of radioactive fission products having accumulated in them during past operation at power. This is called decay-heat. It represents only a few percent of normal reactor power, and reduces to a fraction of a percent within a day. However, a further significant reduction takes a long period of time (weeks to months). As the variation of decay-heat with time is well known, we can calculate the rate of heatup very accurately.

Simple physical considerations indicate the need for prompt coolant replenishment and give us a reliable estimate for the time we have for doing so. We can count on this estimate to be conservative, because in reality there will be a small amount of cooling that, if taken into account, would increase the available time estimated obtained from the "simple" calculation.

In practical accident analysis, much more sophisticated calculations are applied— such calculation models reflect the whole thermal-hydraulic response of the system, keeping track of water and steam contents, pressures, temperatures, and distributions throughout the system, taking into account any coolant injected by the variety of systems designed to provide a timely replenishment of coolant, and in particular keeping track of the cooling history, and hence of

---

* For this and other reasons, accidental power increases (reactivity accidents) are also self-limiting such that nuclear bomb-like explosions may be safely ruled out.

15 • Accident Analysis

the temperature of the fuel. These sophisticated calculations are applied in the loss of coolant accident (LOCA) analyses discussed below.

Other circumstances with fuel overheating could result from an accidental power increase above normal levels, or a reduction in coolant flow caused by failure of pumps, which could also lead to a reduction in cooling capacity. As long as the overheating does not damage the fuel, only coolant boiling will occur, producing vapor bubbles. The resulting decrease in water density will again lead to a shutdown of the fission process. As we have already seen above for LWRs, this is not an unstable situation, for a prompt power reduction will occur as a consequence. Furthermore, at decay-heat levels, natural circulation flow is sufficient for continued cooling.

The situation for LMFBRs is different. Normally, in case of a simultaneous failure of all coolant pumps, there is ample time for *automatic* detection of the fault and for reactor scram: The modern LMFBR designs use pumps with flywheels that extend the "coastdown" times to many minutes, certainly a long enough time for human-initiated scram, in the extremely unlikely case the automatic systems should fail. With the power reduced to decay-heat levels, core cooling is assured by natural circulation flow. Is this an adequate approach?

Failure to "scram" cannot be ruled out completely although built-in redundancies and high reliability make it extremely unlikely. Therefore, the subsequent behavior needs to be investigated, including the possibility of sodium boiling. Thus, the answer to the question above, at least at present, is negative: failure to "scram" cannot be ruled out.

Unfortunately, if there is a transient and a failure to scram, LMFBRs are not necessarily stable. In fact, because in these reactors fissions occur with fast neutrons (hence "fast" reactor), boiling of the sodium coolant would produce less slowing down, which means a corresponding increase in the population of fast neutrons and consequently a higher power level. This is an unstable situation with potential of a snowballing effect. How should we protect against such a circumstance?

Safety engineers and licensing authorities feel compelled to study the consequences of this "what if" question and provide additional protection accordingly. One therefore provides inherent power reduction capabilities in addition to the highly reliable reactor scram system. Then, fuel melting and gross core disruption could occur in only very extreme cases; the resulting accident types are called core disruption accidents.

The snowball effects initiated by sodium boiling would develop on the time scale of seconds. In contrast to an LWR-LOCA situation, which may take a long time and may, therefore, involve several human and system actions (see TMI), a core disruption accident in an LMFBR proceeds on its own. Following a boilout of the coolant out of part of the core, the cladding and fuel would melt and the power would continue to increase. In contrast to LWR fuel, which even upon melting can only produce thermal phenomena, melting the LMFBR fuel could increase the nuclear fission and thus yield very high power levels. This condition is called "recriticality." Then, high vapor pressure could develop with the potential to cause mechanical damage to the reactor vessel.

Recriticality may occur as a result of settling of the fuel mass upon melting; its severity increases with the degree of the resulting compaction of the fuel mass. Again, basic physics come to help us to sort out the real from the imaginary.

First, there is a physical phenomenon known as the "Doppler effect" that helps to limit the excursion in an intrinsically safe way. This Doppler effect is a very basic physical phenomenon. It was discovered by the Austrian physicist Doppler as an increase and subsequent decrease in the pitch of a sound when the source approaches you and then moves away from you. With car horns, this is now a common experience. In astronomy, the Doppler effect gives us the speed of quasars moving away from us.

In nuclear reactors the "pitch" of the thermal motion of the nuclei (relative to the neutrons) increases as the temperature increases. Then the absorption of neutrons in the fuel increases and tends to dominate the fission process. Experiments have shown that the theory is correct and that this nuclear physical effect will slow down the reaction immediately as the fuel heats up.

There is another intrinsic safety feature that is less well determined: any configuration of any materials will expand and eventually "boil up" when sufficiently heated.

It is as simple as opening up a preagitated can of soda or producing vigorous boiling in a kettle of water (or even better put a glass of water in a microwave oven) to see that vapor bubbles can lift up the liquid against gravity. This is called a "boilup." The more intense the boiling, the stronger the boilup. In fact, for intense enough bubbling (vapor production), liquid can be carried away with vapor (gas or steam) just like the expulsion you may observe in the soda-can experiment. This is a real phenomenon, well established experimentally, that would tend to oppose recriticality, and if it does not eliminate it altogether, it at least limits the severity of the recriticality effect provided that the recriticality is occurring slowly. Faster recriticalities could occur in the simple-phase regime (absence of bubbles). The liquid expansion associated with the rapid heating of the fuel will dominate such recriticalities, rendering them energetically benign. Thus the above Doppler and boilup/expansion effects may be viewed as two important *inherent* safety features of LMFBRs. To put it differently, LMFBRs cannot explode as nuclear bombs. Of course, this gives only a glimpse of the situation; in actual accident analysis, all the detailed physical effects and mechanisms have to be considered. The ultimate objective is to obtain realistic estimates of the accident sequence and make the reactor vessels strong enough to withstand the worst mechanical effects that could result.

Finally, we turn to the third and last barrier: the *containment*. As it must maintain continued isolation of fission products from the environment, no fluids are allowed to cross its boundaries. As a minimum, therefore, provision for rejecting in the long term, the decay-heat, through suitable heat exchangers must be made for both LWRs and LMFBRs. In addition, the possible consequences and potential mitigation of core melt events having progressed beyond the reactor vessel boundaries and into the containment environment need also be considered.

As we have seen for LMFBRs, core disruption accidents can be made extremely unlikely but cannot be ruled out in principle. Along the same line, safety analysts argue that even if the reactor vessel maintains its integrity during the core disruption phase, the possibility of failure by thermal attack (melt-through) by the core mass cannot be excluded, although timely and adequate emergency cooling is provided to prevent such an occurrence. Nevertheless, progression of a resulting meltdown event beyond the reactor vessel boundary is also investigated.

The following mechanisms challenging containment integrity are then considered.

1. *Hydrogen explosions.* Interactions of water coolant with strongly overheated cladding in an LWR, and of sodium with water found in the concrete floors of the containment of an LMFBR, give off hydrogen. Containments are typically filled with an atmosphere of air. When the hydrogen content reaches a certain level, this air will spontaneously and rapidly react with the hydrogen to yield a burning at best, or an explosion at worst. Containment pressurization will result. This happened at TMI at 1:30 p.m. on March 28, 1979.

2. *Buildup of noncondensables.* Reaction of the high-temperature molten core mass with the containment concrete would yield concrete decomposition and evolution of carbon dioxide gas, again resulting in containment pressurization. Although the unmitigated progression of such reactions has been assumed in the past, recent research indicates that this position may be unduly conservative.

3. *Sodium fires.* In a core disruption accident in an LMFBR, if the vessel fails, sodium coolant will be released to the containment, reacting with the oxygen of the air and thus burn.

4. *Energetic fuel coolant interactions.* The thermal interaction of the molten core ("corium") with the coolant (water for LWRs, sodium for LMFBRs) may yield rapid buildup of pressure, resembling an explosion. The mechanism of such explosions is due to rapid vapor production (boiling).

There is an extensive literature on vapor explosions resulting from interactions of very hot with comparatively volatile liquids, e.g., steel or aluminum with water, and water with freon-22, propane, or methane, and recently explosions have been obtained with LWR-prototypic materials. However, recent evidence also indicates that owing to geometric constraints on the reactor vessel design, the mixing of the corium with the coolant would be impaired, thus limiting the severity of such explosions to such low levels that they can be of no consequence to containment integrity.

These effects and their analysis have always been considered in the past. However, in the wake of TMI, they have received new attention. In particular, serious thought is given to (1) inerting the containment by filling with nitrogen or inducing controlled (slow) burning and thus avoiding the hydrogen explosion hazard; (2) providing controlled (and filtred) venting of the containment to minimize the hazard caused by overpressurization; and (3) installing devices that will contain the core melt, which would prevent interactions with the concrete and hence the hazard from noncondensable gases at high pressure.

Finally, we note a third general principle that can help reduce the consequences of an accident. The radioactive fission products include noble gases—krypton, xenon—which do not interact chemically, and are released easily in an accident. By the same token, they do not react chemically in the body and are exhaled after being inhaled. Considerable quantities of these noble gases could be released in a serious accident (e.g., 3 million curies were released at TMI). In contrast, radioactive iodine, and various particulates react chemically in the body and are hazardous. However, they also react chemically with metal surfaces and with water, which strongly reduces any escape from the reactor. Only 15 curies of iodine was released at TMI. It is likely that this general principle will prevent any large release of iodine from slowly developing accidents.

3. ACCIDENT ANALYSIS TOOLS AND METHODOLOGY

In Section 2 we emphasized the distinction between physical processes, which are well understood and quantitatively determined, and human or mechanical actions or interferences, which are often random in nature and must be considered probabilistically.

Now let us consider how well we can determine the physical processes.

The reduction of the transient due to the Doppler effect is based on very elementary physical considerations; all the neutron absorption cross-sections needed for its use have been measured in several countries, and the whole effect has been tested in actual power transients in a special reactor (SEFOR). The effect is known to an accuracy of approximately 15%.

The boilup phenomenon, while more complex and less well understood, is nevertheless real.

Let us consider again the soda-can experiment mentioned above and let us attempt to develop a qualitative feel as to the various levels of uncertainty associated with the various assertions we can make about the outcome of this experiment. Many people have opened cans that had been shaken shortly before, and they always made a mess. The *consistency* of the observations is such that we have *no doubt* that a good portion of the liquid contents will be quickly expelled in the form of a froth. Now let us consider a more sophisticated question. If we have a particular can, shaken for a particular time and with a particular intensity opened in a particular way, can we predict what portion of the liquid will eventually remain in the can? Certainly we will find no one naive enough to claim the ability to make such a prediction very precisely; i.e., the prediction of the boilup effect is not very precise, but that does not mean that the boilup does not occur.

The important implication is that "determinism" of physical processes must be viewed in the context of a nonzero uncertainty interval found in practice. This uncertainty interval propagates through sequences of processes, and a variety of scenarios may be created from a seemingly single set of initial conditions and interventions. It follows that as the ranges of uncertainty for individual processes are reduced, the variety of scenarios are also reduced. Thus, the rewards from learning better the characteristics of individual processes consist of reducing the range of scenarios that have to be considered. Judging these uncertainties is probably the most difficult aspect of safety or accident analysis. It helps, in this regard, to break up the entire accident development in several phases.

Just as it would be impossible to think of a symphony as a sequence of simple notes (or even chords), it would be extremely difficult to think of all but the simplest accident scenarios as a straight sequence of physical processes. An intermediate-scale structure is required to provide cohesiveness. Thus, we think of a symphony as a sequence of "movements," each movement made up of a sequence of notes, some essential to the main theme of that movement (and of the composition as a whole), some of only decorative and detailed nature. Similarly, we may think of an accident scenario as a sequence of physical processes. Some of them are very essential in controlling the course of the main events; these are the key physical processes. Others are rather unimportant and affect only details of the accident progression. The advantage of introducing such a structure is that each phase may be studied in great detail as a separate entity. A range of initial conditions, representing perhaps final conditions of a preceding phase, needs to be determined. Any nondeterministic aspects may thus be absorbed in constructing accident scenarios from phase sequencing.

In some cases, phase boundaries may be clearly identified and the above task is rather straightforward. In other cases, neighboring phases show considerable overlap; the transitions are gradual. In some cases, several phases merge together, and in such cases the task of constructing accident scenarios becomes considerably more challenging. Let us recall as an example the loss-of-coolant accident in an LWR. When the leak is large, it only takes around 20 seconds for most of the water coolant to be expelled from the reactor system. This is known as the "blowdown" phase of the LOCA. Although some emergency coolant may be injected during this period, it will be carried out together with the high-velocity outflowing stream of the water–steam mixture. Only near the end of the blowdown phase and when velocities and internal pressures diminish sufficiently will the injected water start accumulating to refill the reactor system.

The "refill" phase will then begin. Now consider a much smaller leak. The blowdown phase extends for much longer times, injected water may be an appreciable fraction of that rushing out, and a merging with the refill phase would be expected.

Now let us go back to our soda-can experiment. Having accumulated a lot of experience with the overflow experiments described previously, would we be in a position to venture a prediction for an "oversized can," say ten times larger than the regular size? We would be foolish to do so, as long as our knowledge from the previous regular-can experiments remained empirical. Suppose on the other hand we were able to identify the key physical processes, and suppose we were able to reduce their physical behavior in mathematical language expressing the controlling fundamental physical laws. A mathematical simulation has thus been effected. This simulation, provided it is successful in predicting the small-can experiments, could then be applied to the new large-scale experiments. Can we be confident of success? No, unless we can obtain some assurances that the new scale does not introduce any new physical phenomena, beyond those already reflected in the modeling. Another potential pitfall arises from the fact that in most cases of practical interest, even fundamentally based mathematical models require certain empirically determined parameters to effect successful predictions. Could it be that the values of these parameters will change in the new application? It would not be unnatural if we felt insecure. Fortunately, however, if we have done a good technical job to this point, little remains to be done to dispel those fears. An indication of scale effects may be obtained by carrying out just a few experiments with different size cans, although the size range may be rather limited and well below the one of actual interest.

The state of affairs in reactor safety, i.e., in accident analysis, is directly analogous to that described above. We will summarize explicitly. It is not practical to conduct simulations of severe accidents with full-size commercial power reactors. Even if we were willing to accept the enormous costs, we could only study a very limited number of scenarios. For a full support of the risk assessment efforts we must rely on mathematical simulations of such accidents supplemented by detailed studies of these accidents (such as TMI) that have actually occurred. The mathematical models (typically in the form of computer codes) will reflect the key physical processes and will provide simulations of one or more accident phases. The heart of this whole analysis-based methodology is in maximizing confidence in the results and in developing a conservative (perhaps even realistic) estimate of the uncertainty interval in the predicted scenario outcomes. The use of experiments as tools is crucial in this respect. In particular, we may identify three classes: (1) "fundamental experiments" usually of small scale and equipped with detailed instrumentation, helping understand each of the key physical processes and thus providing guidance in its mathematical simulation; (2) "full-scale component tests" on the behavior of portions of the full system under

15 • Accident Analysis

simulated portions of the accident, helping to identify any unexpected scale effects; and (3) "system tests" in complete simulations of the reactor system but at smaller scale, helping identify any shortcomings in the mathematical simulation of sequencing and interactions between phases as well as between individual physical processes. In conclusion, certain special experiments may be performed in actual power reactors, simulating aspects of abnormal behavior and then providing an ultimate checkpoint in our predictive capability.

4. CURRENT DEVELOPMENTS AND FUTURE TRENDS

A rational approach for performing accident analysis as a means of predicting the behavior of even the most severe accidents is available. The usefulness of such predictions depends critically on consideration of their uncertainties. At present we find ourselves in the position that, at least for some types of accidents, calculations of a rather detailed nature are necessary for good predictions. The complete and orderly assessment of the uncertainties in such calculations is an ongoing process. A major goal of safety analysts is to continue to reduce these uncertainties, as basis for further improvements of the safety and economics. Because of the existence of uncertainties, vague and rather general but in our opinion unjustified criticisms are made on the adequacy of current safety.

The designer, safety analyst, and government regulator have compensated till now for such uncertainties by following the so-called conservative approach, i.e., the most severe *credible* accidents are identified and the design of the safety system is based on calculations of these severe circumstances. Thus, one takes the pessimistic end of the uncertainty interval. This represents a shortcut approximation to the full accident analysis as described above. Along the line of conservatism, the so-called *single failure criterion* was established and is applied, assuming in effect that if anything can go wrong with active components it will, but simply once on each occasion. This represents a shortcut approximation of the full treatment, consisting of integrating accident scenarios with probabilistic equipment failures. The difficulty with such shortcut approaches is that they tend to emphasize the protection provided in some special and extreme circumstances while the true margin of safety with respect to less severe but perhaps more likely events is not known. This became clear at TMI.

Safety engineers are presently engaged in a broader accident analysis program including the assessment of uncertainties. Some encouraging results are already available. As a consequence of TMI, we will improve on the shortcut to risk assessment sooner than previously anticipated, especially by analyzing consequences of multiple failures. Along the same lines we should expect development of additional protection from accidents more severe than considered heretofore. In particular, the institution of methods for eliminating the hydrogen explosion hazard appears likely. Finally, having learned more about more varied kinds of possible accidents, the operators will be better equipped to respond correctly, thus maximizing the positive aspects of the "human factors."

As we think along these lines, however, it is important to appreciate that even the most meticulous paper studies cannot provide a complete description of the real performance possibilities, and hence of the safety of a power reactor. It is crucial, therefore, that we study in detail the experience with our present LWR power reactors and use them to improve the data base for true safety that can only be achieved through *actual experience* and *technological tradition*. One can never prove nor can one achieve complete safety in advance. Thus, planning for a safe LMFBR program must include construction and operative experience with actual LMFBRs.

5. CONCLUDING REMARKS

Current designs of LWRs and LMFBRs have been developed with special attention to safety and economics. The basic approaches applied in the underlying safety analyses have been illustrated in this essay. The role of conservatism in the design to envelop uncertainty bands was emphasized. Considerable further improvements in form of a much broader accident analysis, including a better assessment of uncertainties, are under way.

The success of these conservative approaches is demonstrated by the performance record of commercial nuclear power reactors. Several incidents and accidents have happened on commercial power reactors, prototypes, or developmental precursors, as it was the case in all similar large-scale technologies. However, no member of the public was ever harmed by these events. About all accidents and incidents provide important information that is thoroughly exploited to further improve the safety of nuclear reactors.

16 The Waste Disposal Risk

Bernard L. Cohen

When fuels are burned to release energy, the fuel materials don't just disappear—as the old adage goes, "matter can neither be created nor destroyed"—but rather they are converted into other forms that we call "wastes." This is true whether we burn uranium in nuclear reactions, or coal, oil, or gas in chemical reactions, and in all of these cases these wastes are responsible for the principal adverse effects of energy generation on our environment. We can therefore gain some useful perspective on the waste problem by starting off with a brief discussion of the wastes from coal burning, which is the principal competitor with nuclear power as a source of electricity.

The principal waste from coal burning is carbon dioxide, produced by a large power plant at a rate of 500 lb. every second. Carbon dioxide is not ordinarily considered to be a pollutant as it occurs abundantly in nature, but there are now very serious worries that the huge quantities of that gas being produced in the burning of coal, oil, and natural gas may be doing irreparable damage to the world's climate.

The most important gaseous pollutant produced by coal burning is sulfur dioxide, which is released from a large power plant at a rate of 10 lb./sec; if it were all inhaled by people, this is enough to kill a million people every few minutes. Of course only a tiny fraction is actually inhaled by people, but this is still enough, according to a National Academy of Sciences study, to kill an average of 25 people per year, cause 60,000 cases of respiratory disease, and do $12 million in property damage. Another gaseous pollutant from coal burning is nitrogen oxide, best known as the principal air pollutant from automobiles—it is the reason we must have catalytic converters in our exhaust lines and use lead-free gasoline; the nitrogen oxide produced by one coal-burning power plant

---

BERNARD L. COHEN • Department of Physics, University of Pittsburgh, Pittsburgh, Pennsylvania 15260.

is equal to that emitted by 200,000 automobiles, and very little is being done about it.

Another pollutant produced in coal burning is smoke and dust, considered by some to be even more harmful than sulfur dioxide. And then there are the solid waste, or ashes, produced at a rate of 1 ton/min, which must be disposed of. A small portion of it can be used in construction applications, but the great majority becomes a heavy burden on our environment.

In addition, coal burning produces a wide variety of organic compounds, a number of which are known to cause cancer. One of these is benzpyrene, which is widely suspected of being the cause of cancer from cigarette smoking. There are also toxic and carcinogenic metals released in coal burning, like arsenic, cadmium, mercury, selenium, etc.

The wastes from burning nuclear fuels are different from these coal-burning wastes in two very spectacular ways. First, they are 5 million times smaller in quantity; the annual wastes from a nuclear power plant that produces the same quantity of electricity as the coal-fired plant we have been discussing would fit under a card table, and after full preparation for disposal would comprise a single truckload. Second, the environmental hazard in this waste is not chemical toxicity as in the case of the waste from coal burning, but rather it derives from the fact that these wastes are *radioactive*.

This last point is a source of considerable difficulty in achieving public acceptance of nuclear power because somehow the public views radioactivity as something very *new,* highly *mysterious,* and extremely *dangerous.* Actually, there is nothing *new* about radioactivity or the radiation it emits. Mankind has always been bombarded by radiation from above in the form of cosmic rays from outer space, from below due to naturally radioactive materials like uranium, thorium, and potassium in the ground, from all sides due to these same materials in the bricks and stones of our buildings, from within largely because of the naturally radioactive potassium that permeates our bodies and is necessary for life, and from the air we breathe due to the naturally radioactive radon gas, which is a product of uranium decay and is an omnipresent constituent of our atmosphere. Each one of these five sources of natural radiation exposes us to at least 100 times as much radiation as we can ever expect to receive from nuclear power installations. Moreover, a new source of radiation—medical X-rays—has been developed in this century, and it now gives us an average radiation exposure comparable to that from natural radiation.

There is also nothing very *mysterious* about radiation. It is easy to detect with good accuracy and extremely high sensitivity, and it has been accorded a great deal of scientific investigation; as a result it is much better understood than such other insults to our environment as air pollution, food additives, insecticides, etc.

16 · The Waste Disposal Risk

Large quantities of radioactivity can be *dangerous* but the public image of this problem is grossly exaggerated. It probably derives from the atomic bomb attacks on Japan, although the great majority of damage done there was by blast and heat. About 25% of the fatalities were due to radiation directly from the bomb itself, but this is analogous to being inside an operating reactor, which no one worries about. The long-lived radioactive waste, which is the principal problem with reactors, did *no* damage in the Japanese bombings.

The proper approach to evaluating dangers from radiation is clearly to deal with it quantitatively rather than to respond emotionally. In this we are fortunate in having available good estimates of the effects of radiation on human health derived from several situations in which people were exposed to large doses as in the A-bomb attacks on Japan, patients treated with X-rays or with radium for medical purposes, etc. As discussed in more detail in Chapter 19, data of this type have been compiled and evaluated by such prestigious groups as the National Academy of Sciences Committee on Biological Effects of Ionizing Radiation (BEIR), the United Nations Scientific Committee on Effects of Atomic Radiation (UNSCEAR), and the International Commission on Radiological Protection (ICRP), and their recommendations will be used here in estimating hazards.

We will confine our attention to the "high-level waste" that is left when spent fuel is chemically reprocessed to remove uranium (U) and plutonium (Pu), which are valuable as fuels. This waste contains the vast majority of the radioactivity produced in reactors, including nearly all of the fission products and the transuranics other than U and Pu, as well as whatever U and Pu are not removed in the reprocessing (typically 0.5% of each). Other wastes generally classed as "low level" include materials from cleanup of effluents, structural components of reactors activated by neutrons, equipment used in handling radioactivity, radioactive isotopes left over from medical and industrial applications, etc. A third category is dispersed into the environment and is discussed elsewhere in this volume as routine emissions of low-level radioactivity (see Chapter 21).

Current plans for handling high-level waste involve converting it into a glass or ceramic, and burying it deep underground in a carefully chosen rock formation. Other disposal plans that have received consideration are rocketing it into outer space, emplacement in the Antarctic and Greenland ice sheets, and burial in deep seabeds. Outer space disposal is feasible but it is prohibitively expensive if the waste is packaged such that, in the event of a rocketry failure, it will return to earth without being dispersed by burn-up on reentry or by impact on the ground and it will be sufficiently shielded so as to be safely approachable by curious bystanders. This expense is drastically reduced if the transuranics are separated from the fission products in reprocessing and only the former are sent out by rocket, as their radiations are much less penetrating and they therefore require far less shielding. However, the cost would still be very substantial and the problem of disposing of the fission products would still remain. Seabed disposal would require complex international agreements and very extensive technical study, but work on it is proceeding at a reasonable pace. We confine our attention here to the land burial option and will show that it is acceptably safe.

The electricity produced in 1 year by a large nuclear power plant sells for about $250 million so we could spend $5 million in disposing of its waste without increasing the cost of electricity by more than 2%. This is a very large sum to spend on the burial of a single truckload of material, but cost is essentially not an issue. We can afford to take nearly any conceivable precaution; that has certainly been the policy and there is every indication that the full $5 million/reactor-year will be spent. There was a ballot issue in a local election resolving that Alpena County (Michigan) not be used as "a nuclear waste dumping ground"; citizens envisioned dump trucks driving in and tilting their carriages back to slide the waste down into a hole. Clearly such an operation does not cost $5 million for disposal of a single truckload.

The present thinking is that the waste would be formed into glass cylinders about 1 foot in diameter by 10 feet long, enclosed in a thick metal casing that has been welded shut—the unit is referred to as a *canister*. The radioactivity emanating from such a canister is rather fierce; without shielding, a person standing 30 feet away would be exposed to a lethal radiation dose in a few minutes. With adequate shielding, however, there are no difficulties in handling the waste. In addition, a great deal of heat is generated by the radioactivity—about 1 kW/canister, the heat produced by an electric iron. A single canister of the type under discussion would contain the wastes from about 30 MWe-year of electricity production, so a 1000-MWe plant would generate about 35 of them each year.

This heat leads to a number of problems that require consideration, and the solutions to these problems place a constraint on how close together the waste canisters can be buried. If the constraint is simply that the temperature in the rock close to the surface of one canister is not increased more than 10–20% by the heat from neighboring canisters, the separation distance must be at least 10 m. Most current thinking favors a spacing of this magnitude, i.e., one canister in about a 100-m^2 area. The annual wastes from a single power plant would therefore occupy an area of 3500 m^2, so a large U.S. nuclear program would require about 1 km^2/year for waste storage.

The heat generation from reprocessed waste decreases by a factor of 10 after 100 years and a factor of 100 after 200 years, after which it decreases very slowly, but by this time the heat generation is a negligible problem. However, the heat flow away from the repository area is rather slow so rock temperatures remain elevated by several tens of degrees for many hundreds of years. The problems caused by these elevated temperatures will be considered later.

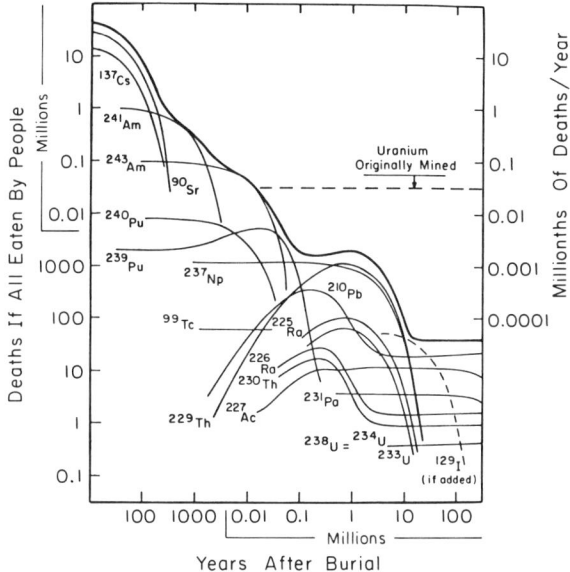

FIGURE 1. Number of cancer-causing doses in the waste from 1 GWe-year of nuclear electricity if all of the material were ingested one time by humans in soluble, digestible form.

Consideration has been given to a wide variety of scenarios under which the waste might be released from its burial area and there is general agreement that the most important is the one in which groundwater comes in contact with the waste, leaches it into solution, and carries the dissolved waste into surface waters (rivers, wells) from which it can get into food and drinking water. An evaluation of the danger from this scenario depends on the toxicity of the waste for *ingestion;* we show this as a function of time in Fig. 1.

Figure 1 is based on the waste from 1 GWe-year (one large power plant operating for 1 year). The ordinate in Fig. 1 is the number of cancer doses in this waste; e.g., the 40 million initial value means that if all of this waste were converted into edible form and fed to billions of people, we would expect 40 million cancers to result. A more credible scenario would be if it were dumped randomly into U.S. rivers; about one part in 10,000 of material in rivers is ingested by people, so we would expect 40 million/10 thousand or 4000 cancers to result. It is thus clear that dumping it into rivers is not an acceptable disposal method. A full nuclear power program in the U.S. might include 400 GWe and thereby produce 16 billion lethal doses each year. It is perhaps worth noting that an annual production of 16 billion lethal doses of a toxic substance is not an unprecedented or even an unusual step. The U.S. produces 400 trillion lethal doses of chlorine gas each year and many trillions of lethal doses of several other gases. If we restrict our attention to oral ingestion, about 100 billion lethal doses of barium compounds and of copper, and 10 billion lethal doses of arsenic are sold each year. Note that the toxicity of the waste decreases 100-fold over a 200-year period, whereas the toxicity of the barium, copper, and arsenic remain forever. It is sometimes argued that the latter materials were here already whereas the nuclear waste is a toxic substance being newly produced, but a large fraction of the barium, copper, and arsenic is imported into this country and is therefore being introduced artificially into our environment. Finally, we point out that the waste is to be buried deep underground whereas the barium, copper, and arsenic remain in our environment; in fact, most of the arsenic is used as a herbicide and is therefore spread over the ground in areas used to grow food!

An alternative representation of the ingestion toxicity in the waste is given in Table I, which lists the amount of the waste (converted to digestible form) one would have to ingest to have a 50% risk of getting cancer. We see that initially the waste is quite lethal—1/100 of an ounce would be fatal—but after 600 years, a full ounce would be required. By this time it is less toxic than some things we keep in our homes such as rat poisons or some insecticides. It is also no more toxic than some naturally occurring rocks that we have never worried about.

From either Fig. 1 or Table I, it seems reasonable to divide the waste problem into two categories, the first few hundred years when it is quite toxic and must be kept out of our environment, and the long-term problem extending for thousands or millions of years during which the waste is not very toxic but it is interesting to explore its effects because of the very long times available to do damage.

When some people hear that the waste must be isolated for hundreds of years, they react in horror; nothing, they say, can be counted on for that long, our political and economic systems may well collapse before then, and almost any structures we build will not last for hundreds of years. The error in this perception is that it is based on experience in our environment here on the surface of the earth. If we were rocks buried 2000 feet underground, we would find the

TABLE I
Cancer Risk from Ingestion of Radioactive Waste

	Years after burial				
	10	100	600	10,000	80,000
Ounces ingested for 50% cancer risk	0.01	0.1	1.0	10	200

environment very different. Things happen very slowly there. If all the rocks between 2000 and 3000 feet below the U.S. were to have a newspaper, it would only come out once in a million years—there would be nothing to report more often than that. Typical times required for changes at that depth are tens of millions of years.

With regard to the very-long-term problem, after a few hundred years the waste becomes very much like the naturally radioactive uranium and thorium in the ground, so it seems reasonable to compare it with them. The ground is full of such radioactivity, and in burying our waste we are adding only a very tiny increment to it. In this connection, Fig. 1 shows that after 15,000 years, the radioactivity in the waste is less harmful than the uranium ore originally mined to produce the fuel from which the waste was generated.

These arguments were meant only to be introductory and qualitative; let us now return to consider in more quantitative detail first the "few hundred years" problem, and then the very-long-term problem. For the first few hundred years there is a great deal of protection from various time delays in the possible release of the waste through groundwater. First and foremost, the waste will be buried in a rock formation in which there is virtually no groundwater, and in which geologists are as certain as possible that there will be no groundwater for a very long time. Four committees of the National Academy of Sciences have studied this problem over the years and have reported "we can be quite certain that there will be no groundwater [in these formations] for at least a thousand years, and probably for very much longer." A few hundred years is a very short time for geologists! A great deal of effort is being expended to assure that the disposal site chosen will be free from groundwater for a very long time.

But what if the geologists should be mistaken and water does get into the rock formation in which the waste is buried? It would first have to leach or dissolve away large fractions of that rock—roughly half of the rock would have to be dissolved before half of the waste is dissolved. This factor might seem to offer minimal protection if the waste is buried in salt as salt is readily dissolved in water. However, in the New Mexico area being used for an experimental repository, if all the water now flowing through the ground were diverted to flow through the salt, it would take 150,000 years to dissolve away the salt enclosing 1 year's waste. The amount of salt there is very substantial—about 2000 feet of solid rock salt—whereas the groundwater in that region is not like a river but rather more like a dampness very slowly working its way through the rock. It takes a long time for dampness to dissolve away such a large quantity of solid rock salt—about 150,000 years.

The next layer of protection is the back-fill material surrounding the waste canisters. This will be a clay that swells up when wet and is thus expected, in most circumstances, to become a tight seal keeping water away from the canister.

If the groundwater should penetrate and dissolve some of the waste, this backfill material also serves to filter the radioactivity out of the water, effectively preventing it from escaping.

But this is getting ahead of our story. If water should penetrate the backfill, before it can reach the waste it must get past the metal casing. Materials for this casing have been developed that give impressive resistance to corrosion. The present favorite is a titanium alloy that has been tested in 350°C brine solution with oxygen bubbling through to maximize corrosion rates—present thinking is that actual temperature rises will be kept below 150°C, and groundwater is generally highly deficient in oxygen—and even under these extreme conditions, corrosion rates were such that water penetration would be delayed for several hundred years. Under more normal groundwater conditions, these casings would retain their integrity for a million years. Thus, the casings alone provide a rather complete protection system even if everything else fails.

The next layer of protection is the waste form, which will probably be glass. We all know from daily experience that glass is not easily dissolved in water. Glass articles from ancient Babylonia have been found in riverbeds where they have been washed over by running river water—not just dampness—for thousands of years without dissolving. A Canadian experiment with waste glass buried in the water table indicated that it will last for a hundred million years, i.e., only about 1/100 millionth dissolves each year.

But suppose that somehow some of the waste did become dissolved in groundwater. Groundwater moves very slowly, typically at less than 1 foot/day—at the New Mexico site it moves something like 1 inch/day. Moreover, groundwater deep underground does not ordinarily travel vertically upward toward the surface; rather it follows the rock layers, which are essentially horizontal, and hence it typically must travel about 50 miles before reaching the surface—in the New Mexico area it must travel several hundred miles. To travel 50 miles at 1 foot/day takes about 1000 years, so this again gives a very substantial protection for the few hundred years that concern us here.

But by no means does the radioactive material move with the velocity of the groundwater. It is constantly filtered out by adsorption on the rock material, and as a result it travels hundreds or thousands of times more slowly than the groundwater. If the groundwater takes a thousand years to reach the surface, the radioactive materials would generally take hundreds of thousands or millions of years.

We thus have seven layers of protection preventing the waste from getting out during the first few hundred years when it is highly toxic—the absence of groundwater, the leach resistance of the surrounding rock, the sealing actions of the back-fill material, the corrosion resistance of the metal casing, the leach resistance of the waste glass itself, the long time required for groundwater to reach the surface, and the filtering action of the rock. This is about as close to

16 • The Waste Disposal Risk

perfect protection as one can get. But what if by some combination of miracles the waste did get out while it is still highly toxic? The increased radioactivity in the water would very rapidly be detected and the water would not be used for drinking or irrigation of food crops, so there would still be no damage to human health.

We now turn to a quantitative evaluation of the long-term hazard from high-level waste. Let us assume for the moment that the probability for an atom of high-level waste to be leached out by groundwater, carried with it into a river, and eventually ingested by a human is the same as that probability for an atom of average rock now submerged in groundwater. We will examine the validity of this assumption later, but we use it here because it allows us to calculate the probability. For the atom of average rock, this can be done as follows.

From measurements of groundwater flow rates and porosity of rock, we can determine the quantity of groundwater per year flowing into a river in a given area. From this and chemical analyses of the groundwater, we can determine how many atoms of each chemical element it is carrying into the river each year. But we know how much rock this water passed through, and from this plus chemical analysis of the rock we can determine the number of atoms of each chemical element it contains. The ratio of the number of atoms of a given chemical element carried into the river each year to the number in the rock then gives the probability per year for escape. The result of this calculation is that the probability for an atom of average rock to be carried by groundwater into a river is one chance in 100 million per year. We therefore assume that this probability applies to an atom of buried radioactive waste.

The next problem is to estimate the probability for an atom in a river to reach a human stomach. For water this is readily calculated as the ratio of water ingested by the U.S. population to water flow in U.S. rivers, which is 1/10,000. If only drinking water were considered, the probability would be less than this for the radioactive material because some of it would be removed by water filtration systems, but there are pathways through the food chain also to be considered, so that the 1/10,000 probability is roughly applicable to the waste. When this probability is multiplied by the 1/100 million chance per year for an atom of radioactive waste to reach a river, the final result is that the probability for an atom of buried radioactive waste to reach a human stomach is about one in a trillion (10^{-12}) per year.

Effectively, then, the vertical scale in Fig. 1 should be multiplied by that factor to determine the number of fatalities expected per year. When this is done and effects are added up over time, the final result is that we may expect about 7 eventual fatalities from each year of all-nuclear electric power in the U.S., or 0.018 eventual fatality per plant-year. This is many thousands of times smaller than the effects of waste from coal burning usually estimated at about 25 fatalities per plant-year.

This estimate was derived under the assumption that buried waste behaves like an average rock submerged in groundwater, so let us consider the validity of such an assumption. One way in which buried waste differs from natural rock is that there must have been a shaft leading to the surface through which the waste was brought in. This shaft would typically be several hundred meters away from the waste as the latter would be transported through long horizontal corridors before emplacement, but it still represents a threat to the security of the waste if this shaft (as well as the bore holes used to explore the geological structure of the area) is not well sealed. This problem has been studied in considerable depth, and there seems to be general agreement that these holes can be sealed so as to be at least as secure as the original rock.

Another difference between buried waste and average rock is the fact that the former contains a source of heat, which raises the temperature of the surrounding rock. There have been some worries that this may lead to cracking of rock formations, which might provide a path for intrusion by groundwater. This question has been under study for over 10 years, and the present conclusion is that rock cracking will not be important unless the temperature rises by several hundred degrees Centigrade. That is the reason for limiting temperature rises to about 150°C. Further studies are in progress, and if they should lead to less optimistic conclusions, the solution would be to spread the waste over a larger area so as to reduce the heat load, or alternatively to delay burial for 50–100 years (heat output is ten times smaller after 100 years). This is then essentially a problem in repository design rather than a question about the safety of buried waste.

Another temperature effect under study is brine migration due to temperature gradients. The solubility of salt in water increases with temperature, so a pocket of brine (i.e., water with dissolved salt) will dissolve away salt on its hotter side and precipitate salt out of solution on its cooler side. This action causes the pocket as a whole to move in the direction of higher temperature, which is toward the buried waste, so we may expect a collection of brine around each waste canister. Brine migration depends on the temperature gradient, the rate at which temperature changes with distance, and this decreases rapidly with time as the heat spreads out through the rock and the heat generated by the waste diminishes. Preliminary studies indicate that brine migration will only occur for the first few decades after emplacement and that the total amount arriving at each canister will be only a few percent of the volume of the canister. It is difficult to see how such a small quantity of liquid can do a great deal of damage. Moreover, during this time, the surface of the canisters is hot enough to boil the brine and cause the steam to escape, provided an escape path is available. Keeping such a path available for this period of time would seem to be a readily soluble problem of repository design, but the entire problem is under study. Temperature gradients could be decreased, if necessary, by changing the size and spacing of the waste canisters.

Still another difference between the buried waste and average rock is that the former is foreign to the environment, so it is not in chemical equilibrium with the surrounding rock–groundwater (if any) regime. However, chemical equilibrium is a surface phenomenon. When a foreign material is emplaced, there is relatively rapid dissolution of a microscopically thin surface film and this is replaced by a surface coating precipitated out of solution. Once this surface coating forms, it is in chemical equilibrium with its surroundings and further dissolution can only occur by diffusion through this coating, a very slow process. Moreover, this further dissolution is replaced by a thickening of the coating, which slows the process still more.

Because we have been discussing ways in which the waste may be *less* secure than the average rock with which it was compared in our analysis, it should be pointed out that there are several ways in which it is *more* secure. It is in a carefully selected rock formation, whereas the average rock is in a formation of only average integrity. It is in a region free of groundwater, whereas our average rock is submerged in groundwater. The latter is not enclosed in the corrosion-resistant casing that adds so much security to the waste, saving the situation even if all else fails. And the last-resort protection of detecting escaping radioactivity is not available to average rock.

In the derivation of our estimate that we may expect 0.018 eventual fatality per plant-year, it should be understood that this total will not be reached for millions of years. But in considering such very-long-term effects, it seems reasonable to take into account the fact that nuclear reactors burn up uranium, and uranium is a source of radiation exposure to humans, largely due to the radon gas emitted from its radioactive daughters. By removing this uranium, it is estimated that we will eventually *save* about 300 lives per plant-year, nearly a million times more than will be lost due to the waste. Thus, on any long time scale, nuclear power is a method of *cleansing* the earth of radioactivity. People living in the distant future will experience *less* radiation exposure as a consequence of our use of nuclear power today.

The reason for this is clear when it is recognized that the radiation emitted in the natural decay of a uranium atom is many hundreds of times more dangerous than the radiation emitted by the fission products into which it is converted in a reactor. Uranium decays by a series of eight α-particle emissions, and each α particle is a hundred times more dangerous than the emissions from a fission product. Furthermore, one of the elements formed in the decay of natural uranium is radon, which is a gas and therefore rises out of the ground to become airborne. As humans are constantly inhaling air, radon and the radioactive materials into which it decays have a very easy pathway into the human body. We repeat— on any long time scale, nuclear power *cleanses* the earth of radioactivity.

Our treatment thus far has dealt with averages, but some people are obsessed by fear of a major catastrophe. As the waste will be spread out over a large area, it is difficult to conceive of a catastrophe that would release more than 1 year's contribution (a circle of ½-mile diameter). The simplest known mechanism for such a release would be the impact of a very large meteorite, which would behave very much like a nuclear bomb explosion, creating a fireball and evaporating an immense quantity of material including the buried waste. The radioactive waste thus released would eventually come down into the air we breathe in a manner similar to nuclear bomb fallout. If such a meteorite impact were to occur shortly after the waste is buried, it would cause several hundred thousand eventual cancers. However, impacts of such large meteorites are extremely rare; one might be expected to strike the area in which a given year's waste is buried only once in a trillion years! Moreover, such a large meteorite could cause far greater harm by falling on a large city, or by falling in a large lake or ocean and thereby creating enormous tidal waves that can do enormous damage. Averaged over time, meteorite impacts are expected to be a million times less damaging than the groundwater-to-ingestion route we discussed above.

Discussion of meteorite impacts immediately raises the question of whether similar effects could be caused willfully by use of nuclear weapons. However, even the largest nuclear bomb ever built (50 megatons) would not damage the rock 2000 feet underground. There is no military application for the much larger bombs that would be required; release of the buried waste would not be an incentive for constructing such a bomb as far greater damage can be done by exploding a much smaller bomb over a city.

Another natural phenomenon that could release the waste would be development of a volcano passing through the repository. However, this would release far less material than the meteorite impact, and is far less probable if the waste is buried in a geologically stable area. For several reasons, geologically unstable areas would not be considered suitable for a repository.

Some consideration of release of buried waste through human intrusion might be in order. Purposeful release from a sealed repository would be an exceedingly difficult, expensive, and dangerous enterprise and would rank far down on the list of available techniques for mass murder. Toxifying food supplies and sources, or use of chemical poison, for example, would be much easier, cheaper, and more effective. However, there is the possibility of inadvertent human intrusion if the site of the repository were forgotten. Studies were made of releases through drilling operations and mining, including the mining of salt, and it was found that these present much less of a hazard than the groundwater-to-ingestion release we have discussed.

We often hear that in producing this waste we are placing a burden on future generations in watching over it. However, it should be recognized that our estimate is based on *no* watching. It was derived from a comparison with average rock and no one is watching to prevent erosion of average rock. Any long-term watching of buried waste would therefore be only to decrease the already very small effects we have calculated.

16 • The Waste Disposal Risk

Moreover, if future generations should decide to watch our buried waste, the burden would be very minimal. The wastes from 300 years of all-nuclear power would lie under an area 10 miles square. The person watching it could arrange to pass over each area about once per month, checking that there is no mining or deep drilling and inspecting warning signs to keep them in good repair. In addition, he should withdraw small water samples from wells and rivers in the area about once each month and measure their radioactivity. This entire job of watching a 300-year waste accumulation would be something less than a full-time job for one person.

In considering burdens on future generations, we should worry far more about the burden we impose on them by consuming, within a few generations, all of the earth's mineral and hydrocarbon resources. We are using up essentially all of the useful deposits of copper, tin, lead, mercury, etc., leaving our progeny very few options for producing materials. And then we are burning up those very valuable hydrocarbons, coal, oil, and gas, each at a rate of millions of tons every day, again depriving our progeny of feedstock for producing plastics and organic chemicals that might serve as substitutes for some of these materials.

In devouring the earth's mineral resources in such a short time, we are impressing a truly tremendous burden on our progeny. The only way we can compensate them would be to leave a technology that will allow them to live in reasonable comfort without these resources. The key to such a technology must be cheap and abundant energy; with it one can devise a substitute for almost anything, but without it there can be no hope—the only option would be to return to a very primitive existence.

In view of these considerations it seems clear that we owe our progeny a technology for producing cheap and abundant energy, and the only technology we can now guarantee is nuclear power from fission.

17 Radon Problems

Bernard L. Cohen

Radon is a naturally occurring radioactive gas derived from the decay of uranium—actually, uranium decays into thorium, which then decays into radium, which in turn decays into radon. As uranium is naturally present in all rock and soil, radon is constantly being generated in them and in materials derived from them like brick, stone, cement, and plaster. Being a gas, this radon diffuses out into the air where it and some of its radioactive decay products are inhaled by humans. If the methods used for estimating health effects of radiation in all other contexts are applied to this exposure, the results indicate that radon inhalation is now causing about 10,000 fatal lung cancers per year in the U.S. This is several times the fatality toll from all other radiation, natural and manmade, combined. Incidentally, insulation of all houses according to government recommended standards would trap the naturally evolving radon inside for longer times and thereby cause an additional 10,000 fatalities/year, so on this basis, energy conservation poses a far greater radiation threat to human health than nuclear power.

The nuclear industry affects these radon health problems through the mining and milling of uranium ore. In the latter process, the uranium is separated out and the residual materials in the ore are pumped into "tailings" ponds that eventually dry out to become large sandlike piles. These residual materials include all the thorium and radium from the original ore, so they continue to emit radon as before, and the emission rate decreases only with the 77,000-year half-life of the thorium. Uranium ore typically contains 500 times as much uranium as average soil, so these mill tailings piles typically release radon at 500 times the rate per square foot of average soil. If no countermeasures are taken, this excess radon emission will eventually cause 300 fatal lung cancers,

---

BERNARD L. COHEN • Department of Physics, University of Pittsburgh, Pittsburgh, Pennsylvania 15260.

half of them within 77,000 years, for each GWe-year of nuclear electricity generation with light-water reactors. This is hundreds of times more than all other radiation health effects from the nuclear industry combined.

Fortunately, however, there is a simple solution to this problem. As radon has a half-life of only 3.8 days, it is only necessary to cover the mill tailings with enough soil for the diffusion time of the gas through it to be many times longer than 3.8 days. Current regulations require that the cover thickness be sufficient to reduce the radon emission 200-fold, which requires 10 to 16 feet of soil with appropriate stabilization measures to assure that it does not undergo abnormally rapid erosion. With these countermeasures, radon emission from mill tailings is expected to cause 1.5 eventual fatalities/GWe-year over the next 100,000 years or so.

While discussing health problems from mill tailings, one should also consider the fact that in mining uranium, a source of radon is removed from the ground. A small fraction of uranium is mined from close enough to the surface to be releasing radon gas into the atmosphere, and the removal of this uranium saves lives; in fact, it saves several times as many lives as are being lost due to radon emissions from covered mill tailings. The net effect of uranium mining and milling, according to the ground rules we have been using, is to *save* about 2 lives/GWe-year.

There are still two aspects that have been left out of our discussion. One is the effect of buildings. Radon levels in buildings are typically 5 times higher than outdoors because radon diffusing up from the ground below or out of bricks, stone, cement, or plaster is trapped inside for a relatively long time. Therefore, if land that is surface mined for uranium is assumed, if not mined, to have had the same probability as average other land to be used for buildings, the saving of life is increased fivefold to about 10 lives/GWe-year eventually saved.

These lives would be saved over hundreds of thousands of years, but there is a widespread feeling that it is meaningless to consider such long time periods. For example, cancer may become a curable disease, or on the contrary civilization may collapse, reducing life expectancy to its pretechnology length of 30 years in which case radiation-induced cancer would be very much less of a threat—it is largely a disease for ages 50 and beyond. Many analysts therefore limit their considerations to about 500 years into the future. If this is done for the radon problem, the net effect is to *save* about 0.05 life/GWe-year. Of course, similar considerations should then also be applied to high-level and low-level radioactive waste, in which case their effects would become completely negligible.

On the other hand, if one opts to consider very long time periods, as has frequently been done by antinuclear activists, we should take into account another new aspect of the problem, erosion of the earth's surface. From the rate at which rivers carry material into the oceans, it is straightforward to calculate that the

North American continent is eroding away at an average rate of about 1 m every 22,000 years. Of course, nearly all of this erosion is occurring in creek- and riverbeds, but rivers and creeks change their courses frequently (if they didn't, the 0.1% of land surface they cover would become 100-m-deep canyons after 22,000 years), and climates change to create new rivers and creeks (e.g., 10,000 years ago, the Arizona desert was a rain forest), so over periods like 100,000 years or more it is reasonable to assume that all land will erode at about the average rate.

This means that *all* uranium mined out of the ground to fuel nuclear reactors would eventually have had its turn on the surface to serve as a source of radon. When this is taken into account (cf. my paper in the Jan. 1981 issue of *Health Physics*), it turns out that mining uranium eventually *saves* about *350* lives/GWe-year. This is thousands of times more than the number of lives lost due to all other health effects of radiation from the nuclear industry combined. If very long time intervals are considered, nuclear power must clearly be credited with being a great *saver* of human life.

Before leaving the subject of radon it should be pointed out that coal contains uranium so that the ash left when coal is burned serves as an important source of radon. Over a 500-year time span, this causes about 0.04 fatal lung cancer/GWe-year of coal-burning electricity, roughly the same as the number *saved* by uranium mining for nuclear electricity. If the effects are added up over very long time periods, the effects of radon from coal burning are to cause 25 fatalities/GWe-year as compared to the 350 lives/GWe-year *saved* by nuclear power. When the radon problem is included, the health advantages of nuclear power over coal burning become truly impressive.

18 Risks in Our Society

Bernard L. Cohen

Our news media are constantly bombarding us with scare stories about radiation, pollution, dangerous chemicals, and other products of our technology. How dangerous are these threats, and how do they compare with other risks we constantly face in our daily lives? In a technical paper published in the June 1979 issue of the journal *Health Physics,* we collected quantitative information on risks from a wide variety of sources and analyzed it in terms of life expectancy reduction (LER), the average amount one's life is shortened by each risk. For example, an average 40-year-old can expect to live another 34.8 years, so if he takes a risk that has a 1% chance of being immediately fatal, his LER is 0.348 year or 127 days. Of course, most risks are with us to varying extents at all ages and the effects must be added up over a lifetime, making the calculations somewhat complex. But the computer has done the work and we will only summarize the results.

One of the greatest risks endured by large numbers of people is remaining unmarried. Statistics show that a *single* white male has a life expectancy of 62.3 years vs. 68.3 years for the *average* white male, corresponding to an LER of 6.0 years. For white females and nonwhite males and females, the LERs are 4.8, 8.8, and 6.0 years, respectively. Note that males suffer much more than females from remaining single and nonwhites suffer more than whites. One might suspect that part of the reason for these differences is that sickly people are less likely to marry but this does not seem to be the case; mortality rates are even higher for widowed and divorced people at every age. For example, the life expectancy for a 55-year-old white male is 19.6 years if he is married, 16.4 years if he is single, 15.7 years if he is widowed, and 13.4 years if he is divorced, assuming in the last three cases that he does not later marry. For white females

---

BERNARD L. COHEN • Department of Physics, University of Pittsburgh, Pittsburgh, Pennsylvania 15260.

these figures are 24.9, 23.0, 22.2, and 22.4 years, respectively, again showing that widowhood and divorce are even more dangerous than staying single.

Next to remaining unmarried, the greatest common risk we encountered was cigarette smoking. For one pack per day, this has an LER of 6.4 years for men and 2.3 years for women—for men this is an LER of 10 min for each cigarette smoked. For noninhalers the lifetime risk from one pack per day is 4.5 years for men and 0.6 year for women, while for those who inhale deeply it is 8.6 years for men and 4.6 years for women. Cigar and pipe smoking are almost harmless if there is no inhalation, but with inhalation the LER is 1.4 years for pipes and 3.2 years for cigars.

Another major risk that we can do something about is being overweight: we lose about 1 month of life expectancy for each pound our weight is above average. For example, the LER for being 20 lb. above average weight is 20 months or 1.7 years. Our weight increases by 7 lb. for every 100-calorie increase in average daily food intake so if an overweight person changes nothing about his eating and exercise habits except for eating one extra slice of bread and butter (100 calories) each day, he will eventually gain 7 lb. and his life expectancy will be reduced by 7 months. This works out to a 15-min LER for each 100 extra calories eaten.

Any discussion of major risks must include the traditional leader, *disease,* which caused life expectancy early in this century to be 20 years less than at present. Among individual diseases, heart disease (LER 5.8 years), cancer (LER 2.7 years), and stroke (LER 1.1 years for men and 1.7 years for women) are numbers one, two, and three. All other diseases combined cost us a little over a year.

Another very important risk is that of being unskilled and/or uneducated. Professional, technical, administrative, and managerial people live 1.5 years longer than those engaged in clerical, sales, skilled, and semiskilled labor, and the latter group lives 2.4 years longer than unskilled laborers. Corporation executives live 3 years longer than average professionals, or a full 7 years longer than unskilled laborers. A similar study in England showed even larger differences, and it found similar differences among wives of workers; the wife of a professional person lives about 4 years longer than the wife of an unskilled laborer. This indicates that the problems are not occupational exposures but rather are socioeconomic.

In seeking to understand the reasons for these differences, it is interesting to consider the causes of death. If we compare unskilled laborers with professional people, their risk of early death from tuberculosis is 4.2 times higher; from accidents, 2.9 times higher; from influenza and pneumonia, 2.8 times higher; from cirrhosis of the liver, 1.8 times higher; and from suicide, 1.7 times higher. These causes are generally associated with unhealthy living conditions and attitudes, and they are largely preventable.

A similar pattern appears in correlations between life expectancy and educational attainment. College-educated people live 2.6 years longer than average, high school graduates live 0.8 year longer, grade school (8 years) graduates live 0.7 year less, and those who dropped out of grade school live 1.7 years less than average. This represents a 4.3-year difference in life expectancy between the most educated and the least educated groups. These differences are about the same for men and women, which again indicates that occupational exposures are not the basic problem here. Dropping out of school at an early age ranks with taking up smoking as one of the most dangerous acts a young person can perform. Even volunteering for combat duty in wartime pales by comparison; the LER from being sent to Vietnam during the war there was 2.0 years in the Marines, 1.1 years in the Army, 0.5 year in the Navy, and 0.28 year in the Air Force.

The most highly publicized risks are those of being killed in accidents—although the actual danger is well below that of the risks we have discussed previously. The LER from all accidents combined is 435 days (1.2 years). Almost half of them involve motor vehicles, which give us an LER of 207 days, 170 days while riding and 37 days as pedestrians. Before the national speed limit was reduced from 65 to 55 mph, the total LER was 40 days higher. On an average, riding 1 mile in an automobile and crossing a street each have an LER of 0.4 min, making them as dangerous as one puff on a cigarette.

We spend most of our time at home and at work, so that is where most of our nontravel-connected accidents occur. The LER for accidents in the home is 95 days, and for occupational accidents it is 74 days. The latter number varies considerably from industry to industry, from about 300 days in mining, quarrying, and construction to 30 days in trade. Nearly half of all workers are in manufacturing and service industries for which the LER is 45 days.

On-the-job accidents are not the only risk, and in fact are not usually the principal risks connected with one's occupation. Estimates of the fraction of all cancer that is due to occupational exposure to chemicals vary from 5 to 20%, which correspond to 50- to 200-day LERs from that source alone for the *average* worker. For those who are directly exposed to chemicals, the LER is presumably considerably higher. There can be little question but that some fraction of heart disease is due to occupational factors; a 1-year LER from this source does not seem unreasonable. Occupational exposures may well play an important role in many other diseases. One area in which a quantitative statement is possible is radiation exposure to workers in the nuclear industry from which the average LER is 30 days.

Some showmanship activities are widely advertised as having very high accident potential, but judging from statistical experience, these dangers are exaggerated in the public mind. For example, professional aerialists—tightrope walkers, trapeze artists, aerial acrobats, and high-pole balancers—get an LER of 5 days per year of participation, or 100 days from a 20-year career, and the risk is similar for automobile and motorcycle racers of various sorts, so the risk of accidental death in these professions is less than in ordinary mining and construction work. The most dangerous profession involving thousands of participants is deep-sea diving with an LER of 40 days per year of participation.

Homicide and suicide are significant risks in our society with LERs of about 135 days each for men, and 43 and 62 days respectively for women. Homicide is more common among the young while suicide becomes several times more important among the elderly. With the toll from homicide so large, one cannot help but question the wisdom of the recent tendency to cut back on police protection and on public lighting.

From the treatment they are given in the news media, one might think that large catastrophes pose an important threat to us, but this is hardly the case. Hurricanes and tornadoes combined give the average American an LER of 1 day, as do airline crashes. Major fires and explosions (those with eight or more fatalities) give us an LER of 0.7 day, and our LER from massive chemical releases is only 0.1 day.

A nuclear reactor meltdown has been portrayed as "the ultimate catastrophe," but even if we had one every 5 years somewhere in the U.S., our LER from them would be only 0.2 day according to government-sponsored scientists, or 2 days according to nuclear critics (this is from a report by the "Union of Concerned Scientists," which acts as scientific advisor to Ralph Nader and buys large newspaper ads calling for a halt to our nuclear power program). The critics estimate that with a full nuclear energy program we may have a meltdown every 5 years whereas the government-sponsored scientists estimate one every 50 years. We have not had the first meltdown yet.

There is a common impression that our prodigious use of energy is a major risk to the average American—how valid is this idea? Burning coal to produce energy causes an LER of 13 days, largely due to air pollution. Accidental electrocution claims about 1200 lives each year in the U.S., and has an LER of 5 days (most of its victims are relatively young). Oil burning causes air pollution and starts fires, which give it an LER of 4 days. Natural gas causes fatalities from asphyxiation, air pollution, explosions, and fires, adding up to an LER of $2\frac{1}{2}$ days. In spite of all the publicity about dangers of nuclear energy, it is generally agreed in the scientific community that all environmental problems other than reactor accidents have much smaller effects than these; a typical estimated LER if all U.S. electricity were nuclear is 0.03 day. For reactor accidents the critics estimate an LER of 2 days while government-sponsored scientists estimate only 0.02 day, so even if we accept the former, the total LER from a full nuclear energy program would be just over 2 days.

When all the effects of energy production and use outlined above are added up, the total LER is about 25 days. Considering the tremendous benefits we derive from them, this does not seem like a high price to pay. For example, it is half the price we pay for not using air bags in automobiles. The most effective improvement we could make in the 25-day LER from our use of energy would be to shift our electricity generation from coal to nuclear—this would cut the LER in half.

Some people worry that there are great environmental risks in our general embracement of technology. A simple test of this viewpoint is the correlation between the life expectancy in various nations and their technological development. In countries with advanced technology like the U.S., western Europe, Australia, and Japan, life expectancy is about 71 years. Life expectancy in a sample of other countries is 68 years in Poland and Rumania, 61 years in Mexico, 55 years in Turkey, 45 years in India, and 32 years in Chad and Ivory Coast. The correlation with technological development is clear, and the complete list strongly reinforces that conclusion. We may infer that technological development brings us about 30 years of increased life expectancy.

This, of course, does not prove that technological development beyond a certain point may not reduce life expectancy. How tenable is that position? As the LER from energy production is 25 days and this uses a substantial fraction of the materials we consume, and produces a substantial fraction of the recognized pollutants in our environment, it seems reasonable to conclude that the LER from all of our technology is not more than 100 days. Counterbalancing this is the fact that technology creates wealth and reduces poverty and we have demonstrated that these change longevity by several years. It therefore seems most probable that even in our present status, the net effect of advancing technology is to increase our average life expectancy.

The media have publicized the dangers of various individual substances from time to time. Coffee is an important cause of bladder cancer, with an LER of 6 days for regular users. There is some evidence that saccharin may cause bladder cancer; the LER from one diet soft drink every day of one's life is 2 days. (Ironically, the weight gain from one extra nondiet soft drink per day causes an LER of 200 days, but still the FDA wants to ban saccharin!) Birth control pills can cause phlebitis, which give its users an LER of 5 days.

Even very tiny risks often receive extensive publicity. Perhaps the best example was the impending fall of our orbiting Sky-Lab, which gave us an LER of 0.002 *sec*. There has been heavy publicity for leaks from radioactive waste burial grounds, although these have not given a single member of the public an LER as large as 10 sec. The Three Mile Island accident gave the average Harrisburg area resident an LER of 1.5 min (0.001 day), and the relatively few people living very close to the plant an LER of up to 2 hr. Our risk of being struck by lightning gives us an LER of 20 hr.

TABLE I
Estimated Life Expectancy Reduction from Risks and Activities

Activity or risk	Days LER
Heart disease	2100
Being unmarried	2000
Cigarette smoking	1600
Cancer	980
Being 30 lb. overweight	900
Grade school dropout	800
Unskilled laborer	700
Stroke	520
Vietnam army duty	400
Mining or construction work (due to accidents only)	300
Motor vehicle accidents	200
Pneumonia, influenza	130
Homicide	90
Drowning	40
Poison + suffocation + asphyxiation	37
Energy production and use	25
Diet drinks	2
Hurricanes, tornadoes	1
Airline crashes	1
All-nuclear electricity	0.04–2[a]
Harrisburg area residents (from TMI accident)	0.001
Radioactive waste burial ground leaks, risk to nearest neighbors	0.0001
Sky-Lab fall	0.00000002

[a] The lower number is the estimate of government-sponsored scientists, and the higher number is the estimate of nuclear critics.

In order to facilitate further discussion, we have collected some of the figures given above in Table I.

This table is abbreviated and includes some simplifications such as averaging between men and women but it will suffice for our purpose of considering measures to reduce risks.

It is clear from Table I that the dominant risks in our society are behavioral, disease related, and socioeconomic, and it is in these that the opportunities lie for important reduction in risks. Research on heart disease, cancer, and stroke can easily give us a life expectancy increase—LEI, which is the negative of LER—of hundreds of days. Even such a simple measure as introducing mobile intensive care units, rapidly responding ambulances with sophisticated emergency equipment and well-trained paramedics, has contributed an LEI of over 100 days to those served by them, largely by saving heart attack victims. This service, by the way, is still available only in larger cities. There are many opportunities for early detection of cancer by screening, which could save tens of thousands of lives at costs in the range of $50,000 per life saved,* again contributing perhaps 100 days of LEI to those serviced by them.

It seems likely that such simple measures as improved dating and counseling services could reduce the effects of being unmarried by 5%, thereby giving 100 days of LEI to those affected. It would seem that cutting back on cigarette advertising could reduce smoking by 6 or 7% and therefore contribute an LEI of 100 days to 40% of our population. Printing calorie content in large type on packaged foods, and instituting publicity campaigns about the mechanics of weight reduction, rather than leaving that field to faddist advertisers after a fast buck, could easily give 100-day LEIs to nearly half of our citizenry.

Combating socioeconomic factors is perhaps more difficult. One possibility would be to take special measures to reduce mortality from pneumonia and influenza, which is largely preventable as evidenced by its threefold lower toll among the more privileged socioeconomic classes. This alone could easily give a 100-day LEI to the underprivileged. However, the real solution to socioeconomic risks is to reduce poverty and menial labor. We have made great strides in this direction during this century, largely through advancing technology, and a strong case could be made for encouraging that process to continue. Some social planners argue that our society would benefit from less technology and more manual labor—Table I speaks powerfully against this idea, as do the statistics on life expectancy in nations now in that situation.

Because they are well suited to media coverage, accidents have always received more than their share of public attention, and important advances have been made in reducing them. Over the past decade, the death rate from accidents in the U.S. has decreased by 22%, which corresponds to an LEI of 110 days for every American. But there is still room for further progress. Air bags in automobiles would give a 50-day LEI to the average American. Reducing the national speed limit from 65 to 55 mph gave us a 40-day LEI, and better enforcement would undoubtedly increase this appreciably as well as helping to save gasoline. There are many improvements in highway safety like impact-absorbing guardrails, breakaway posts on roadside lights and signs, and skid-resistant surfaces at danger spots that could give us several days of LEI. Risks from fires have been substantially reduced by the development and deployment of smoke detectors in homes, giving an LEI of 10 days to those who have them. While accident risks are considerably less important than behavioral, socioeconomic, and disease-related risks, there is still a reasonable incentive for working to reduce them, and they are being reduced, largely through improved technology.

The only other item in Table I large enough to warrant discussion is the

---

* Estimates of dollars expended per life saved are from my paper "Society's valuation of life saving in radiation protection and other contexts," *Health Phys. 38,* 33 (1980).

25-day LER we experience due to energy production. About half of this is due to air pollution, and great efforts are being made to improve on it. Cosmetic effects like removing the visible dirt and foul odors have been successful and they are important as ends in themselves, but they have not greatly affected our LER from air pollution, and it is unlikely that major improvements can be affected as long as coal burning is widely employed. As noted above, the problem could be resolved by conversion to nuclear energy, but the scientific community has been outpoliticked on that possibility. In any case, the LEI one could expect from great progress in air pollution abatement is only about 10 days, which is hardly worthy of great concern in our overall view of risks. It could be completely counterbalanced by simply installing smoke alarms in all homes—only about 25% of homes now have them.

But the most important lesson to be learned from Table I does not deal with techniques for saving lives or extending life expectancy. It is rather a lesson on the shortcomings of information transmittal in our society. Few overweight Harrisburg area residents recognize that their risks from radiation received in the TMI accident is hundreds of thousands of times smaller than their risk from overeating. Few unskilled laborers or their wives realize that the risks inherent in their life-style are hundreds of times greater than their risks from tornadoes, major fires and explosions, airline crashes, and chemical releases combined. The reason for this situation is that the public gets its information through the news media, but providing information to the public is *not* the highest priority of the media people. They are in a highly competitive business where the requirement for survival is to catch and hold the attention of their audience. Unfortunately, stories about a falling satellite, or a town being destroyed by a tornado, or a leak of "deadly radioactivity" are much better suited to this purpose than such mundane matters as overeating and health neglect, so they concentrate on the former. With the media focusing so much attention on sensationalist risks, it is only natural that the public gets the impression that they are what it should worry about.

In our governmental system, public concerns drive government action. There were well-publicized Congressional hearings on the fall of Sky-Lab, on leaks from radioactive waste burial grounds, and on the radiation from the TMI accident, while some of the top items in Table I receive virtually no attention even though they represent millions or billions of times more risk. Our government is spending hundreds of millions of dollars per life saved in its radioactive waste management activities, while it passes up opportunities to save a life for each $50,000 spent in medical and highway safety programs. In a participatory democracy, government must be responsive to public concerns, even if it means assigning a backseat to more pressing public needs. That is the way it is, and that is the way it should be.

But we are still left with the very grave problem of properly educating the public on what it should be concerned about. A heightened sense of responsibility on the part of media people would help, but even the best-intentioned of them must do what is necessary to survive. As long as the public depends on the news media for its information and demands entertainment in the package, misplaced public concern will remain a grave problem, and we will pay heavily both in money and in human suffering as a result. The great power of television has compounded this problem immensely.

Our problem here is basically the age-old problem of "conflict of interest." We seem to have largely solved it in simple cases where personal profit and governmental service are combined, and we have made some progress in situations where a governmental agency both promotes and regulates an industry. There is widespread discussion of the conflict of interest for oil companies in both serving the public and making a profit, and there have been mighty attempts for government to intercede. But there is no more potentially dangerous conflict of interest in our society than the dual role of the media in both informing and entertaining the public especially because their first priority is in the *entertaining,* while the needs of the public are in the *informing,* and even more especially because there is no possible way for government to intercede.

V Special Nuclear Issues, Past and Present

Part V is devoted to the destruction of a large number of myths and misconceptions that have been promulgated by antinuclear spokesmen. Most of the promulgators of the mythology know better; that some of them claim to be scientists makes us particularly bitter.

The first five papers concern myths about radioactivity. The group is led off by two papers by B. L. Cohen: "Health Effects of Low-Level Radiation" and "Routine Releases of Radioactivity from the Nuclear Industry." In the first paper, Cohen explains how our knowledge of radiation effects is derived, and from this knowledge describes how very conservative radiation standards really are. In the second, he considers how the very small releases of radiation from the nuclear industry are generated, what causes the releases, what their pathways are to humans and to the environment, and how the resulting radiation compares with other radiation sources in the environment. The nuclear industry source is trivial in comparison with them and hardly worth the trouble of abating further.

Some well-publicized "studies" have claimed to show large effects of nuclear radiation on people. These studies are scientifically invalid in so many ways that they only find their way into the respectable scientific literature when political pressure is put on journal editors. (Yes, Virginia, it happens!) The next paper, K. O. Ott's "Low-Level Radioactivity and Infant Mortality," briefly documents what is wrong with one of the most well-publicized such works of pseudoscience: Ernest Sternglass's contention that radiation is responsible for a noticeable increase in infant mortality.

Sternglass's work has, of course, been consistently rejected by reputable scientists, so criticizing it has all the sport of shooting fish in a rain barrel. A more difficult myth to detoxify is the persistent myth that plutonium is fantastically poisonous. Yet myth it is, and B. L. Cohen, in "The Myth of Plutonium Toxicity," explains why. Then, in "Myths about High-Level Radioactive Waste," he returns again to this bugaboo, which he already has expanded in Chapter 16.

We are frequently confronted with the question "if high-level radioactive waste is so easy to manage, why hasn't it already been tucked away?" The answer is a sad one. Every attempt to select a facility has been foiled by essentially political interventions by nuclear opponents.

Myths about radiation dominate our perception of what really bothers people, so we have dwelt on them at some length. But there are other myths that need to be demolished as well. In "The Aging Reactor Myth," A. D. Rossin examines the misconception that nuclear power plants wear out very rapidly. Alas, the data that prove this myth have been badly "cooked," and when correct data are substituted we see that nuclear plants follow availability curves that are little different from other types of power plants. If anything, they hold up better, partly because regulatory supervision requires their maintenance to be more meticulous than is the case for coal- or oil-fired plants. In "The Police State Myth," B. I. Spinrad examines the contention that nuclear power jeopardizes civil liberties. Contention is all this is; proofs do not exist. Even our nuclear weapons program has not intruded into the liberties of the citizenry of our country. C. Whipple, in "Insurance and Nuclear Power: The Price–Anderson Act," examines a number of myths about nuclear insurance. He shows that it has been profitable to the government as insurer, less restrictive on claims of injury than normal insurance, and backed up by the clear intent of Congress to extend the coverage in the extraordinarily unlikely event that such is needed. Finally, in "Solar and Nuclear Power are Partners," Rossin debunks the claim that nuclear power stands in the way of solar power development. In fact, these two sources of electricity are synergistic: solar electricity is most economical when used for intermediate (daytime) power demands, while nuclear power is used for base-load (day and night). Of course, solar electricity would be very expensive to install when and if it becomes economical to use it (it is not so now). It could only be afforded if the rest of the electrical bill is low enough so that one can put that installation money—e.g., capital costs—aside. That, of course, is what nuclear power is all about. The bottom line, for both power sources, is keeping consumer costs under control so that we can afford to use electricity, not wastefully, but as an abundant resource.

B.I.S.

19 Health Effects of Low-Level Radiation

Bernard L. Cohen

1. HOW DANGEROUS IS RADIATION?

What is radiation, and how dangerous is it? Radiation consists of subatomic particles that shoot through space at enormous speeds like 100,000 miles per *second*. They easily penetrate deep inside the human body where they can strike and damage biological cells, and thereby cause cancer or genetic defects in later generations.

This sounds very scary, and one can easily get the impression that being struck by one of these particles is a tragic event. However, this cannot be so, because each of us, and every other person who has ever lived or who ever will live on this earth, is struck by about *15,000* of these *particles of radiation every second* of his life *from natural sources,* and when we get a medical X-ray we are struck by about 100 billion of them. The thing that saves us is *not* that it takes some minimum number of these particles to do us harm—as some like to say, no level of radiation is perfectly safe. Any single one of these particles can cause a fatal cancer or a genetic defect in our progeny, but the probability that it will do so is only one in 30 quadrillion (30 million billion).

If we are worried about this risk, there are many things we can do to reduce our exposure. We can wear metal-lined clothing to shield us, like the apron the dentist often lays across us when he takes an X-ray. We can choose the building materials of our homes; brick and stone contain more radioactivity than wood and therefore expose us to more radiation. We can choose to live in areas where natural radiation is lower; in Colorado it is twice the national average whereas in Florida and the Gulf coast area it is well below the average, and there are even important differences between different sections of the same city.

BERNARD L. COHEN • Department of Physics, University of Pittsburgh, Pittsburgh, Pennsylvania 15260.

But most of us don't worry about such things. We recognize that life is a series of risks. Every breath of air may have a germ that will cause fatal pneumonia, but we continue to breathe. Every bite of food may have a chemical that will give us cancer, but still we continue to eat. Every time we get in an automobile we recognize that we may be killed in an accident, but still we drive. We are willing to engage in these games of chance as long as the odds are heavily in our favor. And in the case of radiation we recognize that 30 quadrillion to one is pretty good odds. Unfortunately, however, that point is missed by some sensationalist writers who have produced pages of prose about the horror of being struck by a particle of radiation.

In evaluating dangers of radiation, two rules are paramount. First, it is important to be *quantitative,* not just qualitative. With qualitative reasoning, almost any human activity can be shown to be dangerous; this surely applies to any technology for generating energy, and even more to doing without energy. Only on the basis of quantitative reasoning can rational decisions be made. The second rule is to keep things in perspective—don't focus on one danger while ignoring all others. With these rules in mind, let us proceed.

The usual unit of radiation dose is the millirem. Exposure to 1 mrem of radiation corresponds to being struck by about 5 billion particles. The reason for using the millirem unit rather than the number of particles is that it includes corrections for the type of radiation particle, its energy, the size of the person, etc.

A radiation exposure of 1 mrem has about one chance in 8 million of causing a fatal cancer, and about an equal chance of causing a genetic defect in later generations. These are the estimates by the U.S. National Academy of Sciences Committee on Biological Effects of Ionizing Radiation (known as BEIR), which is composed of about 20 leading American experts in that field, by the United Nations Scientific Committee on Effects of Atomic Radiation (UNSCEAR), which is composed of scientists from 20 countries in 6 continents representing all parts of the political spectrum, and by the International Commission on Radiological Protection (ICRP), which has a Swedish chairman, a British Secretary, and 12 members from 8 different countries. All of these groups agree rather closely on the above risk estimates, and they are used by all national, state, and local organizations charged with responsibilities in the field all over the world, including the U.S. Environmental Protection Agency, the U.S. Nuclear Regulatory Commission, the U.S. National Council on Radiation Protection and Measurements (NCRP), etc.

In quoting a value for the risk per millirem, we are tacitly implying that the risk increases *linearly* with the dose; i.e., if the dose is doubled, the risk is doubled. This is a highly convenient assumption because there is a great deal of information on effects at high doses from a wide variety of sources, including the Japanese A-bomb survivors; British patients treated with high doses of X-rays for ankylosing spondylitis, an arthritis of the spine; German patients injected with an isotope of radium for that same disease; British women given X-ray treatments for menorrhagia, a gynecological malady; American women employed in painting radium numerals on watch dials; infants treated with X-rays for tinea capitis, a ringworm of the scalp; Nova Scotia women in a tuberculosis sanitorium where there was excessive use of diagnostic X-rays; children treated with X-rays for a thymus condition; American women treated with X-rays for mastitis, an inflammation of the breasts following childbirth; American, Czechoslovakian, Swedish, and Canadian miners exposed to high levels of natural radioactivity in the form of radon gas due to poor ventilation in mines; etc. These data are all in generally good agreement on the effects of high-level radiation, so if linearity is accepted, effects of low-level radiation are readily calculated by simple proportionality with dose.

There is some question over whether this simple proportionality with dose can be extended to low levels, with the majority of the community and all of the prestigious committees believing that this is a "conservative" procedure, more likely to overestimate than to underestimate effects of low levels. The evidence that it *over*estimates effects at low doses comes from various sources: (1) well-established repair mechanisms that should usually repair damage from low doses; (2) widely accepted theories of how radiation induces cancer; (3) experiments with animals; (4) data on leukemia among the Japanese A-bomb survivors; (5) data on bone cancer among the radium watch dial painters; (6) a comparison of effects from normal environmental radon and the high levels of radon in some mines; and (7) the fact that cancer takes longer to develop from low doses so death from other causes may occur first.

All of these lines of evidence indicate that low levels of radiation should be less harmful than predicted by linearity. Nevertheless, until recently most prestigious groups condoned the use of linearity to estimate effects of low levels, and we have used it here. However, there has recently been increasing resistance to this path. The NCRP officially declared that it overestimates effects at low doses by a factor of 2 to 10. The National Academy of Sciences BEIR Committee was badly split on the issue, and could not produce a consensus report until the two members representing the extremes of the spectrum of opinions were dropped off. These two wrote minority reports, one espousing strict linearity, and the other declaring that linearity gives a gross overestimate of effects at low levels. The majority report recommended a procedure according to which linearity gives a moderate overestimate. In the low-dose region, this procedure reduces to use of the linear hypothesis, and we use it here.

There has been a tiny minority that contends that linearity underestimates effects of low-level radiation, which would indicate that low-level radiation is *more* dangerous than given by previous estimates, and the media have given tremendous publicity to their viewpoint. The principal evidence they offer is a paper known as the "Mancuso study," which purports to find a number of excess cancers of certain types in a group of radiation workers at the U.S. Government's Hanford (Wash.) Laboratory. The same data have been reanalyzed by three independent groups and the only excess they find is two extra cases each of multiple myeloma and cancer of the pancreas; but there is no evidence of such excesses among the tenfold more numerous Japanese A-bomb survivors who were exposed to the same radiation dose, and these types of cancer were not found to be enhanced relative to other types among any of the highly exposed groups. It therefore seems reasonable to surmise that the four extra cases could have been due to some type of chemical or other exposure.

The Mancuso paper has drawn at least 25 critiques and has been pointedly rejected by the BEIR Committee (unanimously) and the British National Radiological Protection Board (NRPB). It has been ignored by all national and international standard-setting bodies in that they have not altered their standards, although they would be bound to do so if they gave any credence to the Mancuso paper. Nevertheless, the media have continued to shower it with publicity and have virtually ignored the critiques and rejections. The New York Times Information Bank contains 11 entries on Mancuso, but a total of only one on all the critiques. The media have used the dissension in the BEIR Committee to imply that there is support for the Mancuso contention that linearity underestimates effects of low doses. Actually the dissension is on the opposite side of the issue, between those favoring linearity and those who contend that linearity *over*estimates effects of low-level radiation.

2. RADIATION RISKS IN PERSPECTIVE

As 1 mrem is a typical radiation exposure in highly publicized incidents—for example, the average exposure received by nearby citizens in the Three Mile Island accident was 1.2 mrem—let us pause to give some perspective on the dangers of a 1-mrem exposure. As already noted, it has one chance in 8 million of causing a fatal cancer, which corresponds to reducing life expectancy by 1.2 min. This is the amount of life expectancy we would lose in

- Crossing streets four times (due to the risk of being killed by a car)
- Taking three puffs on a cigarette
- Eating 10 extra calories (e.g., one lick on an ice cream cone) if we are overweight
- Raising the national speed limit from 55 to 55.02 mph for 1 year
- Being without a home smoke alarm for 1 week
- Being exposed to typical city air pollution for 1 week

19 • Health Effects of Low-Level Radiation

In addition, this 1 mrem of radiation can cause genetic defects in our progeny. Another source of genetic risk is delaying conception of children, for the risk of genetic disease increases with parental age. We can thus calculate an equivalence between these two risks, and it works out that 1 mrem of radiation has about the same genetic effects as delaying conception of a child by $1\frac{1}{2}$ hr.

With regard to genetic effects, there is a popular notion that they can do long-range harm to the human race. This is absolutely false: genetic selection acts to breed-out bad mutations and breed-in good mutations, so the *long-term* effect of additional radiation is bound to be *favorable* rather than unfavorable. Actually, this long-term effect is negligibly small and the short-term effects do cause human misery so genetic effects of radiation are considered to be bad, but not because they represent a danger to the human race.

Another perspective on genetic effects of radiation may be derived from studies of the Japanese A-bomb survivors. There were about 24,000 in the group who were exposed to an average of 130,000 mrem each, but in the first generation of children born to these people there is no detectable excess of genetic defects.

3. HIGHLY PUBLICIZED INCIDENTS

With this background in understanding the risks from radiation, let's review some of the highly publicized incidents involving radiation to the public. By far the most serious was the TMI accident in which 2 million people living within 50 miles received an average exposure of 1.2 mrem. On an average, each of them has an increased probability of dying of cancer equal to 1.2 chances in 8 million, or about 1 chance in 6 million. With 2 million people having this risk, the total probability that there will some day be a single resulting death is 2 million chances in 6 million, or 33%.

The idea that some day someone may die as a result of releases from an industrial plant seems to be a new one to many people. But 10,000 Americans die each year—more than one every hour, day and night—as a result of toxic chemical releases, better known as air pollution, from coal-burning plants. This means that every time a coal-burning power plant is built, about 1000 people are condemned to an early death from air pollution. Even the antinuclear activists do not claim that there will be more than 100 extra fatalities (on an average) from a nuclear power plant, and most scientists estimate no more than 2.

All of the other highly publicized releases of radioactivity have been far less consequential than the TMI accident. Perhaps the second most publicized radiation scare was the crash of a Russian satellite containing radioactive material in Canada. This was in the headlines for several days, but there were no exposures of more than a few millirems.

There have been highly publicized scare stories about radioactivity leaks from waste burial grounds in Kentucky and New York. The Philadelphia Evening Bulletin, for example, ran a three-part series on these headlined "It's spilling all over the U.S.," "Nuclear grave is haunting Ky.," and "There's no hiding place." But from neither of these leaks were there *any* exposures as large as 0.1 mrem, 12 times smaller than the TMI exposures. Moreover, there were only a handful of people involved, so clearly no cancers are to be expected. Part of the reason is that only a tiny fraction of the buried radioactive material leaked out. But actually even if *all* of the radioactivity in those burial grounds were to leak out of their trenches and spread through the soil, probably there would never be a single fatality (this is demonstrated in my paper in the Dec. 1978 issue of *Nuclear Technology*). The reason for this is that material in the ground does not easily get into people, even if the land is used to grow food crops. A typical atom in the top layers of soil has only about one chance in a billion each year of getting into human food. That is why it is considered acceptably safe to bury short-half-life low-level radioactive materials in shallow trenches. A similar consideration applies to radioactivity released into the air; a particle released *in a city* has only about one chance in 100,000 of ever being breathed in by a person.

There has been wide publicity for some leaks in high-level liquid waste storage tanks in Washington State. These were tanks based on an obsolete technology and used for military waste, but the leaking material has moved only a few feet through the highly absorbing soil, and as it is 40 feet underground with no apparent way to get to the surface or into the river before the radioactivity decays away, it is most difficult to see how anyone can ever be exposed to radiation from those leaks.

There are frequent stories in newspapers about transport incidents involving radioactivity in which a package falls off a truck, or leaks a little liquid onto the road. There have been about 100 such incidents over the last 35 years. But these packages carry only small quantities of weakly radioactive material—high-level waste is carried in special shipping containers that have never broken open or leaked. Typical exposures from these accidents are less than 1 mrem, and there have never been exposures higher than 10 mrem to members of the public. As only a very few people are ever exposed in these accidents, calculations indicate that there is less than a 1% chance that there will ever be a single fatality from all of these accidents combined.

4. MEDIA EXAGGERATION

In spite of this near-perfect record, the public views radiation as a major threat to its safety. How did this peculiar situation come about?

One reason is overcoverage by the media. I did a study of entries on accidents

19 • Health Effects of Low-Level Radiation

in the New York Times Information Bank, which is representative of newspaper coverage, over the 5-year period 1974–78 (before the TMI accident). There was an average of 120 entries per year on accidents with motor vehicles, which kill 50,000 people annually; on industrial accidents, which kill 13,000, the average was 50 entries per year; on asphyxiation accidents, which kill 4500, there were 25 entries per year. But on accidents involving radiation, there were 200 entries per year, more than for the other three types combined, even though there was not a single fatality observed or ever expected from any radiation accident during that time period.

In addition to overcoverage, the media constantly use inflammatory adjectives—"deadly" radiation, "lethal" radioactivity. These adjectives are not used for motor vehicles, which kill 50,000 people each year, for ordinary air pollution, which kills 10,000, or for electricity, which kills well over a thousand by electrocution. With over 100,000 people a year killed in accidents, it is difficult to understand all this attention to a type of accident that isn't responsible for a single one of them.

The media hardly ever compare radiation from the nuclear industry with natural radiation or with medical X-rays, which give the average American hundreds of times more exposure. If all our electricity were nuclear, all radioactivity emissions from the nuclear industry other than from reactor meltdown accidents would expose the average American to about 0.2 mrem/year; this is the conclusion of many impartial studies including those of the American Physical Society, UNSCEAR, the European Economic Community (OECD), and the Environmental Protection Agency. By comparison, natural radiation exposures average about 100 mrem/year, ranging up to 200 mrem/year (in Colorado), and medical X-rays give the average American about 50 mrem/year. In addition, such activities as an airplane trip, a year of typical TV viewing, and carrying a luminous dial watch give people about 1 mrem/year each. Wouldn't it add perspective if media stories were to mention these comparisons? But they hardly ever do.

The media constantly give the impression that health effects of radiation are not well understood. Actually they are far better understood than health effects of air pollution, food additives, industrial chemicals, or a host of other environmental hazards. As a consequence of all this media overcoverage and distortion, the public has been driven "insane" over fear of radiation. It views it as one of the most important hazards in life, although only when it is associated with the nuclear industry and not when it is from natural sources or medical X-rays.

An important consequence of this distorted view on radiation is that the public views nuclear energy as far more dangerous than coal burning although there have been well over a dozen scientific studies of the comparison and they all conclude that nuclear is safer than coal. (This does not include effects of radon from uranium mill tailings as that is a passé issue because the tailings piles are being covered.) Among the groups concurring in this conclusion are committees of the National Academy of Sciences, the American Medical Association, a high-level Ford Foundation Study, and even a report by the antinuclear activist organization Union of Concerned Scientists. Even Ralph Nader has admitted this to me privately.

But the media don't tell the public about these studies, or about the 10,000 fatalities per year from air pollution due to coal burning. It is very difficult to understand the reason for this neglect. The question of nuclear vs. coal involves many thousands of fatalities and tens of billions of dollars per year, so it is surely important. If the public clearly understood that nuclear is the safer of the two, it might well choose nuclear; at least that would be an important element in the decision. Failure to inform the public on this point is surely a grave irresponsibility that, over the years, will cost our country tens of thousands of lives and hundreds of billions of dollars.

5. OCCUPATIONAL EXPOSURE TO RADIATION

Up to this point, we have been discussing radiation exposure to the public, but many people (including myself) are employed in jobs requiring work with radioactive materials and they receive radiation doses hundreds of times larger and hence face hundreds of times higher risks. Are these risks acceptable?

The maximum permissible exposure is 5000 mrem/year, but only a tiny fraction of all radiation workers reach this limit in any one year and the same people would not reach it year after year. About 800,000 Americans work with radiation to the extent that their exposure is routinely monitored, and it averages about 200 mrem/year. The 55,000 monitored workers in nuclear power plants receive an average of 400 mrem/year. We concentrate on the latter group and compare their risk with those of other workers who face other occupational hazards. We do this in terms of loss of life expectancy (LLE) due to these risks.

An exposure of 400 mrem each year from age 18 to 65 increases one's probability of dying of cancer by 1.3% and thereby reduces one's life expectancy by 12 days, so the LLE is 12 days. The simplest comparison is with the risk of being killed in an occupational accident as statistics on this are readily available. For all 80 million American workers, occupational accidents give an average LLE of 74 days. This number varies from industry to industry, and is highest in the mining and construction industries for which the LLE is 330 and 300 days, respectively.

These broad industry categories average over large variations. In Canada, where the average LLE for all workers is 66 days, forestry and fishing are listed as separate categories, with LLE of 540 and 400 days, respectively (in U.S.

statistics these industries are averaged in with broader categories). But within an industry, there are much larger variations. For example, in the Canadian construction industry, demolition workers have an LLE of 1560 days, while plumbers and electricians have an LLE of 38 days; and in Canadian forestry, log fellers have an LLE of 1060 days while sawmill and paper mill workers have an LLE of 50 days. Nuclear power plant workers are part of the U.S. "transportation and public utilities" industry for which the LLE is 165 days. Their 12-day LLE due to radiation might be viewed in perspective with some of the above numbers.

But radiation-induced cancer is a *disease* rather than a sudden fatal injury, so it should really be compared with other *diseases* caused by occupational exposures. Only in a relatively few cases is there a clearly understood link between a particular occupational exposure and death from a particular disease, as in the case of asbestos and mesothelioma, but there are large variations in life expectancy between different occupations even after correcting for socioeconomic factors, and these may be presumed to be due largely to occupational diseases. In British studies, many occupational groups, like gardeners, postal workers, packers, government officials, university teachers, and corporate officials live about 500 days longer than the national average (for their socioeconomic class), while there are many occupational groups for which life expectancy is shorter than average, for example by about 900 days for coal miners, tailors, brewers, printers, and cooks, and by about 1200 days for fishermen, actors, musicians, watchmakers, shoemakers, and pharmacists. U.S. data are only available by industries and are not corrected for socioeconomic factors but similar patterns are found. For example, life expectancy is about 500 days longer than average for employees of the airline, communication, and postal industries, and about 1000 days shorter than average for workers in the coal mining and motor freight industries and for policemen and firemen. From data of this type and other considerations, it can be estimated that the *average* worker loses about 500 days of life expectancy due to occupational disease. The 12 days of life expectancy lost by radiation workers due to their radiation exposure are clearly small compared to many other widely accepted occupational hazards.

The most serious U.S. radiation episode on record is the exposure of uranium miners in the 1945–68 period to high levels of radon, a naturally radioactive gas that can build up to high concentrations where ventilation is poor. Under these conditions, inhalation gives large radiation exposure to the bronchial regions, which can lead to lung cancer. At that time, radon monitoring was sporadic at best, and enforcement was under the jurisdiction of state mine inspectors who had little knowledge or understanding of the problem. Radon levels are expressed in units of WLM (working-level-months), with the allowable exposure 12 WLM/year at that time, but average exposures were several times that figure and tenfold higher exposures were not uncommon. Starting in the early 1960s, evidence of excess lung cancer began to appear; up to 1974, there was an excess of 134 cases. When the problem became apparent, the federal government took jurisdiction, greatly tightened enforcement, and in 1969 lowered the maximum permissible exposure to 4 WLM/year. We here address the question of whether that regulation is providing adequate protection.

If the risk is proportional to the exposure in WLM (linear hypothesis), the U.S. miner data indicate a lung cancer risk of about 1.5 chances in 10,000 for each WLM of exposure. Another way of estimating this risk is by considering environmental radon exposure to nonsmokers (radon is present everywhere so all of us are continuously exposed), which gives a maximum risk/WLM about 4 times lower. We will give results using both of these methods of risk estimation.

Under present working conditions, average uranium miner exposures are about 1.3 WLM/year. With this exposure each year from age 18 to 65, a man's LLE is 11–45 days (the two values represent results of the two risk estimation methods), and his probability of death from lung cancer is increased by 3–12%.

The LLE of 11–45 days must be viewed in perspective with the LLE from other occupational diseases, which for the *average* U.S. worker was given in the last section as about 500 days, and for miners is an additional 1000 days (due to problems other than radon). Miners suffer excessively from many diseases. In a British study, their mortality rates were above average by 119% for nonmalignant respiratory diseases, 84% for infective and parasitic diseases, 60% for diseases of the nervous and genitourinary systems, 55% for accidents, poisoning, and violence, etc. In a study of U.S. coal miners, mortality rates were found to be above average by 57% for nonmalignant respiratory diseases, and 38% for stomach cancer. In California and Washington, mortality rates for all miners were found to be higher than for the general population by 2300% for silicosis and by 350% for tuberculosis.

With regard to the 3–12% increased lung cancer rate expected for uranium miners, it should be recognized that many occupational groups suffer much more excessively from lung cancer. In the British study, lung cancer rates more than 50% above average were reported for foundry workers, riveters, welders, electroplaters, plasterers, paint sprayers, and boiler firemen. In the Los Angeles steel manufacturing industry, lung cancer rates were reported to be 180% above average, and there are many occupational categories in which lung cancer rates in Los Angeles County were reported to be more than 100% above average. Coal miners have about 13% above average lung cancer rates, and miners of all types in California and Washington have about 35% above average lung cancer rates (there is no uranium mining in California, and very little in Washington).

In these perspectives, the lost life expectancy at present levels of radon exposure to uranium miners is small compared to the effects of other occupational diseases.

20 Routine Releases of Radioactivity from the Nuclear Industry

Bernard L. Cohen

1. SOURCES OF RADIOACTIVITY

A light-water reactor consists of long, thin rods (fuel pins) of uranium oxide (UO_2), enriched to about 3% in ^{235}U, submerged in water. In the proper geometry, this arrangement permits a chain reaction in which a neutron striking a ^{235}U nucleus induces a fission reaction, which releases neutrons some of which induce other fission reactions, etc. Each fission reaction releases energy, which is rapidly converted to heat, warming the surrounding water. The reactor therefore serves as a gigantic water heater; as water is pumped through at a rate of hundreds of gallons per second, it is heated to 600°F. The hot water may then be converted to steam either in the reactor [boiling-water reactor (BWR)] or in an external heat exchanger called a steam generator [pressurized-water reactor (PWR)]; the steam then drives a turbine, which turns a generator to produce electric power.

The fuel is in the form of UO_2 ceramic pellets about 1.5 cm long by 1 cm in diameter. About 200 of these pellets are lined up end to end inside a zirconium alloy tube (cladding), which is then sealed by welding. The reactor fuel consists of about 40,000 of these fuel pins, or a total of about 8 million pellets.

When a ^{235}U nucleus is struck by a neutron and undergoes fission, the two pieces into which it splits, called "fission products," are ordinarily radioactive—this is the principal source of radioactivity in the nuclear energy industry. The pieces fly apart with considerable energy, but they are stopped after traveling

---

BERNARD L. COHEN • Department of Physics, University of Pittsburgh, Pittsburgh, Pennsylvania 15260.

only about 0.02 mm (about 1/1000 of an inch), so nearly all the radioactivity remains very close to the original uranium nucleus inside the ceramic fuel pellet. The same is true for the second most important source of radioactivity, the neutrons captured by the uranium to make still heavier radioactive nuclei—neptunium, plutonium, americium, and curium.

Although nearly all the radioactive nuclei remain sealed inside the ceramic pellets, a few of the fission products have a degree of mobility, and some small fraction of them—for example, 10% of the ^{85}Kr, 1% of the ^{131}I, and lesser fractions of many other fission products—eventually diffuse out of the pellets, but they remain contained inside the cladding. However, during the operation of the reactor, about one or two per thousand of the fuel pins develop tiny leaks in the cladding, releasing the radioactive material that has diffused out of the pellets into the water.

An additional source of radioactivity in a reactor is neutrons striking the nuclei of atoms of structural materials, especially the steel plates that hold the fuel pins in place, and the zirconium alloy fuel cladding. When a neutron is captured, the nucleus is changed into a new nucleus, which is usually radioactive, and when these materials are corroded, some of the radioactive nuclei get into the surrounding water.

There are two radioactive isotopes that merit special discussion here. One is carbon-14, which is not a fission product but is produced by neutrons striking nitrogen nuclei followed by proton emission. Nitrogen is present in reactors only as a low-percentage impurity element in the fuel and in the water, but enough ^{14}C is formed from it to warrant significant consideration, especially in view of its 5000-year half-life. Most of it combines with oxygen to form carbon dioxide gas.

The other special isotope is tritium(^{3}H), a radioactive isotope of hydrogen with a 12-year half-life. In one fission reaction in 500, the uranium nucleus splits into three pieces rather than just two, and in 5% of these cases, one fission reaction in 10,000, the third piece is a ^{3}H nucleus. A more important source of ^{3}H in a PWR is neutrons striking boron nuclei, leading to a nuclear reaction in which ^{3}H is released. Boron is a neutron absorber and a small amount of it is dissolved in the water to reduce the number of neutrons when the fuel is fresh and therefore producing an overabundance of neutrons; as the fuel burns out, the boron is withdrawn to keep the neutron production constant. Neutrons captured by deuterium (^{2}H), a rare but omnipresent isotope of hydrogen, serve as a third source of ^{3}H in light-water reactors; in the heavy-water reactors used in Canada, this is a very important source.

As an isotope of hydrogen, ^{3}H readily changes places with the ordinary hydrogen in water, and it is then very difficult to remove. The water in reactors therefore contains a considerable amount of ^{3}H.

2. GASEOUS RELEASES

As the water circulates between the reactor and the steam generators in a PWR, a small fraction of it is diverted on each cycle to a chemical cleanup system in which it is filtered and then passed through a series of demineralizers and activated carbon, which remove nearly all of the radioactive materials except the gases like the isotopes of krypton and xenon, ^{14}C in carbon dioxide, ^3H in water vapor, and small amounts of iodine, a very volatile substance that sometimes tends to behave like a gas. The materials removed in the chemical cleanup are eventually buried in a low-level waste burial ground. The gaseous materials are stripped out of the water by converting the latter to steam and then recondensing it. They are then passed through solutions containing chemicals for scrubbing out as much of the iodine as is practical, and stored in tanks for several months to allow the short-half-life isotopes to decay away. Eventually they are released into the environment from the top of a tall stack in order to avoid large concentrations at any one place on the ground. These gas discharges are one of the principal sources of radioactivity release into the environment from routine operation of nuclear power plants.

In a BWR, the water is converted to steam in the reactor and the steam goes directly to the turbine and then to the condenser where it is condensed back to water, which is returned to the reactor after passage through a large chemical cleanup system. Proper operation of the turbine requires that the condenser be under a partial vacuum, which is obtained with steam jet air ejectors. The radioactive gases come off with the ejected air as do large amounts of nonradioactive hydrogen and oxygen gases formed from decomposition of water by radiation in the reactor. The total quantity of gas coming off is too large to allow tank storage, but its release is delayed for 30 min to several hours by requiring it to percolate through extensive charcoal beds before being filtered and released from a tall stack. To illustrate the importance of the time delay for krypton and xenon, the radioactivity released in curies is reduced by a factor of 50 by 30 min of delay, by another factor of 6 by 10 hr of delay, by another factor of 5 by 3 days of delay (mostly due to the decay of the 9-hr-half-life isotope ^{135}Xe), and by another factor of 100 by 40 days of delay (mostly due to the decay of the 5-day-half-life isotope ^{133}Xe). Clearly, problems from these gases are very much more difficult in BWRs than in PWRs. It should also be noted that a short-half-life isotope does not remain in the environment long enough to do much harm.

There are several sources of escaping gases other than the principal ones we have mentioned. Steam leaking out of the turbine is a source of radioactive gases in a BWR and there are occasional steam blow-offs to relieve excessive pressure. In PWRs, there is generally an appreciable leakage in the steam generators from the reactor water into the steam, so there are some radioactive gas releases from the condenser air ejector and the turbine leakage. Both BWRs and PWRs also release radioactive gases with the air exhausts from the reactor buildings and from gases associated with contaminated water, which will be discussed below. All of these are thoroughly filtered to remove as much suspended dust as possible before being released from the stack, but it is impractical to introduce time delays into their release.

These emissions are subject to a design requirement by the U.S. Nuclear Regulatory Commission that no member of the public, including those living closest to the plant, receive a radiation dose from gaseous emissions larger than 5 mrem/year to the whole body, or 15 mrem/year to the thyroid. In computing these doses, it is assumed that a much higher than average fraction of the fuel pins develop leaks in their cladding, that all milk, cereals, fruits, vegetables, and meat are derived from the most contaminated farm near the plant, and that the hypothetical maximally exposed person spends all of his time at the most exposed location. All pathways to man are considered, including direct radiation from the gaseous materials, inhalation, irradiation from radioactive dust deposited on the ground, deposition on vegetables eaten by people, depositions on grass that is eaten by farm animals transferring the radioactivity into milk or meat, deposition on the ground followed by pickup by plants, etc.

There are well over 100 radioactive isotopes of some importance in gaseous emissions from power plants, but the most important are ^{3}H, ^{14}C, and the isotopes of iodine, krypton, xenon, and cesium. Iodine and cesium isotopes do their damage through deposition on the ground leading to external exposure of passersby, deposition in the lungs when air is inhaled, and the grass → cow → milk pathway. For iodine, deposition on leafy vegetables is also important, and cesium sometimes enters the body via the grass → grazing animal → meat pathway. Krypton and xenon are not retained in the body and therefore do their damage by direct external radiation. ^{3}H and ^{14}C have long enough half-lives to become distributed throughout the earth's biosphere and therefore become incorporated into the human body through all intake mechanisms, especially through food and water.

3. AQUEOUS RELEASES

There are many ways in which some of the reactor water gets out of the system. Pumps and valves leak, as do heat exchangers and the steam from turbines. There are various discharges and leaks in the chemical cleanup systems, and samples are regularly withdrawn for laboratory analysis. All of this water is collected in floor drains, sumps, or tanks. In addition, the ^{3}H content of the reactor water must be kept below certain limits, and when these are exceeded the only recourse is to release some of the water. Less important sources of contaminated water are the drains from the laundry and showers and from decontamination and washing operations, especially in connection with maintenance, repairs, and equipment modifications.

All of these sources are combined into a few waste-water streams, each of which is given different treatment depending on its radioactivity levels and content of nonradioactive materials. In some cases, volumes are reduced by evaporation. Eventually all of the streams are passed through a fine-filtering and a demineralizer system and stored in tanks from which they are released in a controlled way into a cooling water discharge canal, which leads eventually into a nearby river, lake, or ocean. By the time it is released, the radioactivity in the water is at a level that would be reasonably safe for drinking. Once released, they are, of course, diluted by a very large factor.

These releases are subject to a design requirement by the U.S. Nuclear Regulatory Commission that no member of the public be exposed to more than 5 mrem/year from these aqueous releases. In calculating this exposure, very conservative assumptions analogous to those mentioned above for gaseous releases must be used. For example, it is assumed that the hypothetical maximally exposed individual derives all his potable water from the most contaminated supply system, swims 1 hr/day in the plant discharge canal, and eats 40 lb. of fish per year that had spent their lives in that canal.

While most of the radioactivity in the discharged water is from ^{3}H, when that isotope is ingested it passes rapidly out of the body, giving relatively little radiation exposure. The principal exposures are from a mixture of fission products the most important of which are ^{131}I and cesium isotopes, which get into people via drinking water and fish. Some of the corrosion products from structural materials like cobalt, iron, nickel, zinc, and manganese also make a contribution.

There is an abundance of equipment on the market for chemical cleanup of water, so there has been no difficulty in plants meeting requirements on aqueous releases. However, this equipment does contribute significantly to plant costs.

4. SPENT FUEL AND REPROCESSING

When fuel has been consumed to the maximum extent consistent with proper operation of the reactor, the vessel is opened and spent fuel is removed and replaced by fresh fuel. Typically, such a refueling operation takes place once a year, with one-third of the fuel replaced each time and many of the fuel assembly positions shifted in accordance with a pattern carefully calculated to achieve optimum reactor operation. The spent fuel is stored under water in storage pools, and leaking fuel pin claddings contribute radioactive contamination to the water in the pool. This water is cleaned in much the same way as reactor water, further contributing to releases of radioactivity into the environment.

Eventually, it is planned to ship the spent fuel to a reprocessing plant where the fuel pins are cut into short pieces, dissolved in acid, and put through chemical processes to remove the uranium and plutonium for future use as fuel and to convert the remaining materials (wastes) into suitable form for long-term management. When the fuel is dissolved, the gaseous materials along with suspended radioactive dust are released. Past practice has been to filter these gases very thoroughly to remove dust, pass them through processes for iodine removal, and release them from the top of a tall stack.

As processing will be done many months after the fuel is removed from the reactor, all radioactive materials with half-lives of many days or less will have decayed away. This includes all xenon isotopes and all krypton isotopes except ^{85}Kr, which has a 10-year half life. Essentially all ^{131}I (half-life 8.2 days) will be gone, but there is an iodine isotope, ^{129}I, with a 16-million-year half-life. The ^{14}C as carbon dioxide and ^{3}H as water vapor are the other principal components of gaseous releases.

A relatively new requirement to go into effect shortly involves retention of the ^{85}Kr and much of the ^{14}C. Methods for doing this are still to be decided, but a leading candidate is liquifying these gases at low temperatures and bottling them in high-pressure storage tanks for 100 years or so until the ^{85}Kr decays away. The ^{14}C can readily be converted into other chemical forms that are suitable for burial. There is still no available technology for separating the ^{3}H from the vastly greater quantities of normal hydrogen as would be necessary if the former is to be retained and stored. There are several processes under development, including one based on the variation with temperature of the relative affinities of hydrogen and ^{3}H for H_2O and H_2S, one based on slight differences in boiling points between normal and ^{3}H-containing water, and one based on differential adsorption of ^{3}H and normal hydrogen by certain solid materials. Perhaps a more practical solution is to dump the water in oceans where it is much less harmful, or pump it down into the ground from which it would not emerge until long after the ^{3}H has decayed away.

In the past, reprocessing plants have released contaminated water into the environment. This led to much bad publicity in the West Valley, New York, plant where meat from wildlife was found to have measurable radioactivity. As a result, aqueous releases from reprocessing plants are not allowed. All ^{3}H releases must be as water vapor.

Reprocessing plants are subject to the same regulatory requirements as power plants as regards the 5 mrem/year maximum exposure to any member of the public. A modern reprocessing plant should have little difficulty in meeting that standard.

20 • Routine Releases of Radioactivity

5. POPULATION EXPOSURES FROM ROUTINE EMISSIONS

Several groups have developed estimates of the average exposure to members of the public from the routine emissions we have been discussing. The most important contributors are ^{3}H, ^{14}C, and ^{85}Kr, which become distributed around the world because of their long half-lives, so any damage is spread rather uniformly over the world population. The principal sources of these isotopes is reprocessing plants, and estimates are generally based on no retention of ^{85}Kr and ^{14}C.

Because of the long half-lives, the effects are discussed as "dose commitment," the total dose eventually expected from the released material. This leads to a complication in the case of ^{14}C because of its very long half-life, so it is common to consider its effects only over some designated time period. The worldwide dose commitment over a 500-year period from a reprocessing plant serving 50 large power reactors is 0.0009 mrem/year from ^{85}Kr, 0.0009 mrem/year from ^{3}H, and 0.003 mrem/year from ^{14}C for a total of 0.005 mrem/year. If the time period were extended to infinity, the first two would remain unchanged and the ^{14}C dose commitment would be increased to 0.023 mrem/year. The total electricity now used in the world, if it were all generated by nuclear power, would require 14 of these reprocessing plants, giving a 500-year dose commitment of 0.07 mrem/year. Another way of stating this is: if all the world's present electric power usage were all nuclear and remained constant for 500 years, the average person at that time would receive a whole body dose of 0.07 mrem/year from routine releases of low-level radioactivity. If the per capita use of electricity throughout the world were to equal that of the U.S., this would be increased to 0.4 mrem/year. This is still 200 times less than the average person receives from natural radiation, 100 times less than U.S. citizens receive from medical X-rays, 50 times less than the difference in exposure from living in a brick vs. a wood house (due to the natural radioactivity in the brick), and 10 times less than he now receives from nuclear weapons testing fallout. It is also important to remember that these estimates are based on *no* retention of ^{85}Kr and ^{14}C at the reprocessing plants, whereas there are already legal requirements that these be retained starting in the early 1980s.

In Chapter 19, it is estimated that 1 rem causes an average of 1.2×10^{-4} cancers, so 0.4 mrem gives a risk of $1.2 \times 10^{-4} \times 0.4 \times 10^{-3} = 5 \times 10^{-8}$ of cancer per year of exposure, or a loss of life expectancy of less than 1 hr from a lifetime of exposure. Air pollution from coal burning is estimated to reduce average life expectancy by about 12 days, so it is 300 times more dangerous. Loss of life expectancy due to some other risks is shown for perspective in Table I.

The estimates we have given for exposure from routine releases are from the American Physical Society Study Group on Nuclear Fuel Cycles (*Rev. Mod.*

TABLE I
Loss of Life Expectancy (Days) Due to Various Causes

Cigarette smoking (male)	2200
Being 10% overweight	450
Motor vehicle accidents	200
Failure to use seat belts	50
Drowning	40
Fire and burns	27
Suffocation	13
Burning coal (air pollution)	12
Poison gases	7
Electrocution	5
Burning oil (air pollution, fires)	4
Using natural gas (asphyxiation, explosions)	1.5
Hydroelectric dam failures	0.2
Reactor accidents (Rasmussen)	0.02
Routine emissions of radioactivity	0.04[a]
Smoke alarms in homes	−10
Air bags in automobiles	−50
Mobile coronary care units	−125

[a] If world per capita consumption of electricity were equal to present U.S. per capita consumption.

Phys. Vol. 50, 1978). There are other calculations by the U.S. Nuclear Regulatory Commission (NUREG 0002, 1976), by the European Economic Community (E.E. Pochin, Estimated Population Exposure from Nuclear Power Production and Other Radiation Sources, OECD, Paris, 1976), and by the U.S. Environmental Protection Agency (EPA—520/9-73-003D), and all arrive at similar final results.

6. ALARA

The official policy of the U.S. Nuclear Regulatory Commission on routine emissions of radioactivity is that they be kept "as low as reasonably achievable," known as ALARA. This, of course, requires further interpretation and such interpretations have been given. The 5 mrem/year maximum exposure to any member of the public is one such interpretation. Another and more sweeping interpretation is that all measures should be taken to reduce radioactive emissions if their cost is less than $1000 per man-rem of radiation exposure averted (a man-rem is the sum of the exposures in rems of all people exposed; for example, if 1 million people are exposed to 1 mrem, the population exposure is $10^6 \times 10^{-3} = 1000$ man-rem).

This allows a clear determination of what equipment is required in a given installation. For example, at a midwestern reprocessing plant, a caustic scrubber to remove semivolatiles and ^{14}C from the gaseous emission stream would cost $158,000/year including amortizing the initial cost and maintenance, and it would reduce the population dose (relative to that with other equipment but not the caustic scrubber) by 238 man-rem. The cost is then ($158,000/238 = $660/man-rem averted, which is less than $1000/man-rem, so the caustic scrubber is required. On the other hand, a sand filter to remove particulates and semivolatiles would cost $1,459,000/year, and would reduce population dose by 937 man-rem, making the cost $1560/man-rem averted, so this sand filter is not required. Of course, different combinations of equipment can be used, but after the design is finalized, if it can be shown that there is an additional piece that can be added at a cost of less than $1000/man-rem averted, that equipment must be added.

If we divide the ALARA requirement of $1000/man-rem by the cancer induction frequency, 1.2×10^{-4} cancers/man-rem, we find that the former corresponds to about $8 million/cancer averted. If genetic effects are included, the ALARA requirement corresponds to spending $5 million/life saved. This is much more money than society is willing to spend to save lives in other contexts. There are cancer-screening programs that could save lives at a rate of $50,000 each, and emergency coronary care units save a life for every $5000 spent. There are highway safety measures that could save a life for every $80,000 spent. There are even ways that a person can save his own life at an average cost in this range as by installing smoke alarms in his home and air bags in his automobile, and by avoiding riskier jobs in spite of their higher salaries. The ALARA requirement therefore corresponds to placing an exceptionally high value on human life.

21 Low-Level Radioactivity and Infant Mortality

Karl O. Ott

Beginning in 1963, Dr. Sternglass [1] began advancing the suggestion that low-level radiation from radioactive decay of strontium-90 appearing as fallout from atmospheric testing of nuclear weapons has an extremely high effect on human fetuses and on infants. He states [2]:

> I agree that the (dose) levels are extremely low and by all ordinary present fallout calculations should not have produced the effect we observed. But, the point I am trying to make is precisely that strontium-90 appears to have a totally unexpected effect much larger than you would get by comparing it with cosmic radiation or external gamma or X-rays.

A second round of publications started in 1969 with an article in the *Bulletin of the Atomic Scientists* [3] and a presentation at the Annual Meeting of the Health Physics Society [4].

Sternglass's claims have been widely and consistently refuted. As an example, I quote part of a joint statement by the President and the 14 living past Presidents of the Health Physics Society [5]:

> His allegations, made in several forms, have in each instance been analyzed by scientists, physicians, and biostatisticians in the Federal Government, in individual states that have been involved in his reports, and by qualified scientists in other countries.
>
> Without exception, these agencies and scientists have concluded that Dr. Sternglass' arguments are not substantiated by the data he presents. The United States Public Health Service, the Environmental Protection Agency, the States of New York, Pennsylvania, Michigan and Illinois have issued formal reports in rebuttal of Dr. Sternglass' arguments.

KARL O. OTT • School of Nuclear Engineering, Purdue University, West Lafayette, Indiana 47907.

Although there is practically uniform agreement in the scientific community that Sternglass's claims are false (see Ref. 6), they have to be discussed here as one of the famous and widely publicized past myths (e.g., Sternglass, The death of all children, *Esquire* [2]).

Sternglass's claims have been refuted on several grounds (see e.g. Ref. 6). Instead of a review of all the contrary evidence and the scientific analysis of the more complete statistical material, I will only give an account of some of my personal evaluations of the data presented by Sternglass himself.

I first learned of these claims through the article in the *Bulletin* [3]. Somewhat later, I also became aware of the additional material of Ref. 4. I was very surprised that such an effect should exist. However, the subject of infant mortality is so important that I felt that these papers should be presented and discussed in our nuclear engineering colloquium. So I announced a colloquium presentation.

In preparing myself for that presentation, I had to carefully go over Sternglass's data and analysis. In doing so, I became convinced that the entire "evidence" is at least a misinterpretation of statistics, if not a fabrication (whether a deliberate one or not, I have no way of knowing). It would hardly be the first time that a scientist had become convinced of some thesis by superficial analysis of statistical material and subsequently disregarded—more or less subconsciously—contradicting data.

Sternglass's main diagram is presented in Fig. 1 (from Ref. 3). He depicts, on a logarithmic scale, the infant mortality in the U.S. from 1934 through 1968, fits part of the data (1934 to 1951) by a straight line and projects it through 1970. The difference between the actual infant mortality and his projected line he defines as "excess infant mortality." Adding up the total "excess" through 1968, represented by the hatched area, he arrives at 375,000 fatalities. He then claims that this—in his opinion real—"excess" is due to the low-level radiation from the decay of strontium-90, contained in the fallout.

In preparing my colloqium presentation, I made the following observations:

The *reduction* of infant mortality results from a number of efforts and factors, such as from better hygienic conditions and medical care, and application of added medical knowledge (e.g., antibiotics). The contribution of each factor to the mortality *reduction* should have its own variation with time; some factors would tend to saturate, others would act on a longer period of time. There is no reason whatsoever that all the efforts and factors in reducing infant mortality should compound in a way that amounts to the *same percentage decrease* year after year, as it corresponds to the straight-line extrapolation in Fig. 1.

As there is no justification for an exponential trend of the infant mortality data, the points of Fig. 1 are plotted on a normal, i.e., linear, scale (see Fig. 2). The straight line of Fig. 1 appears in Fig. 2 as a curve (dashed line). There are many curves (e.g., the heavy dotted one in Fig. 2) that represent a smooth fit to the statistically fluctuating data points. This leads to the following *conclusion*:

21 • Low-Level Radioactivity and Infant Mortality

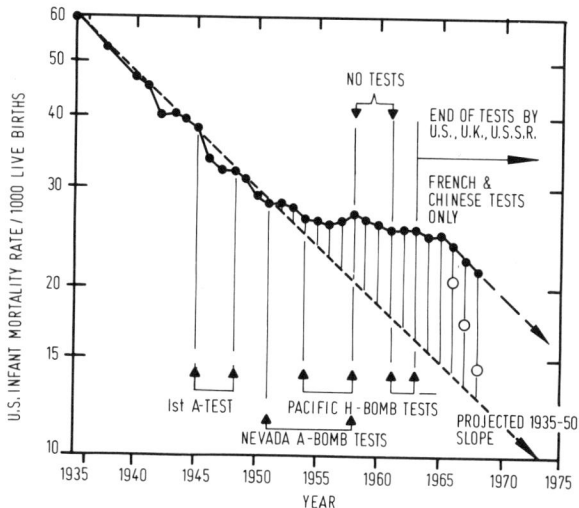

FIGURE 1. Mortality rate in the U.S. for infants 0–1 year. Data from "Vital Statistics of the United States," National Center for Health Statistics; logarithmic scale.

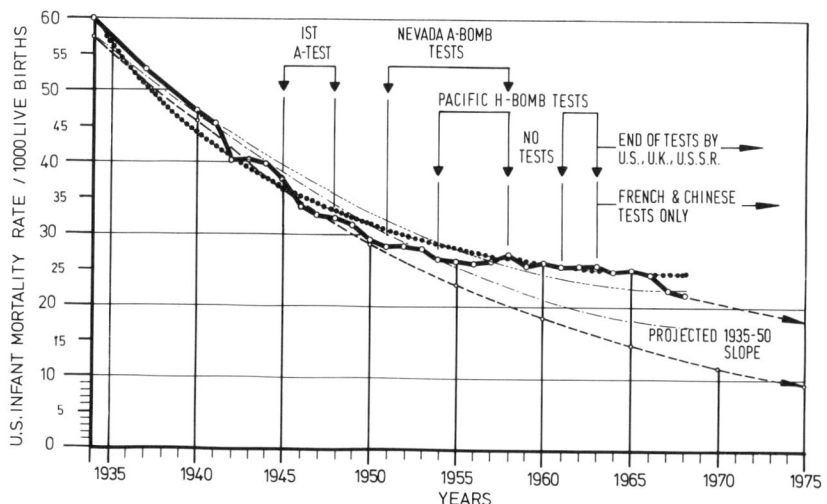

FIGURE 2. Mortality rate in the U.S. for infants 0–1 year. Data from "Vital Statistics of the United States," National Center for Health Statistics; linear scale.

The magnitude, as well as the existence of an "excess" is totally dependent on how one draws a curve through the data points; none of these smooth curves has a sufficient inherent validity as to justify the claim that a deviation from it represents an actual "excess" mortality.

The next question was: if data fitting does not establish the existence of an excess, how about the claimed causal connection, the strontium-90 from fallout?

The bomb test periods are indicated in Fig. 1, ending in 1963. If the excess were due to strontium-90, one would expect that some years after the last test-explosions, after the strontium-90 has come down and eventually disappeared, the "excess" effect would have to disappear too. Unbiased observation indicates that the effect (the "excess") must disappear with its alleged cause; otherwise the claimed causal relation cannot exist. However, Sternglass argues differently. He makes the following observation concerning the last few points in Fig. 1: ". . . within two years following the test-ban of 1963, the infant mortality rate resumed its decline at a rate approaching that prevailing prior to the onset of large-scale atmospheric testing" [3]. He indicates this conclusion in Fig. 1 by an arrow.

What Sternglass employs here to argue for a causal connection between fallout and his "excess" mortality is a grave *logical mistake:* After the alleged cause (the strontium-90) has disappeared, the logically correct conclusion is that the excess as such must disappear; it does not suffice that only the rate is affected. If his causal connection were correct, the curve would have to approach his lower projection about as indicated by the three circles in Fig. 1, in order to make the "excess" disappear together with its alleged cause.

While the "discovery" of his "excess" mortality was based on medical implausibility and on qualitative and quantitative mishandling of the statistical data, Sternglass's "proof" of his alleged causal connection with fallout is based on a logical mistake. Applying common sense or logical rigor to the interpretation of the last few points in Fig. 1 actually *disproves* his claim with his own data, for the points do not go down as they should after the alleged cause has disappeared.

In this short essay I do not want to discuss more of Sternglass's analyses. It would only be more of the same. Common sense gets offended by effects appearing 2 or 3 years before the "cause." Standard scientific methodology gets offended by his arbitrary data selection and plotting practices, and by deriving, from single-digit numbers, three-digit values for rates of decline or rise, as he did in a similar type of analysis of infant mortality around some nuclear test reactors.

Needless to say, the colloquium presentation I gave after a thorough reading of Sternglass's papers turned out to be quite different than originally intended. Instead of a review of scientific results—though improbable—it became a lecture on the fallacy of statistical analysis if scientific rigor and common sense are disregarded.

What I couldn't know at that time was that about 2 years later, there would be a revealing epilogue in another colloquium presentation at Purdue University, given by Sternglass himself.

In January 1972, I published my observations on Sternglass's analysis [7]. In that paper I emphasized that Sternglass suggested a logical mistake (the change of the slope indicated by the last three points of Fig. 1) as proof for his claim.

Later in 1972, Sternglass gave a colloquium presentation at Purdue University on a technical subject. Though unrelated to this technical subject, he promoted his claims of an "excess" infant mortality and its alleged causal relation to strontium-90 from fallout. He also presented Fig. 1, at least so it appeared to me at first glance. But the last three points did not stay up, they were moving down in large steps toward the lower projected slope about as indicated by the circles in Fig. 1. (Remember, for a logically correct support of his claim, these three points *have* to go down.) I was stunned, wondering if I was witnessing the presentation of a forgery, wondering if he had read my January 1972 paper and altered his data and his argumentation.

But just before his logically corrected "proof" disappeared from the screen, I noticed a word in the upper right corner: MAINE. Apparently, Sternglass had discontinued to justify his claim by his original logical mistake. But then he needed other data, for the U.S. data readily disprove his claim. He found "suitable data" for his altered argument in one of the smaller states, Maine. The data for Maine appeared to be quite similar to the U.S. data, except for the last three points.

This episode not only totally supports the uniform rejection of Sternglass's claims in the scientific community, it may also have provided a revealing insight into the modern methodology of myth creating.

REFERENCES

1. E. J. Sternglass, Ionizing radiation in the prenatal stage and the development of childhood cancer, *Science 140*, 1102 (1963).
2. E. J. Sternglass, The death of all children, *Esquire* (Sept. 1969).
3. E. J. Sternglass, Infant mortality and nuclear tests, *Bull. At. Sci. 25*, 18 (1969).
4. E. J. Sternglass, D. Sashin, and R. Rocchio, Strontium-90: Evidence for a Possible Genetic Effect in Man, presented at the 14th Annual Meeting of the Health Physics Society, Pittsburgh, June 8–12, 1969. Proceedings published by the Health Physics Society.
5. Statement *en-toto* made by the President and Past Presidents of the Health Physics Society with Regard to Presentation by Dr. Ernest Sternglass, 16th Annual Meeting of the Health Physics Society, New York, July 15, 1971.
6. "Low Level Radiation: A Summary of Responses to Ten Years of Allegations by Dr. Ernest Sternglass," compiled and edited by C. B. Yulich *et al.*, Annex No. 1 of a paper presented at the Fifth International Conference on Science and Society, Herceg-Novi, Yugoslavia, July 9, 1973.
7. K. O. Ott, Das Sternglass-Phänomen und die Gofman-Tamplin-Kontroverse, *Atomwirtschaft 17*, 25 (1972).

22 The Myth of Plutonium Toxicity

Bernard L. Cohen

Plutonium is constantly referred to by the news media as "the most toxic substance known to man." Ralph Nader has said that a pound of plutomium could cause 8 billion cancers, and former Senator Ribicoff has said that a single particle of plutonium inhaled into the lung can cause cancer. There is no scientific basis for any of these statements as I have shown in a paper in the refereed scientific journal *Health Physics* (Vol. 32, pp. 359–379, 1977). Nader asked the Nuclear Regulatory Commission to evaluate my paper, which they did in considerable depth and detail, but when they gave it a "clean bill of health" he ignored their report. When he accused me of "trying to detoxify plutonium with a pen," I offered to eat as much plutonium as he would eat of caffeine, which my paper shows is comparably dangerous, or given reasonable TV coverage, to personally inhale 1000 times as much plutonium as he says would be fatal, or in response to former Senator Ribicoff's statement to inhale 1000 particles of plutonium of any size that can be suspended in air. My offer was made to all major TV networks but there has never been a reply beyond a request for a copy of my paper. Yet the false statements continue in the news media and surely 95% of the public accept them as fact although virtually no one in the radiation health scientific community gives them credence. We have here a complete breakdown in communication between the scientific community and the news media, and an unprecedented display of irresponsibility by the latter. One must also question the ethics of Nader and Ribicoff; I have sent them my papers and written them personal letters, but I have never received a reply.

Let's get at the truth here about plutonium toxicity. We begin by outlining a calculation of the cancer risk from intake of plutonium (we refer to it by its chemical symbol, Pu) based on standard procedures recommended by all national

---

BERNARD L. COHEN • Department of Physics, University of Pittsburgh, Pittsburgh, Pennsylvania 15260.

1. ESTIMATE OF PLUTONIUM TOXICITY FROM STANDARD PROCEDURES

The first step is to calculate the radiation dose in rem (the unit of dose) to each organ of the human body per gram of Pu intake. According to ICRP (International Commission on Radiation Protection) Publication No. 19, about 25% of inhaled particles of the size of interest (0.5–5 μm in diameter) deposit in the lung, and 60% of this is eliminated only with a 500-day half-life. From this information and the known rate and energy of α-particle emission, we can calculate the radiation energy deposited in the lung, which is directly convertible to dose in rem.

According to ICRP Publication 19, 5% of inhaled Pu gets into the bloodstream from which 45% gets into the bone and an equal amount collects in the liver; the times required for elimination from these are 70 and 35 years, respectively. This is all the information needed to calculate doses to bone and liver in rem per gram of Pu inhaled.

If Pu is ingested with food or water in soluble form, the ICRP estimates that 3×10^{-5} (30 parts per million) gets through the intestine walls into the bloodstream. From this and the information given above, calculation of rem to the bone and liver per gram of Pu ingested is straightforward. In addition, there is dosage to the gastrointestinal tract calculable by ICRP prescriptions.

Once the dose in rem is calculated, the next step is to convert this to cancer risk using the BEIR Report, the standard reference in this area produced by the National Academy of Sciences Committee on Biological Effects of Ionizing Radiation. It recommends a model in which there is a 15-year latent period following exposure during which there are no effects, followed by a 30-year "plateau" period during which there is a constant risk of 1.3×10^{-6} (1.3 chances per million) per year per rem for lung cancer and 0.2, 1.0, and 0.3×10^{-6} per year per rem for bone, gastrointestinal tract, and liver* cancer, respectively.

For children less than 10 years old, these are divided by 5, and for an older person, there is a calculable probability that death will result from other causes before the cancer develops. With this information we can calculate the cancer risk as a function of age at intake. Averaging over ages, we obtain the average cancer risk per gram of Pu intake.

---

* In the BEIR Report, liver cancer is included among "all other" for which the risk is 1.0×10^{-6}; the value used here is based partly on other information.

22 • The Myth of Plutonium Toxicity

TABLE I
Cancer Doses in Micrograms (Defined as the Inverse of Risk per Microgram)

Entrance mode	^{239}Pu	Reactor-Pu
Inhalation (dust in air)	1300	200
Ingestion with food or water	6.5×10^6	1×10^6

The results are given in Table I for the most important isotope of Pu, ^{239}Pu, which contains 1 curie of radioactivity for each 16g, and for the mixture of Pu isotopes that would be commonly found in power reactors, which is 6 times more intensely radioactive (1 curie in each 2.5 g). We refer to the latter as "reactor-Pu" and use it in our discussions where appropriate.

Table I shows the inverse of the risk, which we call the "cancer dose." For example, we see that the risk of inhaling reactor-Pu is 1/200 per µg, so if one inhales 10 µg, he has one chance in 20 of developing cancer as a result. Another application is that in a large population we may expect one cancer for every 200 µg inhaled, so if a total of 1000 µg is inhaled by people, we may expect 5 cancers (regardless of the number of people involved).

Estimates of cancer doses of Pu have also been derived using different methods by the British Medical Research Council in its report "The Toxicity of Plutonium," and by Dr. C. W. Mays (who developed some of the important basic information in his experiments on dogs) in a report published by the International Atomic Energy Agency (IAEA-SM-202/806), and they agree closely with Table I. We see from Table I that Pu is dangerous principally when inhaled as a fine dust. It is not very toxic when ingested with food or drink because of its very small probability of passing through the intestine walls into the bloodstream. Pu forms large molecules, which have great difficulty in passing through membranes.

In addition to causing cancer, intake of plutonium can also cause genetic defects among progeny in the next 5–10 generations, but the total number of eventual genetic defects before they are bred out is only 20% of the number of cancers. For simplicity we will restrict our discussion to cancers, but the genetic effects can always be included by applying the 20% addition.

The estimates in Table I are based on data from radiation effects on humans as analyzed in the BEIR Report. These include the Japanese A-bomb survivors, miners exposed to radon gas, people treated for various maladies with radium or with X-rays, etc. None of these effects were from Pu—there is no evidence for any injury to humans from Pu toxicity. However, there is a considerable amount of data from animal studies with Pu, and this is summarized for lung

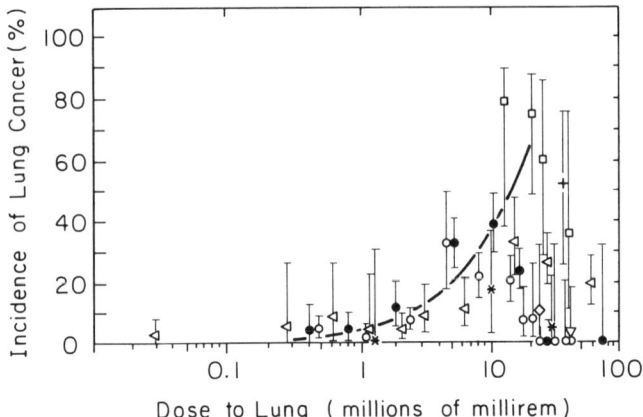

FIGURE 1. Data from animal studies with Pu, summarized for lung cancer. □, PuO$_2$, dogs; ▽, PuO$_2$, mice; ◇, PuO$_2$, mice; ○, Pu citrate, rats; ●, Pu pentacarbonate, rats; ◁, Pu nitrate, rats; +, Pu nitrate, rabbits; *, Pu pentacarbonate, rabbits.

cancer in Fig. 1 where the line shows the estimate from our calculation. In general the agreement is quite reasonable.

There has been a great deal of publicity about the high point for beagle dogs (the highest point in Fig. 1), but we see that our curve passes within the error bars given by the authors. One aspect of that experiment that is frequently overlooked is that the latent period for development of the cancers increased with decreasing dose, and in fact the dogs contributing to the point under discussion developed cancer rather late in life. If this effect is extrapolated to lower doses, the latent period for most doses usually considered would greatly exceed life expectancy, so the effects we derive in this paper would be substantially reduced.

2. CRITICISMS OF STANDARD PROCEDURES

There have been several criticisms of treatments like the one we have given. The best known of these is the "hot-particle" theory, which gives greatly increased effects (by a factor of 100,000) due to the fact that the Pu is not evenly distributed over the lung but is concentrated in particles, which give much higher than average doses to a few cells. This theory has been studied and rejected by the following groups:

- A Committee of the U.S. National Academy of Sciences especially assembled for this study in a report entitled "Health Effects of Alpha-emitting Particles in the Respiratory Tract

- U.S. National Council on Radiation Protection and Measurements (NCRP), a very distinguished group composed of about 70 in our nation's leading radiobiomedical research scientists, in NCRP Publication No. 46
- British Medical Research Council in "The Toxicity of Plutonium"
- U.K. National Radiological Protection Board in its Report R-29 and Bulletin No. 8 (1974)
- U.S. AEC in a very elaborate study, WASH-1320, authored by three of the world's leading researchers on Pu toxicity
- U.S. NRC in Federal Register, Vol. 41, No. 76
- U.K. Royal Commission on Environmental Pollution—Sixth Report—Nuclear Power and the Environment

One easily understood aspect of these criticisms is that there were about 25 workers at Los Alamos who inhaled varying amounts of Pu about 30 years ago, and according to the "hot-particle" theory each should have experienced an average of 200 lung cancers, whereas there have been *no* lung cancers as yet among them. According to our estimates in Table I, there is a 40% chance that one of them would have had a lung cancer, so this is experimental evidence that Table I does not grossly underestimate the cancer risk from Pu intake.

Another criticism of the "hot-particle" theory is that there are experiments on animals in which two groups were exposed to the same total amount of Pu but in one of them it was much more in the form of hot particles—an that group experienced fewer cancers. It was also pointed out that particles in the lung do not stay in one place but are constantly moving about so that their exposure does not fall on only a few cells.

After these rejections of the "hot-particle" theory appeared, John Gofman, a former research scientist who has spent the past several years as the full-time leader of an antinuclear organization, came out with a new theory ascribing enhanced toxicity to Pu. His paper was not written for a scientific journal but was inserted in the Congressional Record by Senator Gravel. His basic premise was that smoking destroys the cilia, the fine hairs that stop dust particles from entering the bronchial region—this much was well established—and that Pu particles therefore remain in that region for a very long time, allowing their radiation to cause bronchial cancers. This allows him to ignore the animal data as animals do not smoke. He also manages to explain the lack of lung cancers among the 25 Los Alamos workers by a combination of four improbable hypotheses, the failure of any one of which would destroy his theory.

There have been at least seven individual critiques of the Gofman theory. Perhaps the most telling criticism is that there was a series of experiments at New York University in which a number of graduate students inhaled a controlled amount of radioactive dust and the rate at which this dust was cleared from the bronchial region was directly determined by placing radiation detectors over their chests and measuring the radiation intensity as a function of time. It was found that there was no difference between smokers and nonsmokers, and the experimenters concluded that smokers do more coughing and have increased mucous flow, which compensates for their lack of cilia. In fact, if dust accumulated in the bronchial region of smokers in the manner postulated by Gofman, their bronchial tubes would be completely closed and they would die by suffocation.

There were many more weak points in the details of the Gofman paper. He misuses the BEIR Report, he miscalculates the area of the bronchial region by a factor of 17 and thereby incorrectly increases the toxicity by that factor, he misuses the ICRP lung model, etc. He even suggested that the great increase in lung cancer in recent years may be due to Pu, but this increase has been steady since the 1930s whereas Pu-induced cancers should not have occurred until 1960. Moreover, the lung cancer increases have been in areas with chemical industry and high air pollution, and there has been no increase in areas downwind from the Nevada test site where Pu would have its maximum effect.

A relatively less publicized attack on the conventional approach to evaluating Pu toxicity is the "warm-particle" theory of Edward Martell. He hypothesizes that natural radiation is one of the principal causes of lung cancer, but this idea has not been accepted by the cancer research community.

K. Z. Morgan has proposed that the relative biological effectiveness (RBE) for Pu in bone might be 250 times larger than the usual value. C. W. Mays, on whose experiments much of Morgan's hypothesis is based, reanalyzed Morgan's work and concluded that if his approach is correct, the increase should be only by a factor of 10. There is experimental information on this from some supposedly "terminally ill" patients injected with Pu in 1945–46 to study Pu metabolism. Four of these are still alive and one who was injected with a rather large quantity died of unrelated causes only in 1968. If the RBE of Pu were 10 times the present value, there is a better than even chance that one of these five would have gotten bone cancer, but none did. As our calculated inhalation effects are dominated by lung cancer, a factor of 10 increase in bone cancer risk would only double the total inhalation risk.

S. M. Wolfe, an employee of a Nader-sponsored group, drew far-reaching conclusions from the fact that 11 of the first 30 deaths in the US Transuranic Registry (a registry of people who have worked with plutonium) revealed cancers on autopsy, whereas based on listed cause-of-death for all U.S. males, only 6.2 of each 30 deaths is from cancer. His paper, which was never published in the scientific literature, received very wide publicity in the news media. However, it turned out that autopsies were done preferentially on people who had died of cancer, and that explained the entire effect. In addition, it was pointed out that Pu is expected to cause cancers of the lung, bone, and liver, whereas among the 11 cases there were no bone or liver cancers, and less than the expected number of lung cancers for a normal population. Needless to say, the news media never bothered to report that the Wolfe paper was based on an incorrect premise.

22 • The Myth of Plutonium Toxicity

In evaluating all of the criticisms outlined above, it is important to realize that they are actively considered every year by a committee of the ICRP and they have repeatedly been rejected. Likewise, the EPA, which has jurisdiction in the U.S., studied the matter and decided not to modify its standards. No standard-setting or official study group in any country has given credence to any of these criticisms of the standard procedures we used in deriving Table I.

3. CONSEQUENCES OF PLUTONIUM DISPERSAL

It is clear from Table I that Pu is dangerous principally as an inhalant, so we now consider the consequences of a dispersal of Pu powder in a populated area. The calculations are done with a standard meteorological model, in which the dust cloud moves with the wind dispersing in the downwind, crosswind, and vertical directions. Meteorologists have determined the extent of dispersal as a function of wind velocity and atmospheric stability. Figure 2 shows the results of calculations assigning the atmospheric stability most characteristic of each wind velocity. This is different between day and night, so separate curves are given for each.

These curves give the area within which various fractions, q/Q, of the dispersed Pu are taken in by a person inhaling at an average rate. For example, we see from Fig. 2 that for a typical daytime 8 m/sec wind velocity, only in an area of 500 m^2 is as much as 10^{-6} (one-millionth) of the dispersed Pu inhaled. A typical city population is 10^{-2} people/m^2, so there would typically be about 5 people in this area. Similarly, from Fig. 2, about 60 people would inhale 10^{-7}, 700 people would inhale 10^{-8}, etc. of the dispersed Pu.

As we know the cancer risk per microgram of Pu inhaled from Table I, it is straightforward to calculate the total number of cancers expected per gram of Pu dispersed. When corrections are applied for the fraction of typical Pu powders that are in particles of respirable size, the efficiency in dispersal, the protection afforded by being inside buildings, and decreased breathing rates at night, the result is that we may expect about one eventual cancer for every 24 g of Pu dispersed, or about 19 fatalities per pound.

If there is a warning, as in a blackmail scenario, people can be instructed to breathe through a folded handkerchief or a thick article of clothing, with a resulting decrease in fatalities to 3 per pound dispersed.

Eventually, the Pu settles to the ground but it may then be blown up by winds. Meteorologists have also developed methods for calculating these effects ("deposition" and "resuspension"). Within the first few months, this causes about one-third as many cancers as inhalation from the initial cloud. Beyond this time period, resuspension is of much less and continually decreasing importance as the Pu becomes part of the soil.

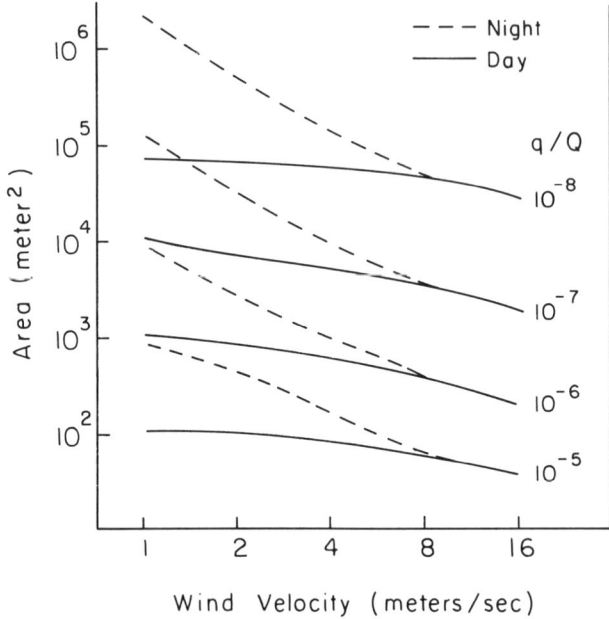

FIGURE 2. Area over which the ratio of inhaled to dispersed Pu has values shown for q/Q versus wind velocity under typical day and night atmospheric conditions.

Of course, ^{239}Pu lasts for tens of thousands of years, so let us consider its effects over this time period. We know the amount of *uranium* in soil and we know now how much there is as dust in the air, so we can estimate now much is inhaled per year—it calculates out to be 1.3×10^{-11} of that in the top 20 cm of soil. If this factor is applied to the Pu after it becomes part of the soil, we find that over the 25,000-year half-life there will eventually be about one fatality per 2500 g of Pu dispersed. Thus, we see that the long half-life is almost irrelevant; nearly all of the damage eventually done occurs very soon after dispersal.

A summary of all these effects of Pu dispersal is given in Table II. It also includes plant uptake into food. There is a great deal of information on uptake of Pu by plants both from laboratory experiments and from several areas where an appreciable amount of Pu has gotten into the soil from bomb tests or from various research activities. Plant uptake is small for the same reason that Pu does not easily pass through the walls of the intestines—it forms large molecules, which do not easily pass through membranes. From Table II we see that the total eventual effect of Pu dispersal in a city is one fatality per 18 g dispersed without warning, or 25 fatalities per pound.

TABLE II
Summary of Fatalities per Gram
of Reactor-Pu Dispersed

Inhalation from cloud	0.042 (1/24)
Resuspension	0.014
Long term	0.0004 (1/2500)
Plant uptake into food	0.002
Total	0.058 (1/18)

4. DANGERS OF PLUTONIUM DISPERSAL

The fear is sometimes expressed that the world may become "contaminated" with ^{239}Pu. To evaluate this potentiality, we calculate that if all of the world's present electric power were produced by fast breeder reactors in an equilibrium situation where Pu is consumed as fast as it is produced, the total amount of ^{239}Pu existing in the world would be 2×10^8 curies.

By comparison, the radium (^{226}Ra) in each meter of depth of the earth's crust is 1.2×10^9 curies, so there is as much Ra in each 17 cm of depth as there would be ^{239}Pu in the whole world. For ingestion, Ra is 40 times more toxic than Pu as it passes through the intestine walls much more easily. For direct inhalation, Ra is less hazardous than Pu, but it serves as a source of radon gas, which comes up out of the ground and mixes with the air we breathe, and therefore is a serious inhalation hazard, so as material on the ground, Ra is a 40-fold greater inhalation hazard than Pu.

Thus, as a long-term hazard either for ingestion or for inhalation, Ra is 40 times worse than Pu; the total Pu in existence for an all-breeder power system would then be as dangerous as the Ra in each 4 mm of our soil. Of course, nearly all of this Pu would be in reactors or in other parts of the nuclear industry, well isolated from the environment.

There is now a legal requirement on the allowable releases of Pu from nuclear plants, which is such that if all U.S. power were nuclear and derived from fast breeder reactors (they use the most Pu), the total releases would be about 0.6 g/year. If we use Table II, this would predict an average of 0.03 fatality/year, but that would be valid only if nuclear plants were in cities; as they are not, the expected effects are about 10 times less, or one fatality in 300 years.

Some perspective on this problem may be obtained by comparing the 0.6 g/year tht may some day be released by the nuclear industry with the amount of Pu that has been dispersed in the atmosphere in nuclear bomb tests, which is 5 million g. Estimates on the same basis that we have been using predict about

200 U.S. fatalities to date from Pu releases in bomb tests, and 4000 in the world. It also predicts about 200 fatalities worldwide from the reentry burn-up in 1964 of a space vehicle carrying a SNAP-9A ^{238}Pu-powered energy source. It is important to keep in mind that all of these estimates are theoretical. There is no direct evidence for Pu toxicity having caused serious injury to any human being, *anywhere, ever.*

The reason why the legal requirement on plutonium releases is so stringent is not because Pu is so dangerous, but because the technology is available for keeping the releases that low, and in fact this technology is very close to present practice. Pu dust particles tend to stick to each other and to their containers, so Pu is not easily dispersed. It is also very readily collected on filters; anywhere Pu powder is used, the air is exhausted through filters, which catch all but about one part per billion of the dust suspended in air.

Of course, these control measures are expensive and they increase the cost of nuclear electricity. As previously noted, the reason they are required is not because Pu is so dangerous—one fatality every 300 years is surely a trivial problem when burning coal, our only viable alternative to nuclear energy, is killing 10,000 people every year with its air pollution—but because the public is afraid of plutonium. Ralph Nader, former Senator Ribicoff, John Gofman, and their like have done their work well, and the public is paying the price in its electric bills.

One often hears that in large-scale production of Pu we will be creating unprecedented quantities of a poisonous material. Because Pu is dangerous principally as an inhalant, we compare it in Table III with quantities of other poisonous inhalants produced in the U.S. We see that it is relatively trivial by comparison. Moreover, it should be noted that Pu is not easily dispersed whereas the others are gases and hence readily dispersible. Of course, Pu released to the environment will last far longer than these gases, which would be decomposed chemically, but recall from our earlier discussion that nearly all of the damage done in Pu dispersal is by the initial cloud of dust; all of the later resuspension and the thousands of years spent in the soil do far less damage. It is thus not

TABLE III
Lethal Inhalation Doses Produced Annually in the U.S. ($\times 10^{12}$)

Chlorine	400
Phosgene	18
Ammonia	6
Hydrogen cyanide	6
Pu if all U.S. power were from fast breeder reactors	1

unfair to compare Pu with the poison gases, and we see from Table III that it will always be far less of a hazard.

It is often argued that there is a great deal we do not know about Pu toxicity. While this may be true, one would be hard-pressed to name another public health issue that is as well understood and controlled. Surely it would not be air pollution from burning coal, which is a million times more serious a problem. Surely it is not food additives or insecticides or such that may well be doing real harm to our health. Pu hazards are far better understood than any of these, and the one fatality per 300 years they may some day cause is truly trivial by comparison.

In spite of the facts we have cited here, facts well known in the scientific community, the myth of Pu toxicity lingers on. The news media ignore us, and prefer to continue scaring the public at every opportunity. They don't recognize the difference between political issues on which everyone is equally entitled to an opinion, and scientific issues, which are susceptible to scientific investigation and proof. The myth may linger forever.

23 Myths about High-Level Radioactive Waste

Bernard L. Cohen

The subject of high-level radioactive waste was discussed in detail in Chapter 16. Here we discuss some of the distortions and myths about that problem that have received wide publicity.

Disposal of Radioactive Waste Is an Unsolved Problem. This statement depends on one's definition of "unsolved problem." If the problem is to dispose of radioactive waste in a way that is *perfectly* safe, all waste disposal problems are unsolved. For example, the wastes from coal burning are largely discharged into the environment with minimal control; they are responsible for air pollution that is probably killing 10,000 Americans each year. Is that a solved problem? It has been shown that we could simply dump our radioactive waste in the ocean, and it would be hundreds of times safer than that (and it would not harm ocean ecology, either).

But we have much better solutions than ocean dumping. Burial deep underground or under the ocean floor would be many times safer still.

Radioactive Waste Remains Lethal for Hundreds of Thousands of Years. After 600 years, the waste loses 99% of its radioactivity. A lethal dose at that stage is about 1 ounce eaten in digestible form. This is ten times less than a lethal dose of arsenic which is sprayed on agricultural land, and less than a lethal dose of caffeine or copper. These things are often disposed of without great care, whereas the nuclear waste would be buried one-third of a mile deep underground in a secure rock formation.

If Groundwater Gets into the Repository, There Is Great Danger. A great deal of the rock beneath us is saturated with ground water, but yet it is dissolved

---

BERNARD L. COHEN • Department of Physics, University of Pittsburgh, Pittsburgh, Pennsylvania 15260.

away only very slowly over a period of hundreds of millions of years. If the waste is dissolved away at such a rate, calculations indicate that if all our power were nuclear for many millions of years, the millions of years of accumulated waste would cause less than ten deaths per year in the U.S.

Escape of Buried Radioactive Waste from a Repository Could Result in Large Loss of Life. Even if everything should go wrong and all safeguards should somehow be bypassed, allowing the waste to escape into groundwater and rivers, it is difficult to envision a situation resulting in appreciable loss of life. Radioactivity is very easily detected, even in amounts thousands of times too small to be harmful, so the escape would readily be detected and it would be easy to warn people not to drink water that is excessively contaminated and not to use it for irrigation of food crops.

In Burying Radioactive Waste, We Are Introducing Large Quantities of a New Pollutant into the Earth. The earth is already full of natural radioactivity—it always has been and always will be. The principal naturally radioactive materials in the earth are uranium, potassium, thorium, and radium, but there are many others. In adding our waste to the earth's crust, we increase the total radioactivity within the top half mile of U.S. soil by only about one part per million each year. Moreover, the natural radioactivity in the earth, with the exception of that within a few feet of the surface, is doing essentially no harm; it can be shown that it is causing far less than one fatality per year in the U.S. There is every reason to believe that our buried radioactive waste will be comparably innocuous if even minimal precaution is exercised in burying it.

Large Quantities of High-Level Radioactive Waste Are Being Generated. The high-level waste produced by one power plant in 1 year would fit under a typical card table, and after full preparation for burial would comprise a single truckload.

We Are Placing a Great Burden on Future Generations in Requiring Them to Guard Our Waste. Very little if any guarding would be necessary once the repositories are sealed. Guarding a thousand years of accumulated waste would be a part-time job for one person. His only tasks would be to make periodic checks on water samples, and to check that no mining or drilling takes place, although even that would not be very dangerous.

By contrast, we are placing a tremendous burden on future generations in using up all the coal, oil, and gas (and most other valuable minerals) especially if we do not leave them with an alternative technology for producing energy cheaply and abundantly. Nuclear energy is the only such technology we can now guarantee.

24 The Aging Reactor Myth

A. David Rossin

Many attacks on nuclear power still include the argument that the LWRs now in commercial service encounter serious operating problems at the age of 4 or 5 years. They claim that this indicates a deterioration and it will inevitably happen to new plants. These claims are still made despite the fact that the "analysis" on which the original charge was based was neither sound nor conclusive, and that operating experience on many more reactors has not only failed to follow the prediction, but has shown the opposite! After the first refueling, LWRs in the U.S. have shown a statistically significant improvement in capacity factors [1].

Large electricity-generating plants are built with the expectation that they will operate successfully for 30 years or more. On most utility systems it is likely that plants in the latter half of their lifetime will be operating less than newer and more efficient units. But it was surprising to the utility industry that so many critics seized on the charges made by the late David D. Comey in his testimony to the FEA Project Independence hearings in September 1974. Comey subsequently published the work in November 1974 [2], but in neither instance had it been subjected to any form of peer review.

Margen and Linde of Studsvik, Sweden, plotted the same data Comey had used, but they showed all the points where they belonged [3]. Comey had plotted all the data for 7 years and over (for only three older plants) as a single point, and this gave his curve the appearance of a dropoff in performance after the first several years. The Swedish researchers showed that there was no statistical effect at all.

Rossin shows the Swedish curve and also argues that the approach has no validity because undue emphasis is placed on short-term data that give a misleading impression when plotted as Comey plots his points [4]. Data for $\frac{1}{2}$ year

---

A. DAVID ROSSIN • Electric Power Research Institute, Palo Alto, California 94303.

of operation during which a refueling outage took place look bad, but refuelings are on 12- or 18-month schedules. Actually, so few data points were used that statistical validity cannot even be claimed.

Since 1974 many nuclear units have passed through the 4- to 5-year age at which Comey claims deterioration will occur. Their performance has not matched his predictions at all. In fact, efforts by others to verify Comey's claims with a bigger data base have given the opposite result [5]. Performance holds up well and even improves on average for some types of LWRs.

There will be operating problems in any type of power plant, and nuclear units will be no exception. Several generic problems, such as pipe cracks in BWRs, sparger cracking and replacement, and steam generator tube denting (PWRs), have all caused lengthy shutdowns. Nevertheless, overall nuclear plant performance has been as good or better than coal-fired units, with better fuel economics and more prospects for improvement in the future.

REFERENCES

1. M. E. Lapides, EPRI (July 1977).
2. D. D. Comey, Will idle capacity kill nuclear power?, *Bull. At. Sci. 30*, 23–28 (Nov. 1974).
3. G. Margen and M. Linde, unpublished observations.
4. A. D. Rossin, Reliability and Economics of Light Water Reactors, ANS White Paper, American Nuclear Society, LaGrange, Ill. (May 1976).
5. C. Komanoff, CEP Report (Dec. 1976).

25 The Police State Myth

Bernard I. Spinrad

An obscure but often repeated assertion that is fashionable in antinuclear circles is that the use of nuclear energy would lead to a suppression of civil liberties in the U.S. The apparent relationship is that nuclear materials could be stolen, and nuclear systems sabotaged, and that in the minds of the accusers only a police state could effectively protect against these possibilities.

It is a cheering thing to be able to report that this is not what the critics really mean. After all, there are many precious and dangerous things that could be stolen and misused and many systems whose sabotage would hurt us. The potential for harm, at greater levels than anything contributed by nuclear power to social risk, has existed for a long time, and we have gotten by safely without damage to civil liberties. Legitimate policing power and security systems are acceptable provided they function within public trust (and deserve that trust). They have proven workable.

On further inquiry, it turns out that the "police state" argument is one more example of unethical forensics by antinuclear activists. By using the term, they win the support of the nihilistic fringe, to whom all police activities are an anathema. However, careful questioning about the basis of the accusation brings out two specific charges:

1. The management of nuclear materials would require investigation of the stability and loyalties of persons chosen to manage. Such investigation is incompatible with some people's concepts of civil rights.
2. In case of theft of nuclear weapons material, police countermeasures might have to include violations of due process with regard to entry and search.

BERNARD I. SPINRAD • Department of Nuclear Engineering, Iowa State University, Ames, Iowa 50011.

Both of the charges have validity, but they do not add up to a police state. As to the first charge—the requirement that some background investigations be made of responsible personnel—we have long precedents in private industry (the references and files of the "confidential secretary," the investigation of the corporate treasurer by the bonding agency, etc.) and in public life (the investigations into the financial dealings of Bert Lance and into the opinions of Kent Hansen, for example). *Proper* investigation of a person proposed for a responsible position is common prudence, does not violate that person's civil liberties, and is nondiscriminatory under due process. In brief, it should be routine to investigate and verify the credentials of persons chosen to manage nuclear materials and systems, but this is not a "police state" action.

The second charge—that theft of nuclear weapons material would provoke police countermeasures in violation of Bill of Rights guarantees—may also be true. Such countermeasures are justified by a "clear and present danger" determination, which is the basis of national or local emergency proclamations and the imposition of martial law. Acts of God bring about this type of situation every year, when public order is threatened after hurricanes, floods, or blackouts, or when public safety requires evacuations, relocation of people, or destruction of their property (as, for example, in the setting of backfires to control forest fires).

We do not expect nuclear materials theft to be commonplace. The record, which shows no known theft in over 30 years, speaks for itself. Thus, this possibility does not materially alter the frequency with which suspension of liberties would be required by circumstances.

Nuclear materials theft may not, in fact, require such measures, however. The material is radioactive, and radioactivity is easily sensed at some distance. This fact would so simplify police investigations that there might not be need for close searches.

26 Insurance and Nuclear Power

The Price–Anderson Act

Chris Whipple

The Price–Anderson Act, which establishes procedures for insuring nuclear facilities (including nuclear power plants), was enacted with the dual purpose of protecting the public and encouraging the development of a private nuclear energy industry. As with most aspects of commercial nuclear power, this insurance system has become a target of criticism in recent years, criticisms that can generally be grouped into four categories: that the Price–Anderson Act provides a federal subsidy to the nuclear power industry [1], that the public would be inadequately compensated in the event of a nuclear accident [2–4], that the exclusion from coverage of damage caused by nuclear accidents from all homeowners insurance policies represents a lack of confidence in nuclear power by private insurers [1], and that the Price–Anderson Act removes incentives for safe operation of nuclear power plants [5]. These criticisms have been reviewed in several publications [6–8], and a brief synopsis of these reviews, coupled with the most recent and authoritative review available, that of the Supreme Court (see Note 4), will be given following a description of nuclear insurance as practiced under the Price–Anderson Act.

NUCLEAR POWER PLANT INSURANCE

The most controversial aspect of the Price–Anderson Act is that should an accident occur, the aggregate liability of the reactor operator, the NRC, or any others who might be at fault (such as equipment manufacturers) is limited to $560 million. Lawsuits for amounts in excess of $560 million are prohibited. If

---

CHRIS WHIPPLE • Electric Power Research Institute, Palo Alto, California 94303.

damages exceed that amount, and no additional money were available, the $560 million would be divided on a prorated basis. However, the 1975 renewal of the Price–Anderson Act provides [9]:

> That in the event of a nuclear incident involving damages in excess of that amount of aggregate liability, the Congress will thoroughly review the particular incident and will take whatever action is deemed necessary and appropriate to protect the public from the consequences of a disaster of such magnitude. . . .

The method of obtaining $560 million of coverage under the original form of the Price–Anderson Act was through a combination of private and government insurance. Nuclear power plant operators are required to purchase as much conventional liability insurance as is available from private sources (originally $60 million) and the additional amount necessary to reach $560 million from the government. When the Act was renewed in 1975, this method was modified to include a layer of coverage above that provided by the insurance industry but below that sold by the government through assessments of all reactor operators in the event of a reactor accident producing damages in excess of the privately available insurance [10]. The limit upon the amount of these assessments was to be set by the NRC between $2 million and $5 million per reactor; the NRC has chosen $5 million for all commercial reactors. Thus, the $560 million liability coverage is presently achieved through private insurance (currently $160 million), then assessments of reactor operators (currently $360 million based upon 72 plants), and then $40 million of government insurance.

In time, as the sum of private and assessment coverage rises to $560 million, either through increases in the availability of private coverage or the start-up of additional plants, the government will no longer provide insurance, and beyond that point, the limit will automatically rise above $560 million to the sum of that available through the first two sources. Under the Price–Anderson Act, the reactor operator assumes all public liability (subject to the $560 million limitation), including that that might otherwise fall upon the manufacturers of the plant or its equipment.

These features provide the basis under which the private development of nuclear power was encouraged, because it was felt that many manufacturers would be unwilling to enter a new market of this type without protection from ruinous liability losses, albeit at extremely low probability. These features are also the basis for the controversy, because critics of nuclear power view these specific provisions as undermining public protection.

To offset the features that are favorable to industry (from the viewpoint of the public) is a requirement that the reactor operator is strictly liable for damages regardless of cause. The operator must agree to waive the use of a defense that argues that he was not negligent, that the claimants implicitly or explicitly accepted the risk, that injured parties contributed to the accident by their own negligence, that the cause was due to acts of God, or that liability rests with some other party (such as the manufacturer of the plant or its equipment). The operator can deny a causal chain between the accident and the claimants' losses.

This "no fault" approach offers significant advantages to claimants following a nuclear accident when compared to conventional accidents. Following the Three Mile Island accident of May 28, 1979, representatives of American Nuclear Insurers and the Mutual Atomic Energy Liability Underwriters opened an office on March 30 to provide reimbursement for living expenses to those who had evacuated. About $2 million was paid as a result of claims made. At the peak of this operation, 51 claims representatives were on the scene processing applications [11]. In a general insurance case, at a time when they have suffered a loss, claimants may be faced with mounting a legal suit against a company with greater legal resources and little incentive to seek a rapid settlement. The further requirement that the claimant prove that the company is at fault can add to these difficulties, particularly when there are extremely complex technical issues involved in establishing fault.

There is also the situation in which the facility causing the damage represents the total assets of a company—when the accident occurs no assets are left to pursue. This situation offers the ownership little if any incentive for carrying liability insurance. Examples of this are reportedly common in the shipping industry, where various holding companies, each owning one ship, provide liability protection for a parent company.

Clearly the Price–Anderson Act specifies insurance mechanisms that, when compared to conventional insurance, are different from those of conventional insurance, and some of the differences have been sources of controversy. Of the four criticisms listed initially, two (inadequate protection of the public and limited safety incentives for reactor operators) are directly attributable to the limitation on liability. The other two (the exclusion from homeowners policies of damages caused by nuclear accidents and a federal subsidy through government-sold insurance) are not related directly to the liability limitation. These last two criticisms will be briefly discussed first.

The claim that nuclear power is subsidized by the rates charged for the insurance sold by the federal government is rapidly becoming moot, because in several years it is anticipated that the sum of privately available insurance and retroactive assessments will exceed $560 million, and at that time the government will no longer sell this insurance.

Prior to the retroactive assessment policy, the government sold nuclear insurance for $30 per thermal megawatt, or roughly $100,000 per reactor per year. The fairness of this charge can only be estimated roughly, but based upon the estimates of the Reactor Safety Study (WASH-1400) the expected costs to the government per reactor year (from losses exceeding $160 million up to $560 million) is less than one-tenth of the cost of the premium. In actual experience no claims have been made or paid by the government (including those arising from the TMI accident). In fact, the government has been turning a tidy profit on this business. Although this period is too short to provide a statistical verification of the reasonableness of the premium, this rate of $100,000 has virtually no impact upon the economics of a power plant that produces over $100 million worth of electricity per year.

The second criticism, that the nuclear exclusion clause in homeowners policies indicates a lack of confidence by the private insurance industry, is difficult to understand in view of the coverage provided to reactor operators by private insurance pools. In addition to the $160 million liability coverage per reactor, an additional $300 million of property insurance (to partially cover the loss that a utility would incur in the event of a serious accident) is available. The private insurance industry is therefore providing $460 million insurance per power plant, among the highest amounts provided for any activity [6, 8]. Although this indicates strong confidence by the insurance industry, it does not explain the reason for the homeowners exclusion clause. One explanation, offered by the pools that sell the nuclear insurance, is that individual coverage is unnecessary because it is already provided by the insurance purchases of the reactor operator [12]. An additional explanation was offered during testimony before the Supreme Court (see Note 4; testimony is quoted in footnote 9 of the ruling) that the "nuclear industry has essentially absorbed the entire capacity of the private insurance market in their need for property and liability insurance."

The remaining two criticisms of the Price–Anderson Act are that the limitation on liability removes a significant safety incentive and that the public would not be protected in the event of accident damages exceeding $560 million. The safety incentive issue was considered by the Supreme Court—their opinion perhaps the most recent and certainly likely to be the most unbiased on the subject. They wrote:

> This District Court's further conclusion that the Price–Anderson Act tends to encourage irresponsibility . . . on the part of builders and owners of the nuclear power plants, 431 F. Supp. 222, simply cannot withstand careful scrutiny. We recently outlined the multitude of detailed steps involved in the review of any application for a license to construct or to operate a nuclear power plant, Vermont Yankee Nuclear Power Corp v. NRDC, No. 76-419, slip op., at 4-5 (April 2, 1978); nothing in the liability limitation provision undermines or alters in any respect the rigor and integrity of that process. Moreover, in the event of a nuclear accident the utility itself would suffer perhaps the largest damages. While obviously not to be compared with the loss of human life and injury to health, the risk of financial loss and possible bankruptcy to the utility is in itself no small incentive to avoid the kind of irresponsibility and cavalier conduct implicitly attributed to licensees by the District Court.

The TMI accident demonstrated that the Supreme Court was quite accurate in its assessment of exposure to losses a utility faces.

The question of the degree of public protection afforded by Price–Anderson was central to the Supreme Court's ruling. In discussing limited versus unlimited liability, one point, often misunderstood, is the distinction between the amount of liability and the amount of damages paid. Under the circumstances in which liability is unlimited, recoveries are limited to the assets and insurance of the liable party. In most cases this is less than $560 million [7, 8]. This point was addressed directly by the Court:

> The expert testimony before the District Court indicated that Duke Power, one of the largest utilities in the country, could not be expected to accumulate more than $200 million for damage claims before reaching the point of insolvency. App. 393-397. This amount, even when coupled with the amount of available private insurance, would be less than the $560 million provided by the Act. Moreover, if the liability were of sufficient magnitude to force the utility or component manufacturer into bankruptcy or reorganization, recovery would likely be further reduced and delayed.

Under Price–Anderson, the limitation on liability at $560 million is not intended to be absolute. The intent of Congress to provide additional relief is quite clear [13]:

> This limitation does not, as a practical matter, detract from the public protection afforded by this legislation. In the first place, the likelihood of an accident occurring which would result in claims exceeding the sum of the financial protection required and the government indemnity is exceedingly remote, albeit theoretically possible. Perhaps more important, in the event of national disaster of this magnitude, it is obvious that Congress would have to review the problem and take appropriate action. The history of other natural or man-made disasters, such as the Texas City incident, bears this out. The limitation of liability serves primarily as a device for facilitating further congressional review of such a situation, rather than as an ultimate bar to further relief of the public.

In reviewing this Congressional policy of providing additional relief, the Supreme Court found:

> In the past Congress has provided emergency assistance for victims of catastrophic accidents even in the absence of a prior statutory commitment to do so. For example, in 1955, Congress passed the Texas City Disaster Relief Act, 69 Stat. 707, to provide relief for victims of the explosion of ammonium nitrate fertilizer in 1947. Congress took this action despite the decision in Dalehite v. United States, 346 U.S. 15 (1953), holding the United States free from any liability under the Federal Tort Claims Act for the damages incurred and injuries suffered. More recently Congress enacted legislation to provide relief for victims of the flood resulting from the collapse of the Teton Dam in Idaho. Pub. L 94-400, 90 Stat. 1211. Under the Act, the Secretary of the Interior was authorized to provide full compensation for any deaths, personal injuries or property damage caused by the failure of the dam.*
>
> The Price–Anderson Act is, of course, a significant improvement on these prior relief efforts because it provides an advance guarantee of recovery up to $560 million

---

* Payments totaled about $400 million.

plus an express commitment by Congress to take whatever further steps are necessary to aid the victims of a nuclear incident.

The final question is whether the protection afforded the public under Price–Anderson is equivalent to that that would be available under common law. On this subject the Court was quite direct:

> We view the congressional assurance of a $560 million fund for recovery, accompanied by an express statutory commitment, to 'take whatever action is deemed necessary and appropriate to protect the public from the consequences of' a nuclear accident, 42 U.S.C. 2210(e), to be a fair and reasonable substitute for the uncertain recovery of damages of this magnitude from a utility or component manufacturer, whose resources might well be exhausted at an early stage. The record in this case raises serious questions about the ability of a utility or component manufacturer to satisfy a judgment approaching $560 million—the amount guaranteed under the Price–Anderson Act. Nor are we persuaded that the mandatory waiver of defenses required by the Act is of no benefit to potential claimants. Since there has never been, to our knowledge, a case arising out of a nuclear incident like those covered by the Price–Anderson Act, any discussion of the standard of liability that state courts will apply is necessarily speculative. At the minimum, the statutorily mandated waiver of defenses establishes at the threshold the right of injured parties to compensation without proof of fault and eliminates the burden of delay and uncertainty which would follow from the need to litigate the question of liability after an accident. Further, even if strict liability were routinely applied, the common-law doctrine is subject to exceptions for acts of God or of third parties (see Prosser, Law of Torts, supra, at 520-521)— two of the very factors which appellees emphasized in the District Court in the course of arguing that the risks of a nuclear accident are greater than generally admitted. All of these considerations belie the suggestion that the Act leaves the potential victims of a nuclear disaster in a more disadvantageous position than they would be in if left to their common-law remedies—not known in modern times for either their speed or economy.
>
> . . . The Price–Anderson Act not only provides a reasonable, prompt and equitable mechanism for compensating victims of a catastrophic nuclear incident, it also guarantees a level of net compensation generally exceeding that recoverable in private litigation. Moreover, the Act contains an explicit congressional commitment to take further action to aid victims of a nuclear accident in the event that the $560 million ceiling on liability is exceeded. This panoply of remedies and guarantees is at least a reasonably just substitute for the common-law rights replaced by the Price–Anderson Act.

This chapter was completed in July 1978 and revised in February 1981.

REFERENCES AND FOOTNOTES

1. Public Interest Report, The Price–Anderson Act and the Nuclear Industry: The Attempt to Insure the Uninsurable, a joint project of the Environmental Alert Group (Los Angeles) and the Environmental Education Group, undated paper, likely published sometime in 1975. The paper opens with a reprint of a standard nuclear exclusion clause from a homeowners insurance policy.

2. H. S. Denenberg, Nuclear power: Uninsurable, *Progressive Magazine* (Nov. 1974).
3. The California Nuclear Initiative (Proposition 15, June 8, 1976 ballot, defeated) contained, under section 67503(a), the requirement that "after one year from the date of passage of this measure, the liability limits assured, either by law or waiver, as determined by a California court of competent jurisdiction. . . ."
4. The Carolina Environmental Study Group, the Catawba Central Labor Union, and 40 individuals sued Duke Power and the Nuclear Regulatory Commission. This action, which commenced in 1973, sought among other things a declaration that the Price–Anderson Act was unconstitutional. On April 5, 1977, the U.S. District Court for the Western District of North Carolina ruled that the Act was unconstitutional in two respects: (1) it violated the Due Process Clause of the Fifth Amendment because it allowed injuries to occur without assuring adequate compensation to the victims; (2) the Act offended the equal protection component of the Fifth Amendment by forcing the victims of nuclear incidents to bear the burden of injury, whereas society as a whole benefits from the existence and development of nuclear power. This ruling was overturned by a unanimous vote of the Supreme Court on June 26, 1978.
5. J. M. Marshall and L. I. Lieb, Liability and Safety in Nuclear Power Plants, UCLA Report UCLA-ENG-7724 (Feb. 1977).
6. C. Whipple, Insurance of Nuclear Power Plants, EPRI Report (June, 1976).
7. K. Solomon and D. Okrent, Catastrophic Events Leading to De Facto Limits on Liability, UCLA Report UCLA-ENG-7732 (May 1977). [See also Some Comments on De Facto Limits on Liability, Rand P-5885 (June 1977) by the same authors.]
8. C. Whipple, K. Solomon, and D. Okrent, Insurance and Catastrophic Events: Can We Expect De Facto Limits on Liability Recoveries?, Rand P-5940 (March 1978). Also published in Proceedings of the ANS Topical Meeting on Probabilistic Analysis of Nuclear Reactor Safety, Los Angeles, May 8–10, 1978.
9. Public Law 94-197, 94th Congress, H.R. 8631, December 31, 1975, 89 STAT.1113.
10. Public Law 94-197, 94th Congress, H.R. 8631, December 31, 1975, 89 STAT.1112.
11. American Nuclear Insurers Letter, Vol. 3, No. 6 (June 1979); available from American Nuclear Insurers, The Exchange, Building 3, Farmington Avenue, Farmington, Conn. 06032.
12. NEL–PIA Reports, Report No. 3, Revised 9/77. NEL–PIA is the Nuclear Energy Liability–Property Insurance Association, a pool of liability and property insurance companies that combine their resources to insure nuclear facilities and supplies.
13. H.R. Rep. No. 89-883, 89th Congress, 1st Session 6-7 (1965).

27 Solar and Nuclear Power Are Partners

A. David Rossin

The words come in the same breath: "Solar energy could do it but nuclear power stands in the way!" Some solar energy proponents state their case as a demand for an end to nuclear power to open the future for the sun. The claim is voiced that continued reliance on electric grids with coal-fired and nuclear plants is almost a conspiracy against alternative, soft, decentralized (or solar) energy sources.

The argument is worth testing. The test reveals that:

- Solar and nuclear power do not compete with one another.
- No energy source can do the job alone. In fact, solar and nuclear are partners.
- Solar energy's future is brightest only if nuclear power succeeds.

Why are these facts so?

The first test is answered by examining utility electric power supply and demand patterns. Utilities with nuclear plants use them day and night to supply what is known as the *base load*. In most regions, utilities have to generate almost twice as much energy during the day as they do in the middle of the night. No utility would consider using nuclear plants for all its generating capacity right up to peak loads. The base-load power must be supplemented by more coal power (usually from older and less efficient coal-burning units) and, if necessary, by oil or gas.

The sun shines during the day! That's the only time when solar energy can be generated! That's also when the need is greatest. Solar energy could be used to decrease the amount of electricity the utility plants must supply at the very time when utilities have to use their peaking units, the ones that generate the most expensive power and the ones that use the scarcest fuels: oil and gas.

A. DAVID ROSSIN • Electric Power Research Institute, Palo Alto, California 94303.

Solar energy actually competes with oil-fired power, not with base-load nuclear and coal. Electricity generated by burning oil costs three to five times as much as nuclear power (and the money for quite a bit of that goes directly to OPEC).

As an example of costs, in 1979 Commonwealth Edison's nuclear power plants produced electricity at 1.6 cents/kW-hr all costs included; its large coal plants at 3.0 cents/kW-hr while just the *fuel oil cost alone* for oil-fired plants (not including the manpower or investment costs) ranged from 3.6 to 6.0 cents/kW-hr.

Peak load generation costs are so high that it pays to store some electricity generated at night in a pumped storage facility 300 miles away. By the time the energy is transmitted up and back and the water pumped up and returned to Lake Michigan through turbines, it costs $1\frac{1}{2}$ times its original generation cost. It is still a bargain for the people of northern Illinois.

At this time there is no source of solar electricity that can even approach the price of oil-fired power. If it could, it would be a great help, but the best current estimates range from 30 cents/kW-hr on up.

There are opportunities for homeowners and businesses to design passive solar buildings that reduce energy requirements. Effective passive designs would reduce demand for utility power during the day. Again, the benefits are shared.

Nuclear advocates have never argued that nuclear power would do the job alone. The key to our energy solution is diversity, with each form of energy supply doing the thing that it can do best and having them all work together so that the system can meet the legitimate demands of the people that depend on it with a minimum of cost and environmental impact.

The appropriate technologies for base-load central-station electric generation, at least for the next several decades, are coal and nuclear power. Solar energy's contribution will not determine these requirements, but it can help reduce peak power demand and save oil.

So solar energy and nuclear power are *partners, not competitors*. This should raise some questons in people's minds when a person who claims to be an advocate of solar energy also claims that we have to stop nuclear power or stop building new nuclear plants as if they compete with solar energy.

Because nuclear plants produce energy at less cost to the consumer than the alternatives, a new nuclear plant will save the customer money on his electric bill in the years to come; stopping those plants means *higher* electric bills.

This leaves less money available to the homeowner or to the business firm, money that could be invested in solar energy if the technology appears promising.

There are a lot of issues in the energy debate today, but the real solar advocate does not attack nuclear power. The test is whether the proposed actions really help save energy and oil. With even a quick look at the realities, it becomes clear that nuclear and solar power are partners in the task of meeting legitimate energy needs.

Index

Accelerator breeders, 217–218
Accident phases, 295
Advocacy, xii
AEC, 11
Air pollution, 299
ALARA, 346–347
Alternate breeder fuel cycles, 210–213
Alternative breeder reactors, 213–214
Alternative energy, 250
American Nuclear Insurers, 375
American Physical Society, 11
A. Philips Randolph Institute, 113
Arab/OPEC, 5
Average capacity factor, 128

Best Available Control Technology, 116
Biomass, 59
Bomb, nuclear, 266
Breeder reactor development programs, 117
Breeder reactors, 69–70, 154
Bulletin of the Atomic Scientists, 349
Burnup, fuel, 267

California Environmental Study Group, 379
California Nuclear Initiative, 379
California Plan, 114
California Public Resources Code, 115
Capacity factor, U.S. LWR units, 127
Capital investment, 85
Carter, 117, 252
Catawba Central Labor Union, 379

Cats paw, terrorists, 269
CIVEX process, 259–262
Civil disobedience, ix
Civil liberties, 371–372
Clandestine groups, 269
Clean Air Act Amendments of 1977, 116
Clean Air Acts, 117
Clean Water Acts, 117
Clinch River Breeder Reactor Demonstration Plant, 117
Coal, 61–64, 299, 315, 335–336
Cogeneration, 98
Completeness of accident scenarios, 283
Conditional probabilities, 278
Consequence analyses, accidents, 280
Consequence indicator, accidents, 275
Conservation, 53–55, 111–112
Conservation potential, 81
Constraints on energy systems, 22–24
Containment, nuclear reactor, 289
Contingency measures, 114
Conversion, nuclear plants to coal, 149
Core disruptive accidents, 291
Cost factors, 142
Council on Environmental Quality, 116

Decentralized energy sources, 121
Declaration of emergency, 115
Deductive logic, 279
Defense-in-depth, 273
Deindustrialization, vii

383

Delays and cancellations, 140
Denenberg, H.S., 379
Department of Energy, 117
Distribution function
  cumulative, 285
  risk, 284
Diversion resistance, 254–258
DOE, 11
Doppler-effect feedback, 292
Downside risks, 119
Duke Power, 379

EBR-2, 258
Economic control, x
Edison Electric Institute, 113
Eightfold way to nuclear weapons, 224–225
Electricité de France, 140
Electricity use, 97
Energetic fuel-coolant interactions, 293
Energy
  alternatives, 149
  consumption, 49–50, 75
  crisis, 113
  demand, 50–53
  growth, 86
  shortage, 115
  supply, 33, 77
  use, 77
Energy Commission, 115
Energy-deficient society, 118
Energy Plan, 117
Energy Policy, 106, 117
Energy Shortage Contingency Plan, 114
England, 117
Enrichment plants, 245
EPA, 10
ERDA, 11
Event trees, 274, 277, 278

Fallout, 349
Fast breeder reactor, 154
Fault trees, 274, 278
Federal Power Commission, 113
Finite supply of energy resources, 111
Ford, 252
France, 117
French breeder program, 157
Fuel cycles, 254
Fuel-efficient practices, 111
Full scale component tests, 296

Fundamental experiments, 296
Fusion-fission hybrid reactors, 217–218

Georgine, Robert A. 113
Geothermal energy, 57–58
GNP, 111
Gofman, 359–360
Growth vs. no growth, 113

Hard and soft paths, 91
Health care, x
Heavy-Water Reactor (HWR), 236
Homemade bomb, 265–269
Horizontal proliferation, 223–227
Hot particle theory, 359
Hydrogen explosions, 293
Hydropower, 55–57

IAEA, 239, 254
Indifference costs, 198
Industrial revolution, xiii
Infant mortality, 350
Inherent safety, 290
Insurance, 373
International fuel cycle evaluation, 252
International Institute for Applied Systems
    Analysis, 17
International Nuclear Fuel Cycle Evaluation,
    193–194
Interruption of electricity, 115
Isotope separation, 224–225, 234–235

Japan, 117
Jobs, 112

King, Llewellyn, 7, 8

Labor unions, 113
Large coal-fired plant performance, 128
Liability, 373
Lieb, L. I., 379
Life expectancy, 317
Light-Water Reactor (LWR), 154, 274, 288
Linear hypothesis, 331
Liquid-Metal Fast Breeder Reactor (LMFBR),
    154, 289
Load curtailment, 115
Load reductions, 115
Load shedding, 113
Loss of coolant accidents, 291

Index

Lovins, Amory, 8, 76
Low-energy-use society, 111, 116
Lowest Achievable Emission Rate, 116

Mancuso, 332
Manhattan Project, 230
Margin for error, 117
Market penetration of energy systems, 25
Marshall, J. M., 379
Mathematical simulation, 296
Media, 325, 334
Meteorites, 310
Moral issues, xii
Multinational nuclear undertakings, 186–187
Mutual Atomic Energy Liability Underwriters, 375
Myth creating, 353

NAACP, 112
National Academy of Sciences, 11
National Electrical Reliability Council, 113
National Energy Policy, 112
National Environmental Policy Act, 117
Natural gas, 64–65
Natural uranium reactors, 239
NEPA, 116
NNPA, 252
NNW, 239
No-growth society, 111, 116
Nonattainment Areas, 116
Norwegian reactor project, 181–183
NPT, 227, 239, 241, 252
NRC, 10, 116
Nuclear and coal-fired generation costs, 126
Nuclear energy scenarios, 28–29
Nuclear
  freeze, 245, 254
  fuel breeding, 209–210
  power, 100
  safety, 66–67
  waste, 10–12
  waste radioisotopes, 10
  weapons, 223, 243–244, 310, 349
Nuclear Isolationism, 183–184, 185

Occupational exposure, 336–337
Oil, 64–66
  costs, viii, ix
  glut, viii

Oil (cont'd)
  overdependence, vii, viii
  shocks, ix, x

Per capita consumption of energy, 111
Persian Gulf, x
Phenix, 156, 157
Plutonium, 12–13, 267, 355
  weapons grade, 266
Pollard, 13
Power generation, cost comparison of LMFBR vs. LWR, 177
Power reactors, 224, 230–231, 236, 237
Premature detonation, 266
Pressurized water reactor (PWR), 177
Prevention of Significant Deterioration, 116
Price-Anderson Act, 373–379
Production reactors, 224–225, 234, 237
Proliferation, 223
Protection of the environment, 113

Radiation risks, 300, 329, 345, 355
Radioactive waste, 67–68
Radioactivity releases, 339, 363–364
Radon, 313, 337
Rapsodie, 157
Reactor grade plutonium, 238
Reactor meltdown, 320
Reactor Safety Study, 280
Recriticality, 292
Reduced lifestyle, 111
Remote handling, 258
Renewable energy sources, 31
Reprocessing, 255, 258, 343
  overseas, 268
  spent nuclear fuel, 197–198
  U.S., 268
Research reactors, 230–231
Resource Conservation and Recovery Acts, 117
Revolution of rising expectations, 113
Rising aspirations, 112
Rising expectations, ix
Risk, 273, 317, 336–338
  assessment, 275
  total, 275
Risks
  social, vii
  wars, viii, x
Rolling blackouts, 114, 115

Rotating blackouts, 115
Russians, 118
Rustin, B., 113

Sabotage, 68–69
Safeguarding recycled plutonium, 203–205
Safeguards, 243, 244, 249
Sanctions, 242
Shortage index, 115
Social implications, 105
Soft energy strategy, 92
Soft technologies, 75
Solar energy, 59–61
Spent fuel, 239
Sternglass, 349, 350
Strontium-90, 349
Structured nuclear world, 250
Subnational groups, 269
Subnational proliferation, 223, 243
Super-Phenix, 156, 157
Super-Phenix II, 157
Super-Phenix Mark I, 157
Supreme Court, 373
Symbiotic reactor systems, 215–216
System tests, 297

Terrorists, 258, 265
Theft, nuclear material, 267
Thorium, 256
Three Mile Island, 1, 6, 7, 333, 375
Tidal power, 57

Total capital costs, 102
Toxic Substance Acts, 117
Transport accidents, 334
Treaties, 226

Umbrella, nuclear, 227, 245
Unstructured nuclear world, 250
Uranium enrichment, 190–193
Uranium mill tailings, 314
Uranium supply, 228, 229
Uranium trade, 189–190
Uranium-233, 256
U.S. General Accounting Office, 11
U.S. policy, 117

Vertical proliferation, 223–226
Volcanos, 310
Voltage reductions, 113, 115

Waste, 10–12, 299, 334, 367
Waste heat, 53–54
Weapons
  design, 232–233
  location of, 267
  plutonium, 224, 236
Western Europe, 117
Wind power, 58
World energy, 19–20, 33–40
World energy supplies, vii
World's energy, 111
Worldwide consumption of energy, 111

Muirhead Library
Michigan Christian College
Rochester, Michigan

ENNIS AND NANCY HAM LIBRARY
ROCHESTER COLLEGE
800 WEST AVON ROAD
ROCHESTER HILLS, MI 48307

MICHIGAN
CHRISTIAN
COLLEGE
LIBRARY
ROCHESTER, MICH.